OCCUPATIONAL HAZARDS
OF PESTICIDE EXPOSURE

OCCUPATIONAL HAZARDS OF PESTICIDE EXPOSURE

Sampling, Monitoring, Measuring

Edited by

Donald J. Ecobichon

Queen's University
Kingston, Ontario, Canada

CRC Press
Taylor & Francis Group
Boca Raton London New York

CRC Press is an imprint of the
Taylor & Francis Group, an **informa** business

USA	Publishing Office:	Taylor & Francis 325 Chestnut Street Philadelphia, PA 19106 Tel: (215) 625-8900 Fax: (215) 625-2940
	Distribution Center:	Taylor & Francis 47 Runway Road Levittown, PA 19057 Tel: (215) 269-0400 Fax: (215) 269-0363
UK		Taylor & Francis 1 Gunpowder Square London EC4A 3DE Tel: +44 171 583 0490 Fax: +44 171 583 0581

OCCUPATIONAL HAZARDS OF PESTICIDE EXPOSURE: Sampling, Monitoring, Measuring

A CIP catalog record for this book is available from the British Library.
⊛ The paper in this publication meets the requirements of the ANSI Standard Z39.48-1984 (Permanence of Paper).

Library of Congress Cataloging-in-Publication Data

Occupational hazards of pesticide exposure: sampling, monitoring, and
   measuring / [edited by] Donald J. Ecobichon.
     P.   cm.
   Includes bibliographical references and index.
   ISBN 1-56032-706-5 (cloth : alk. paper). —ISBN 1-56032-707-3
(paper : alk. paper)
     1. Pesticides—Toxicology. 2. Pesticide applicators (Persons)—
Health and hygiene. 3. Pesticides—Application—Safety measures.
I. Ecobichon, Donald J.
RA1270.P40226   1998
615.9′51—DC21   98-21964
                                        CIP

ISBN 1-56032-706-5 (cloth)
ISBN 1-56032-707-3 (paper)

CONTENTS

PREFACE

Pesticides are used extensively in agriculture, forestry, and public health. Despite the obvious benefits, such agents must be considered toxic to all who handle them when these agents are not used according to recommended application technology, dosage levels, reentry time intervals for crops, application-harvest time intervals, appropriate personal protective equipment, and so forth. The misuse of chemical pesticides is a problem of global proportions, recognized by the World Health Organization and other international and national agencies.

The operative or "key" word with respect to toxicity is "exposure." In 1985, Graham Turnbull edited *Occupational Hazards to Pesticide Use* (Taylor & Francis), presenting a scientific approach to occupational pesticide exposure, primarily considering exposure in relation to residues on clothing patches and skin under various conditions of application. In the 12 years since then, exposure has taken on new meaning, relating residues deposited on the skin with systemically absorbed amounts of toxicant measured by sophisticated, quantitative analyses of the pesticide, metabolites, degradation products, and surrogate tracer chemicals in biological fluids. This has been the subject of many published papers.

This book approaches pesticide exposure systematically, examining both occupational and bystander exposure from aspects including aerial transport with on- and off-target loss of material, transferable or dislodgeable residues from surfaces, epidermal deposition and absorption through the skin, biomonitoring of systemically absorbed residues, and the assessment of adverse health effects including neurological and neurobehavioral end points of toxicity. It is hoped that the reader will find this approach useful in gaining insight into the problems of exposure characterization or assessment that face anyone designing such experiments to monitor biological effects of exposure or developing regulatory guidelines for application practices or to protect human health.

In seeing the need for such a text, I approached individuals known as experts in their respective fields who, although busy, were coerced into contributing chapters. As editor, I have learned much from reading what they wrote, the techniques described and their application, the science of exposure estimation, where exposure assessment is today and where it is going, all supported by extensive collections of references. My colleagues have done an excellent job of communicating their concepts, knowledge, and ideas. I am confident that the book will fulfill the visualized need and I thank them for their participation in this endeavour.

Donald J. Ecobichon, Ph.D.

LIST OF CONTRIBUTORS

BRUCE A. ARCHIBALD
Food Laboratory Services Division
University of Guelph, 95 Stone Road West
Guelph, Ontario, Canada N1H 8J7

DONALD J. ECOBICHON
Department of Pharmacology
and Toxicology, Queen's University
Kingston, Ontario, Canada K7L 3N6

DONNA HOUGHTON
Novartis Crop Protection Canada Inc.
140 Research Lane, Research Park
University of Guelph
Guelph, Ontario, Canada N1G 4Z3

ROBERT I. KRIEGER
Department of Entomology
University of California, Riverside
Riverside, CA 92521-0314

RICHARD P. MOODY
Health Canada, Health Protection Branch
Product Safety Laboratory
1800 Walkley Road
Postal Locator # 6402A2
Ottawa, Ontario, Canada K1A 0L2

HERBERT N. NIGG
Citrus Research and Education Center
University of Florida
700 Experiment Station Road
Lake Alfred, FL 33850

CHRISTOPHER M. RILEY
Research and Productivity Council
921 College Hill Road
Federicton, New Brunswick
Canada E3B 6Z9

GWEN M. RITCEY
Department of Environmental Biology
University of Guelph
Guelph, Ontario
Canada N1G ZW1

KEITH R. SOLOMON
Center for Toxicology
University of Guelph
Guelph, Ontario
Canada N1G 1Y4

GERALD R. STEPHENSON
Department of Environmental Biology
University of Guelph
Guelph, Ontario
Canada N1G ZW1

CHARLES J. WIESNER
Wiesner and Associates
23 Kentville Road
Durham Bridge, New Brunswick
Canada E6C 1K7

ONE

INTRODUCTION

Donald J. Ecobichon

Department of Pharmacology and Toxicology, Queen's University, Kingston, Ontario, Canada K7L 3N6

Pesticides, by the very nature of their use and pattern classification (e.g., suffix "icide" — insecticides, herbicides) are toxic to some form of life; most, at some dose level, have the potential to be toxic to humans. Although the efficacy of pesticides for the protection or preservation of health, food, and fiber cannot be disputed, these benefits are offset by the terrible costs of an estimated 3 million cases annually of acute, life-threatening intoxications worldwide, with perhaps as many or more unreported cases and some 220,000 deaths (WHO 1990). There exists a long history of pesticide poisonings among workers in agriculture, residential application, and in public health (Hayes 1982). However, the numbers quoted above do not take into account persistent or delayed effects arising from either acute high-level or prolonged, low-level exposures. Specific target-organ toxicity and diseases, adverse reproductive outcomes (infertility, spontaneous abortions, birth defects), and even cancer are poorly studied or go unreported except as anecdotal cases in the literature. The incidence of acute poisoning in emerging nations is 13-fold higher than in highly industrialized countries, whereas the extent of other, more subtle, effects remains undocumented (Forget 1989).

The incidences of acute pesticide poisonings, accidental or intentional, have been well documented, providing some details about the biological effects as well as reasonably good estimates of the amount of agent ingested (Ecobichon and Joy 1994; Hayes 1982; Hayes and Laws 1991; Kaloyanova and El Batawi 1991). Accidental or occupational exposures via other routes (dermal, inhalation) have been less well defined, frequently being associated with job classifications (e.g., mixer/loader, applicator, flagger, harvester, handler), with poor resolution of the actual exposure. Early attempts to assess worker exposure focused on determining the amount of agent deposited on the skin or clothing; this approach developed into the patch test, with analysis of residues deposited on a gauze square pinned to the clothing at various sites (Turnbull 1985).

Deposition of the agent on the surface (skin, clothing, or surrogate patch) was considered to represent *direct exposure*, the assumption made being that what was deposited was absorbed (Franklin et al. 1981). By contrast, measurement of biological effects in the body and quantitation of metabolite levels in biological fluids were considered to represent *indirect exposure* (Franklin et al. 1981). In reality, direct exposure should be attributed only to pesticide residues gaining entry into the body via systemic absorption following ingestion, inhalation, or transdermal penetration. This concept is shown schematically in Fig. 1-1, where the individual has absorbed much of the deposited agent through the skin and, as a consequence, shows strong biological effects along with the excretion of both parent compound and metabolites or byproducts. The transdermal absorption varies considerably from chemical to chemical.

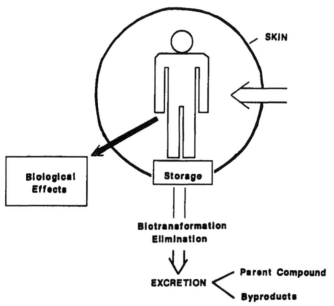

Figure 1-1 Penetration of a pesticide through the unprotected skin of a worker and systemic bioavailability to elicit adverse biological effects as well as for storage, biotransformation, and elimination of the parent compound and metabolic byproducts.

Based on the toxicity of the organophosphorus ester insecticides over that observed among earlier organochlorine compounds, it became evident that dermal absorption was the major route of pesticide acquisition. The early studies by Batchelor and Walker (1954) revealed that, for a casually dressed worker, a significant area of skin was available for exposure, including the face, hands, forearms, back of neck, front of neck and "v" of the chest, approximately 2,930 cm^2 (Ecobichon 1996). The importance of dermal acquisition of pesticides has been confirmed by many studies (Bonsall 1985; Durham, Wolfe, and Elliot 1972; Fenske and Elkner 1990; Wojeck et al. 1981; Wolfe, Armstrong, and Durham 1966, 1972). Estimation of dermal exposure became more complex with the simple but elegant experiments performed by Maibach et al. (1971), which demonstrated large regional differences in the penetration of parathion and malathion on the basis of urinary excretion of radiolabeled carbon. Penetration through the forearm and urinary excretion of a number of pesticides or byproducts revealed a wide range of results for different chemicals, the differences being associated with the physical and chemical properties of the agents (Feldman and Maibach 1974). Such studies raised concerns about the need for greater protection of exposed workers because levels of deposited agent on exposed skin were 20–1,700 times the amount reaching the respiratory tract (Feldman and Maibach 1974).

Extensive studies, such as those of Brady et al. (1991), Fenske and Elkner (1990), Frank, Campbell, and Sirons (1985), Franklin et al. (1981), and Methner and Fenske (1994), have readily illustrated the measures required to reduce exposure. The use of personal protective equipment in the form of a cap or broad-brimmed hat, coveralls with long sleeves, rubber boots, and solvent-resistant gloves would result, as shown in Fig. 1-2, in a reduction in the exposed surface area of skin and entrapment of pesticide residues in or on the equipment. With less agent penetrating through to the skin, a lower body burden of potentially toxic agent would be absorbed, a more efficient biotransformation and excretion of the parent compound would occur with elevated levels of byproducts, and less severe biological effects would be observed. The need for better protective equipment has led to the development of appropriate, chemical-resistant coveralls, solvent-impermeable gloves and footwear, safety promotions with decals to clearly mark "glove zones" on spraying equipment (where hands should be protected

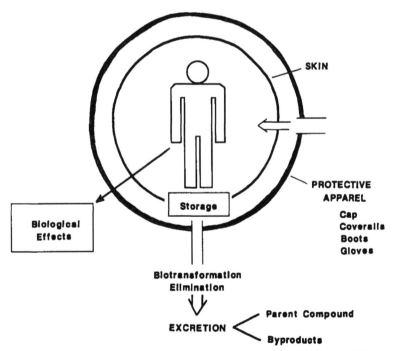

Figure 1-2 Effect of protective apparel on the reduction of systemic bioavailability of a pesticide in an exposed worker, reduced adverse biological effects (smaller arrow), and availability for more efficient biotransformation and the elimination of parent compound and metabolic byproducts.

at all times), a plethora of pictographic pamphlets illustrating safety precautions, and occupational legislation mandating the use of protective equipment. Such measures have been reflected in fewer pesticide poisonings in many industrialized countries but, unfortunately, these safety promotions have had little impact in emerging nations, where there is inadequate legislation and little personal protective equipment available except at exorbitant cost.

The additional sophisticated measures shown in Fig. 1-3 erect another protective barrier to exposure, further reducing the in vivo acquisition of pesticide residues, reducing the body burden of agent. Because exposure is reduced, biotransformation of the pesticide and excretion of its byproducts usually is more efficient, and overt toxicity is shifted to biological effects of a more covert or subtle nature. Protective measures, even of the simplest kind, have necessitated the development of advanced techniques for both low-level residue analysis and the quantification of subtle biomarkers representative of biological effects. This is partly the focus of this book.

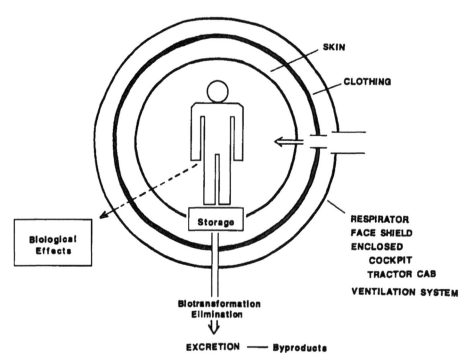

Figure 1-3 Effect of additional protective measures concomitant with protective apparel in reducing the systemic bioavailability of a pesticide in an exposed worker, minimizing adverse biological effects (broken arrow), and the availability for storage, efficient biotransformation, and elimination of metabolic byproducts.

Except under special circumstances, dermal absorption is the major route by which humans acquire pesticides. The estimation of human exposure has been carried out by many investigators, most methods requiring that sweeping assumptions be made to progress from a field application rate to an estimation of the biologically available dose acquired by a worker, a bystander, or another individual (Batchelor and Walker 1954;

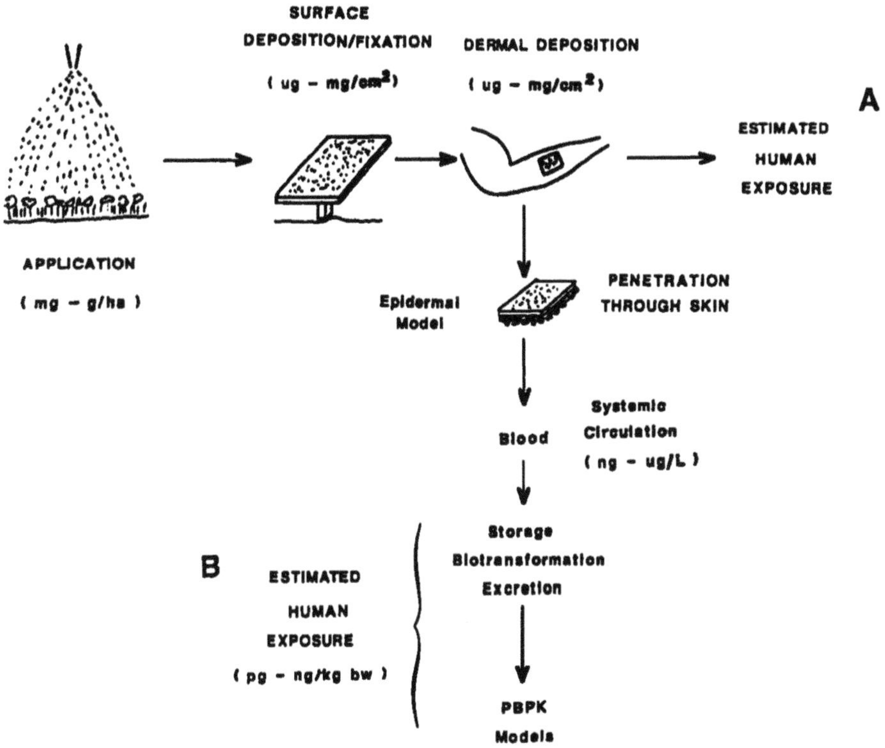

Figure 1-4 Schematic diagram depicting two approaches to the estimation of human exposure to pesticides: A) Exposure is estimated from field-deposit concentrations extrapolated to the surface area of exposed skin (face, hands, forearms, front and back of neck, "v" of chest) of a casually dressed worker, with assumptions being made concerning the percentage of the deposited agent absorbed. B) Estimation of exposure is based on the systemic, bioavailable levels of agent and/or breakdown products measured in tissues and/or fluids, making use of suitable PBPK models.

Byers, Kamble, and Nitkowski 1993; Draper and Street 1981; Wolfe et al. 1972). Figure 1-4 is a schematic diagram showing the steps taken for many assessments of human exposure, moving from an estimated application rate to surface deposition or fixation, equating this measurement with dermal deposition, and making assumptions of either total absorption of the deposited agent or of some fraction thereof derived from experimental models. More recent and more accurate estimates of human exposure have entailed additional measurements of (1) the dermal penetration of the agent, (2) the systemic blood levels of the agent, (3) the stored agent, and/or (4) the quantity biotransformed and excreted in biological fluids, these latter parameters being quantitated temporally using suitable physiologically based pharmacokinetic (PBPK) models.

For a number of years, I have followed what my colleagues have been attempting to do and a new(er) definition of exposure has evolved. In a few situations, I have played a part in the generation of the data whereas, at other times, I simply have enjoyed stimulating discussions while attending poster presentations at toxicological meetings. The purpose of this book is to examine newer approaches involved in quantifying pesticide exposure in occupational settings, examining techniques and methodologies for sampling, monitoring, and measuring for risk assessment purposes. This book is an attempt to pull together some of the most important factors related to worker exposure to pesticides, reviewing what has been done previously, discussing the advantages of newer techniques and what they offer in the way of improved risk assessment and risk management, and considering what pathways might be followed profitably in the future. In what I hope is a logical progression, my colleagues and I present chapters dealing with the following topics.

- On-target deposition of pesticides following ground or aerial application is discussed in the context of concerns about exposures to the owner of the crop, the applicators, and the workers involved in harvesting or handling the produce in the field. The converse of the problem, the off-target loss of agent, is discussed in the context again of concerns about the owner and the workers, as well as bystander populations living and working in rural or semirural agricultural regions.
- Deposition or fixation of the applied agent on foliage (e.g., plant leaves, crops, grasses) and the time-related dislodging of residues and the subsequent transfer to other surfaces (e.g., clothing, skin) are addressed in the context of concerns about exposures to those involved in hand harvesting field crops such as cabbage, cauliflower, broccoli, or tomatoes, or various tree fruits and berries.
- The classical approach to occupational monitoring of agricultural workers (e.g., mixer/loader, applicator, harvester, handler) is discussed as well as personal protection. This chapter provides a stepping stone to newer approaches.
- The efficacy of dermal penetration is explored in terms of the fraction of the pesticide that, for toxicologists, becomes the real or systemic dose capable of eliciting signs or symptoms. Epidermal models, both in vitro and in vivo, designed to quantitate dermal penetration, are described for a range of chemicals as well as how these models may replace human volunteer studies and assist in extrapolating highly variable, species-specific results to humans.
- Applications of pesticides and assessment of exposure in enclosed spaces (e.g., greenhouses, mushroom barns), where personnel are exposed to high temperatures and humidity and where aersosols do not behave as they do out-of-doors, are discussed. New and sophisticated techniques that provide better estimates of exposure are described.
- Biological monitoring of exposed laboratory and field workers is accomplished by use of methods and biomarkers that are relatively noninvasive but still reflect target-organ toxicity.
- Techniques are explored for assessing long-term neurological or neurobehavioral toxicity resulting from acute high-level (accidental) or chronic (seasonal, annual) low-level exposures. These studies, although difficult to conduct because of their subjective nature, are being accomplished by testing of workers whose exposure levels are known.

REFERENCES

Batchelor, G. S., and K. C. Walker. 1954. Health hazards involved in use of parathion in fruit orchards of North Central Washington. *AMA Arch. Ind. Hyg. Occup. Med.* 10:522–29.

Bonsall, T. L. 1985. Measurement of occupational exposure to pesticides. In *Occupational hazards of pesticide use*, ed. G. J. Turnbull, 13–33. London: Taylor and Francis.

Brady, U. E., R. Tippins, J. Perry, J. R. Young, and R. D. Wauchope. 1991. *Bull. Environ. Contam. Toxicol.* 46:343–50.

Byers, M. E., S. T. Kamble, and J. F. Witkowski. 1993. Drift during center-pivot chemigration of chlorpyrifos with and without crop oil. *Bull. Environ. Contam. Toxicol.* 51:60–67.

Draper, W. M., and J. C. Street. 1981. Drift from a commercial aerial application of methyl and ethyl parathion: An estimation of potential human exposure. *Bull. Environ. Contam. Toxicol.* 26:530–36.

Durham, W. F., H. R. Wolfe, and J. W. Elliot. 1972. Absorption and excretion of parathion by spraymen. *Arch. Environ. Health* 24:381–87.

Ecobichon, D. J. 1996. Pesticides. In *Casarett and Doull's toxicology. The basic science of poisons*. 5th ed., 643–89. New York: McGraw-Hill.

Ecobichon, D. J., and R. M. Joy. 1994. *Pesticides and neurological diseases*, 2nd ed. Boca Raton, Fla.: CRC Press.

Feldman, R. J., and H. I. Maibach. 1974. Percutaneous penetration of some pesticides and herbicides in man. *Toxicol. Appl. Pharmacol.* 28:126–32.

Fenske, R. A., and K. P. Elkner. 1990. Multi-route exposure assessment and biological monitoring of urban pesticide applicators during structural control treatments with chlorpyrifos. *Toxicol. Ind. Health* 6:349–71.

Forget, G. 1989. Pesticides: Necessary but dangerous poisons. International Development Research Council, Ottawa, Canada. *Report* 18:4–5.

Frank, R., R. A. Campbell, and G. J. Sirons. 1985. Forestry workers involved in aerial application of 2,4-dichlorophenoxyacetic acid (2,4-D) exposure and urinary excretion. *Arch. Environ. Contam. Toxicol.* 14:427–35.

Franklin, C. A., R. A. Fenske, R. Greenhalgh, L. Mathieu, H. V. Denley, J. J. Leffingwell, and R. C. Spear. 1981. Correlation of urinary pesticide metabolite excretion with estimated dermal contact in the course of occupational exposure to guthion. *J. Toxicol. Environ. Health* 1:715–31.

Hayes, W. J., Jr. 1982. *Pesticides studied in man*. Baltimore: Williams and Wilkins.

Hayes, W. J. Jr. and E. R. Laws, Jr. 1991. *Handbook of pesticide toxocology*, Vols. 1–3. New York: Academic Press.

Kaloyanova, F. P., and M. A. El Batawi. 1991. *Human toxicology of pesticides*. Boca Raton, Fla.: CRC Press.

Maibach, H. I., R. J. Feldman, T. H. Milby, and W. F. Serat. 1971. Regional variation in percutaneous penetration in man. *Arch. Environ. Health* 23:208–11.

Methner, M. M., and R. A. Fenske. 1994. Pesticide exposure during greenhouse applications, II. Chemical permeation through protective clothing in contact with treated foliage. *Appl. Occup. Environ. Hyg.* 9:567–74.

Turnbull, G. J., ed. 1995. *Occupational hazards of pesticide use*. London: Taylor and Francis.

Wojeck, G. A., H. N. Nigg, J. H. Stamper, and D. E. Bradway. 1981. Worker exposure to ethion in Florida citrus. *Arch. Environ. Contam. Toxicol.* 10:725–35.

Wolfe, H. R., J. F. Armstrong, and W. F. Durham. 1996. Pesticide exposure from concentrate spraying. *Arch. Environ. Health* 13:340–44.

Wolfe, H. R., J. F. Armstrong, D. C. Staiff, and S. W. Comer. 1972. Exposure of spraymen to pesticides. *Arch. Environ. Health.* 25:29–31.

World Health Organization (WHO). 1990. *Public impact of pesticides used in agriculture*. Geneva.

.

ON-TARGET AND OFF-TARGET DEPOSITION[†]

Christopher M. Riley

Research and Productivity Council, Fredericton, New Brunswick, Canada E3B 6Z9

Charles J. Wiesner

Wiesner and Associates, Durham Bridge, New Brunswick, Canada E6C 1K7

2-1 INTRODUCTION

It is the transference, i.e., the delivery, of the pesticide from the application equipment to the intended biological target that results in the initial widespread distribution of the pesticide into the environment. Consequently, pesticide exposure during mixing and loading operations typically is limited to only a few specific individuals, but the pesticide application process can be viewed as the first step in the use of pesticides that may result in the exposure of the general population, be they agricultural workers who may have to reenter and work amongst a treated crop or local residents who may come into contact with pesticide drift or off-target pesticide deposits.

This chapter first reviews and explains the factors and processes that affect the dispersal and deposition of pesticides. This is followed by a description of the various techniques that can be used to measure spray deposit and drift. The third and fourth components comprise a review of information from a number of spray-drift and bystander-exposure studies. The chapter concludes with a description and discussion of mathematical models that can be used to predict spray deposit, spray drift, and bystander exposure.

2-2 SPRAY TRANSPORT, DROPLET DISPERSAL, AND SPRAY DEPOSITION

2-2-1 Spray Transport and Spray Dispersal

The behavior, transport, and deposition of pesticide sprays is influenced and determined by a number of factors including gravity, ambient meteorological conditions, and, in

---

[†] Portions of this text are taken from contract reports prepared by the authors and are reproduced with the kind permission of Environment Canada.

case of aerially applied pesticides, the aircraft wake. The extent to which these factors determine the fate of sprays is largely dependent upon droplet size and the emitted droplet spectrum.

2-2-2 Droplet Size and Atomization

Every atomizer produces a range or spectrum of droplet sizes, and initial droplet size or emitted droplet spectrum is determined by formulation characteristics, the type of atomizer, and how the atomizer is operated.

At this point it may be useful to the reader to describe how droplet size information usually is presented. Figure 2-1 shows droplet size distribution data obtained for 29% isopropanol in water using the TeeJet D10-45 disc and core-type hydraulic nozzle operating at a flow rate of 3.4 l/min. This nozzle configuration is used for the aerial application of silvicultural herbicides and the data were obtained using a wind speed of 22 m/s with a nozzle orientation perpendicular to the wind direction.

Data are presented in terms of droplet numbers and volume fraction in a range of different droplet size categories. A usual feature of most commercial atomizers is the high frequency of small droplets. In the example presented, over 70% of all droplets produced by the atomizer were less than 47 μm in diameter. However, in terms of volume, these droplets represented only 1% of the emitted material.

The two types of data presented in Figure 2-1 typically are used to calculate cumulative droplet distributions which can be used to establish characteristic diameters of the emitted droplet spectrum. The most common characteristic diameter is perhaps the $D_{V_{0.5}}$ or volume median diameter (vmd). The $D_{V_{0.5}}$ divides the droplet spectrum into two equal halves in terms of volume; that is, 50% of the spray volume is emitted as droplets smaller than the $D_{V_{0.5}}$ and 50% of the spray volume is emitted as droplets greater than the $D_{V_{0.5}}$. Other characteristic diameters include the $D_{V_{0.1}}$ and $D_{V_{0.9}}$ where 10% of the emitted volume is in droplets smaller than the $D_{V_{0.1}}$ value and 90% of the emitted volume is in droplets less than the $D_{V_{0.9}}$ value. Similar characteristics of the droplet spectra in terms of droplet numbers are described using $D_{n_{0.1}}$, $D_{n_{0.5}}$, and $D_{n_{0.9}}$ values.

The most commonly used atomizer is the hydraulic nozzle. The characteristics of the nozzle tip as well as the operating pressure determine the size and uniformity of the emitted droplet spectrum. Table 2-1 illustrates the effect of nozzle type, spray pressure, and flow rate on the emitted droplet spectrum. Flat-fan, flood, and hollow-cone nozzles produce similar-size droplets and a similar volume of small droplets when compared at equal spray pressure. In the context of ground-based agricultural sprays, drift-prone droplets are considered to be below 100-μm diameter (Maybank 1992; Lagerfelt 1988). The flood nozzle tends to produce slightly larger droplets than the flat fan, whereas the flat fan produces slightly larger droplets than the hollow cone.

Nozzles with larger orifices and hence greater emission rates generally produce a coarse spray and a reduced drift-prone fraction (Table 2-2). Increasing the hydraulic pressure increases the flow rate through the nozzle tip and produces smaller droplets. Conversely, reducing the pressure increases the droplet size.

The orientation of nozzles relative to direction of travel can influence drift from aerial and airblast applications. Because of greater wind shearing when nozzles are pointed into the wind, nozzles pointed toward the direction of travel will produce smaller droplets

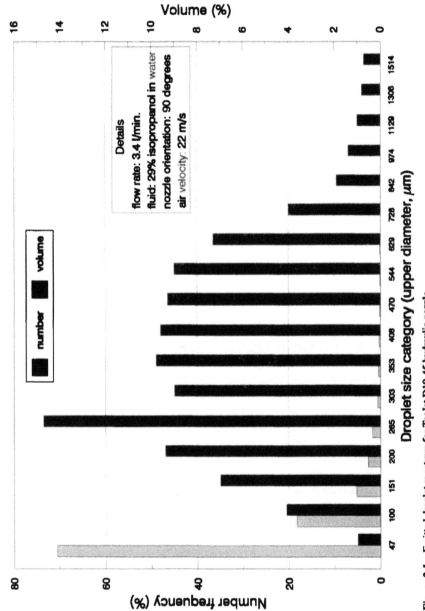

Figure 2-1 Emitted droplet spectrum for TeeJet D10-45 hydraulic nozzle.
Source: Picot and Kristmanson (1997).

Table 2-1 Influence of nozzle type and spray pressure on droplet size

Nozzle type	Delivery rates (gal./min)	Spray pressure (lb/in.^{2})	Spray angle (deg)	Volume median diameter (μm)	Volume with 100-μm diameter (%)
Flat fan (LF-2)	0.12	15	65	239	
	0.17	30	76	194	
	0.20	40	80	178	17.5
Flood (D-1)	0.12	15	90	289	
	0.17	30	115	210	
	0.20	40	125	185	15.5
Hollow cone (HC-12)	0.12	15		228	
	0.17	30		185	
	0.20	40	70	170	19.0
Whirl chamber (WRW-2)	0.12	15		195	
	0.17	30		158	
	0.20	40	120	145	23.0
Raindrop (RD-1)	0.11	15		506	
	0.16	30		358	
	0.18	40	90	310	0.8

Source: Delavan Manufacturing Co., Lexington, Tennessee.

than nozzles that are pointed straight back (Parkin and Wyatt 1981; Yates, Cowden, and Akesson 1985). With rotary atomizers the rotational speed of the atomizer together with the flow rate of the liquid typically control the emitted droplet spectrum for a given formulation. Increased rotational speeds and reduced flow rates decrease the size of the emitted droplet spectrum and increase the volume fraction of drift-prone droplets.

The type of formulation and the physical properties of the spray liquid, such as viscosity and surface tension, also affect the performance and droplet spectrum produced by different atomizers. Reductions in surface tension and viscosity typically reduce droplet size. Spray additives, usually viscosity modifiers or thickeners, have been introduced to reduce the drift-prone volume fraction of the emitted droplet spectrum. Their effects generally are not achieved by narrowing the droplet spectrum but by shifting the droplet spectrum upward to produce a spray with a vmd in the range of 300 to 700 μm. In

Table 2-2 Typical droplet size values for different nozzles and pressures

Nozzle type	Pressure (kPa)	$D_{V_{0.5}}(\mu m)$	Small droplet volume fraction (%)
650067	280	200	15
8001	280	300	5
8002	280	380	3
11001	280	210	8
11002	280	270	4
8002	210	500	2
8002LP	100	700	<1

Source: Maybank (1992).

doing so, these materials may adversely affect the spray pattern produced by hydraulic nozzle tips.

2-2-3 Droplet Physics

If an object, or droplet, is released from rest into still air, it will fall and accelerate until the aerodynamic drag equals the gravitational forces acting on it, at which point the object is said to have reached terminal sedimentation velocity. Terminal sedimentation velocity varies tremendously with the size of the droplet; for example, for a droplet of 50 μm, it is 8 cm/s, whereas for droplets of 100 μm, 250 μm, and 500 μm, terminal velocities are approximately 27, 94, and 208 cm/s, respectively.

In practice, droplets are not released from rest but have an initial vertical component that is imparted during the atomization process and a horizontal velocity component resulting from the forward speed of the delivery vehicle. These initial velocities are reduced quickly by aerodynamic drag and a droplet will be projected only a short distance into the air before it loses any initial velocity and acquires terminal sedimentation velocity.

It can be shown that the stopping distance of a droplet, that is, the distance over which a droplet loses its initial momentum, is equal to $(V_s V_0)/g$, where V_s is the sedimentation velocity (m/s), V_0 is the initial droplet velocity (m/s), and g is the acceleration due to gravity (9.8 m/s^2). Because V_s decreases rapidly with a decrease in droplet size, so does the stopping distance and thus the initial velocity only has a significant effect on large droplets, for example, of 250 μm or more. Droplets that are projected downward will decelerate to their terminal sedimentation velocities and, in the case of droplets >200 μm, are likely to be projected into the crop canopy or onto the soil. Smaller droplets have less momentum and are reduced by aerodynamic forces to their sedimentation velocities much more rapidly.

Following atomization, droplet size and hence droplet sedimentation velocities are reduced by evaporation (Goering, Bode, and Gebhardt 1972; Williamson and Threadgill 1974). This is particularly true of spray mixes that use water as the carrier but even undiluted ultralow-volume (ULV) sprays have a volatile fraction that is subject to evaporation. Evaporation reduces the size of droplets and makes them more prone to drift. The rate of evaporation of water-based pesticide tank mixes is largely independent of formulation and increases with increasing temperature and decreasing relative humidity. Figure 2-2 shows the effects of temperature and relative humidity on the evaporation of water droplets. The lower limit on droplet evaporation is determined by the nonvolatile fraction. Most pesticide sprays are applied in dilution with water and have only a small fraction of material that is nonvolatile. In the case of spray that contains over 95% water, a droplet with an initial diameter of 100 μm could evaporate to less than 36 μm. Small droplets have a larger surface area to volume ratio and hence smaller droplets evaporate more quickly than larger droplets. Larger droplets fall relatively quickly and are unlikely to be affected by evaporation unless emitted from a high altitude.

Elliott and Wilson (1983) suggested that the droplet spectrum may be divided into three different size categories based on the effects of evaporation. The first category consists of droplets for which evaporation makes no significant reduction in size before they reach the target. The second category consists of the small droplets that are initially drift

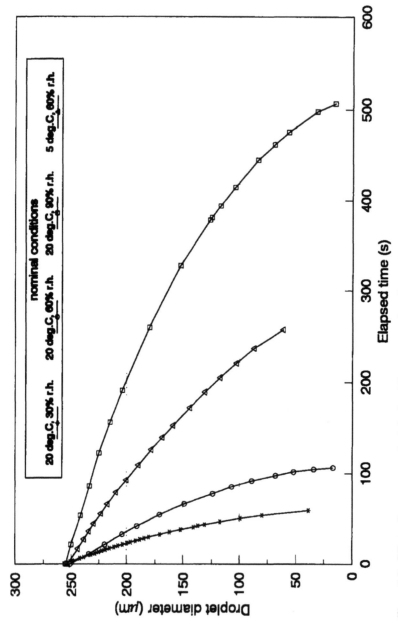

Figure 2-2 Effects of temperature and relative humidity on evaporation of water droplets suspended in an airstream.
Source: C. M. Riley (unpublished data).

prone and evaporate to residual cores of nonvolatile material with very low deposition efficiencies. The third category is the intermediate droplet range in which evaporation reduces the size of the droplets and changes the mechanism of deposition and impaction from one largely of sedimentation to one largely of turbulence. For agricultural boom spraying, the first two size categories may be considered as being greater than 150 μm and less than 50 μm, respectively. The drift potential for the intermediate size category is highly dependent upon application conditions.

Formulations and additives that do not influence droplet size at the time of atomization may reduce drift by reducing evaporation of the carrier, thereby maintaining droplet size above the critical drift-prone size.

In the absence of other effects, all spray droplets would fall under the influence of gravity and deposit within the target area. In practice, meteorological conditions and, in the case of aerial applications, the aircraft wake have an important influence on droplet behavior.

2-2-4 Meteorological Effects

In the presence of wind, droplets are displaced downwind by an amount depending on the strength of the wind and the time over which the droplet is exposed to this effect which, in turn, depends on the height of release and the sedimentation velocity of the droplet. Wind velocity typically increases with height above the ground; consequently, spray droplets released closer to the ground are affected by a lower wind velocity.

For droplets that have a large sedimentation velocity relative to the local air velocity, $S = (HU)/V_s$, where S is the position of the peak deposit (m) relative to the point of emission, H is the release height (m), U is the mean wind speed (in m/s), and V_s is the sedimentation velocity of the droplet size being considered (m/s) (Gunn et al. 1948).

Turbulence and atmospheric stability are important factors in that they modify the descent speed of droplets relative to the ground. This is particularly true for the smaller drift-prone droplets which rapidly assume the motion of local air currents. When air moves across the surface of the land, it experiences a frictional drag from the ground. This slows the air close to the ground and typically produces a logarithmic, wind-speed profile. The shearing stresses associated with the increase in wind speed with height produce eddies or frictional turbulence in the airflow. For any given wind speed the depth of the air over which this effect takes place and the magnitude of the turbulence increases with the roughness of the surface. In unstable conditions, thermally induced buoyancy increases the vertical velocity component of the turbulence (Thom 1975).

When spraying, it is important to consider the vertical velocity within the turbulence eddies. An indication of the turbulence can be obtained from the friction velocity. The friction velocity is defined as the square root of the correlation of the horizontal turbulent velocity with the vertical turbulent velocity and is designated U^*. If U^* is much larger than the terminal velocity of the droplet being sprayed, then turbulent forces will determine the motion of the droplet. Over typical agricultural surfaces under neutral conditions, U^* is approximately 10% of the mean wind speed measured at a height of 3 m (Elliott and Wilson 1983).

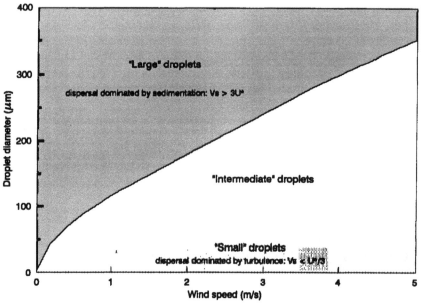

Figure 2-3 Importance of sedimentation and turbulence as dispersal mechanisms for different droplet sizes as affected by wind speed in neutral atmospheric stability conditions ($U^*/u = 0.1$).
Source: After Craymer and Boyle (1973).

The behavior of turbulent eddies is modified by atmospheric stability. In neutral conditions, the normal "lapse" situation, cooler air gently sinks, displacing lower warm air and causing vertical mixing of the atmosphere.

Under these conditions, eddies can be regarded as circular and the eddy velocity is constant at all points on the eddy. In unstable conditions, vertical movement is enhanced by buoyancy and there is increased vertical mixing of the atmosphere. Eddies are stretched vertically and there is an increase in vertical turbulent velocity relative to the neutral condition; that is, U^* is greater than the typical value of 0.1. Under stable or inversion conditions, vertical movement is dampened and may be suppressed completely, resulting in U^* values that are much less than the typical value of 0.1.

The degree to which turbulence dominates the motion of droplets is dependent upon mean wind speed. Figure 2-3 considers the deposit mechanism of different droplet sizes under different neutral wind-speed conditions. Because the sedimentation velocities of small droplets are negligible, their transport is dependent upon turbulence. Some droplets are brought to the ground earlier than would be the case by sedimentation alone whereas others are carried much farther downwind than would be the case in a no-turbulence situation. The effect of atmospheric stability is particularly important when considering the potential for drift of aerially applied pesticide sprays. Figure 2-4 demonstrates the effects of atmospheric stability on the behavior of the drift-prone fraction of the spray cloud. The effect of spraying in an inversion layer, which may be several hundreds of meters deep, is to produce a thin concentrated layer of drifting spray (Fig. 2-4A). With little or no atmospheric turbulence and only the turbulence created by the aircraft wake,

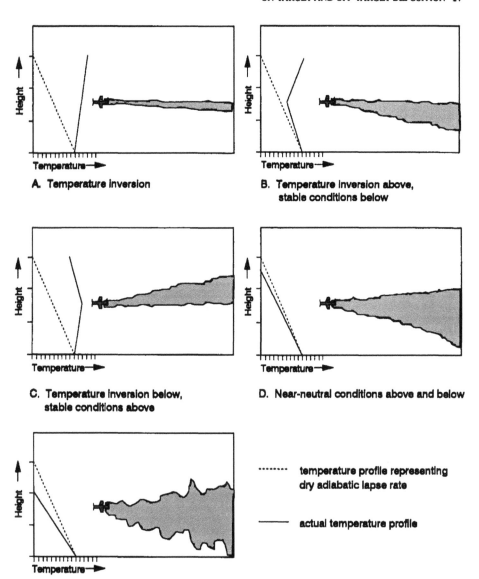

A. Temperature inversion

B. Temperature inversion above,
stable conditions below

C. Temperature inversion below,
stable conditions above

D. Near-neutral conditions above and below

E. Unstable conditions above and below

········ temperature profile representing
dry adiabatic lapse rate

——— actual temperature profile

Figure 2-4 Effects of atmospheric stability on transport of drift-prone sprays from an aircraft.
Source: Adapted from Bierly and Hewson (1962).

there is little chance of turbulent deposition and a good potential for long-range off-target drift into environmentally sensitive areas. Spraying beneath an inversion (Fig. 2-4B) is perhaps the best application condition in that the stable air above the aircraft acts as a lid, capping the spray cloud and reducing spray dispersion above the height of release, but still facilitates turbulent transport of the spray cloud down to the target area.

An inversion layer also can act as a relatively impervious barrier through which drifting spray clouds released above the layer cannot penetrate (Fig. 2-4C). Turbulence above the inversion layer mixes the spray cloud vertically but very stable air below the layer suppresses the mixing action and the spray cloud is prevented from reaching the target. In this case, there is a high potential for drift but, because the cloud is well above the ground and can disperse vertically, the chances of damage due to the drift is much less than that shown in Fig. 2-4A. The behavior of sprays released into a near-neutral or unstable atmosphere is shown in Fig. 2-4D and E. Dispersion in unstable conditions is enhanced by upward and downward motions created by thermal eddies and is more rapid in the stronger eddies of the very unstable condition. The concentration of droplets within the spray cloud released into an unstable air mass is further reduced by turbulent impaction of droplets on the ground or vegetation canopy.

In general terms, as wind speed and temperature increase and the atmosphere becomes more unstable, swath displacement and off-target deposition increases. Turbulent deposition of small droplets increases and drifting clouds of small droplets and vapor are diluted by the action of vertical diffusion. This results in a rapid decrease in drift concentrations downwind of the source. In stable, low-wind-speed conditions, vertical diffusion of the spray cloud is suppressed, droplet impaction is largely by sedimentation, and droplets too small to deposit effectively by sedimentation tend to remain airborne. In these conditions, swath displacement of pesticide sprays is minimal; however, small droplets and pesticide vapors may drift in the form of a concentrated layer which may have a high potential for damage when the drifting cloud eventually makes contact with a susceptible environment.

In some cases, the spray equipment itself can create air turbulence that entrains small droplets and worsens the drift problem. This is particularly the case with ground boom sprayers and self-propelled and truck-mounted sprayers at high speeds and also may be the reason for the increased drift of small droplets observed with some shrouded sprayers. Boom bounce caused by excessive speeds over rough terrain also increases drift and results in poor spray distribution.

The behavior and drift of aerially applied pesticide sprays are influenced by the vortices generated by the aircraft wing, the propeller, or the rotor of the helicopter (Lawson and Uk 1980; Atias and Weihs 1985; Mickle 1996). The vortices are generated as a result of the variation in lift produced across the wing of the aircraft or the rotor. The lift is greatest at the aircraft centerline and, consequently, the strongest vortices are shed at the wing tip. Individual vortices shed from the trailing edge of the wing combine or roll up to form the characteristic double-core aircraft wake (Fig. 2-5). Roll-up occurs at a distance of a few wingspans behind the wing. This corresponds to about 1 s in time (Lawson and Uk 1980). The direction of the circulation within the vortices is determined by the pressure distribution on the upper and lower surfaces of the wing and results in vortices that have a downward motion on the side closest to the aircraft centerline.

As the distance from the aircraft increases, the two contrarotating vortices interact, pushing each other toward the ground. When the vortices get to within half a wingspan or less above the ground, the vortices can bounce off the ground and the cores can move upward and outward. Farther away from the aircraft, the wake begins to decay because of atmospheric turbulence. The life span and rate of decay of the wake are dependent on

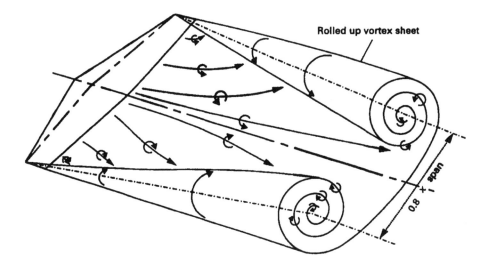

A. Fixed-wing aircraft.

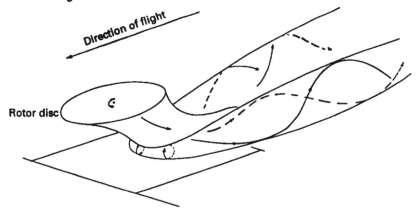

B. Rotary-wing aircraft.

Figure 2-5 Trailing vortex systems behind fixed- and rotary-wing aircraft.
Source: Elliott and Wilson (1983) after Spillman (1977).
Source: Quantick (1985) after Wooley (1962).

the strength of the wake, the stability of the atmosphere, the wind speed, and the wind direction.

The behavior of droplets released into the wake of the aircraft will depend on droplet size and the release position relative to the trailing vortex system. Vortex strength is related to the weight, the wing loading, and the speed of the aircraft but typical agricultural aircraft have vortex circulation velocities of up to 1 to 2 m/s (Atias and Weihs 1985).

Droplets with sedimentation velocities greater than the velocities in the vortex flow, that is, >450 μm, will be relatively unaffected (Wickens 1977; Trayford and Welsh 1977). However, the trajectories of smaller droplets will be dominated by the vortex flow and the spiraling vortices may move spray particles aloft with the outboard updrafts and/or down with the inboard downdrafts. Droplets that initially are not pushed down onto the target or that do not deposit by sedimentation may become entrained within the vortex system. If this happens, droplets may be carried down to, or away from, the target area until the wake decays and the droplets are released from the vortex system to come under the influence of wind speed and atmospheric turbulence.

2-2-5 Spray Deposition

Spray deposition results from inertial impaction, drop sedimentation and electrostatic attraction. These processes play different roles in the deposition of large and small drops. Electrostatic attraction is effective only over short distances but is used by some electrostatic sprayers to improve spray deposition (Law 1980).

Inertial impaction of a droplet occurs when it is carried by air motion up to an obstacle in the airflow. The air deviates around the obstacle, but the droplet may or may not be deposited there, depending on its momentum and size. The size of the obstacle also affects the collection efficiency of droplets. As the droplet size and the speed of approaching drops increase, there are proportionately more droplets deposited by inertial impaction. Also, as the size of the obstacle decreases, a larger proportion of approaching droplets are deposited.

Drop sedimentation is the gravitational settling of a droplet onto an obstacle. This occurs as a droplet descends onto an obstacle, and becomes more efficient as the droplet descent speed and obstacle size increase. The relative importance of these processes is different for large and small droplets. Generally speaking, inertial impaction is a more important deposition process for small droplets than for large ones, because of the low descent speed of small droplets. For large droplets, the reverse is usually true, and deposition by settling onto an obstacle is more important than inertial impaction.

2-2-6 Volatilization of Pesticide Sprays

Pesticide volatilization and the drift of pesticide vapors is a potentially significant source of environmental contamination and bystander exposure. Volatilization may occur from the droplet at the time of application or subsequently from spray deposits on plant and soil surfaces. Pesticide volatilization is a complex process that is largely dependent upon the vapor pressure of the pesticide and the formulation. Volatilization during air transport is promoted by small droplets, low relative humidity, and high air temperatures. Postdeposit volatility also is enhanced by high temperatures at the treated surface and by local air movement.

Volatilization during atomization and air transport is considered less important than volatilization from plant and soil surfaces. For example, Crabbe, Krzymien et al. (1980) found that, in the near field, a drifting spray cloud of aerially applied fenitrothion contained essentially no vapor-phase fenitrothion. However, following spray deposition, considerable vapor flux was detected. White et al. (1977) found that vapor losses during the application of trifluralin, a relatively volatile preemergence soil-incorporated herbicide, amounted to only 3.5% of the total applied, whereas seasonal volatilization losses from the soil were estimated at more than 22%.

Once a pesticide has volatilized, it can move with local air currents away from the target area. Unlike the drift of spray droplets, which often can be determined visually, vapor drift cannot be seen and will drift farther and over a much longer period of time than spray droplets. The subject of pesticide volatilization has received much attention in the scientific literature and those requiring more information are directed to the many extensive reviews of the subject, such as those by Grover (1991), Spencer et al. (1973), and Taylor and Glotfelty (1988).

2-3 TECHNIQUES FOR MEASUREMENT OF PESTICIDE SPRAY DEPOSIT AND DRIFT

2-3-1 Analytical Techniques

Before considering the various sampling techniques that can be used, it is appropriate to consider the range of analytical techniques that are available for the assessment of spray deposit.

The most preferable analytical techniques are those that enable a direct measurement of the active ingredient in the spray mix, i.e., gas chromatography (GC) or liquid chromatography (LC). These methods are typically very accurate and highly sensitive but can be quite costly if elaborate sample preparation techniques are required.

These costs sometimes can be reduced by making use of the stable organic tracer that can be added to the pesticide tank mix or formulated in a simulated tank mix. Nevertheless, spray deposit assessment by GC and HPLC requires specialized equipment and a high degree of technical expertise.

Inorganic tracers such as copper salts also can be used as trace chemicals. These materials can be analyzed by simple organic analytical techniques such as titration and the result is a cheaper, less sophisticated, but generally less sensitive, drift assessment technique.

Purists might argue that spray drift studies should be carried out only with the actual pesticide spray mix because the addition of any material or a tracer might alter the physical properties of the spray formulation and hence alter the emitted droplet spectrum, or the behavior and fate of the trace material might be different from that of the active ingredient in the spray mix. Despite these concerns, trace materials and particularly dyes have been used extensively in spray-drift studies. Photostable dyes or fluorescent tracers can be added to the pesticide spray mix, washed off the sampling substrates and analyzed by colorimetry or fluorometry. These techniques are suitable for the rapid, sensitive, and relatively cheap analysis of many samples.

Other analytical methods include semiquantitative immunoassay techniques that are available for a number of pesticides and formulation components. A bioassay, that is, measurement of the exposure and response of biological test organisms, represents another indirect technique that can be used with dose-response data to quantify off-target

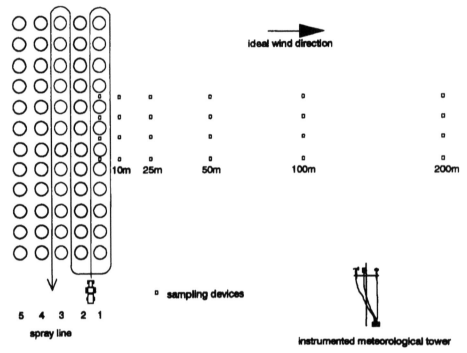

Figure 2-6 Example of a spray-drift study site.

deposit and drift. Microscopic deposit assessment techniques, such as the sizing and counting of droplet stains, typically are used to obtain qualitative rather than quantitative data.

2-3-2 Sampling Techniques for Evaluating Spray Deposit on the Ground

In general, the most preferable surface on which to assess spray deposit is the natural surface itself. Soil and ground cover such as leaf litter can be collected from a known area and the spray deposit analyzed by GC or HPLC. Typically, this requires fairly elaborate sampling, sample extraction, and cleanup procedures, and so, artificial samplers often are used. A variety of plates and sheets, such as stainless steel plates (Riley, Wiesner, and Ernst 1989), glass plates (Payne, Feng, and Reynolds 1990), Mylar sheets (Bode, and Zain 1987), polyethylene sheets (Payne et al. 1988), filter papers (Ernst et al. 1991), and aluminum foil (Crossland, Shires, and Bennett 1982) have been reported in the literature. These sampling surfaces are placed on, or just above, the ground (to prevent contamination) and typically are deployed horizontally at various distances along replicated, parallel sampling transects (Fig. 2-6). An added advantage of artificial sampling surfaces is that they enable multiple spray trials to be carried out on the same site. Once contaminated, natural surfaces on a site cannot be used for subsequent spray trials.

Spray drift is often a concern with respect to aquatic environments. Off-target spray deposits on the surface of the water can be evaluated using water samples (Crossland,

Shires, and Bennett 1982). However, interpretation of the data from such an analysis will be complicated in flowing water and by vertical mixing within the body of water. Such problems can be addressed using sequential sampling techniques and by isolating areas of the water surface using partially submerged pens or buckets. Flat artificial sampling surfaces also can be deployed on floating support structures such as anchored styrofoam sheets.

Water- or oil-sensitive cards can be deployed to provide information on the density and spectrum of the deposited droplet. If the spray formulation has been dyed, then white glossy cards can be used (Barry et al. 1996). The percentage of spray coverage, a relative measure of spray deposit, can be determined quickly and easily using image analysis equipment (Kim et al. 1988). However, if this is not available, relative estimates of deposit density by deposit class can be assigned by visual inspection (Courshee and Ireson 1961). Deposit stains can be measured and/or counted microscopically and the corresponding droplet size calculated using appropriate spread factors.

Spread factors are typically formulation and droplet-size specific. With water-based spray mixes, droplet size and composition will change with time as the droplet evaporates. Ideally, different spread factors would be determined for use at various sampling distances downwind of the target area. Glass slides covered in magnesium oxide do away with the need for elaborate spread-factor determination because impacting droplets typically form craters that have a diameter equal to approximately 1.16 times the diameter of the impacting droplet (Waite 1978).

2-3-3 Sampling Techniques for Evaluating Spray Deposit in a Plant Canopy

There are many publications that describe techniques for determining deposit within the plant canopy (e.g., de Batista et al. 1985; Cooke et al. 1990; Southcombe 1985). These techniques vary greatly depending upon the type of information required and the nature of the study, for example, sprayer performance evaluations, dose acquisition studies, deposit and efficacy studies, and spray retention experiments. In choosing techniques for determining pesticide exposure, those yielding absolute qualitative or quantitative data are of greater value than those yielding relative information.

As stated previously, the most preferable sampling surface to use is the natural surface itself. Spray deposits on nonwoody plant material can be analyzed readily by GC, HPLC, or inorganic tracer analysis. Colorimetry and/or fluorometry may be possible but dye recoveries can be low or inconsistent.

Vegetation corresponding to a known area on the ground, for example, 1 m^2, can be collected and analyzed as a bulk sample (MacNeil and Hikichi, 1986). By collecting samples from various downwind locations, an off-target deposit profile can be determined. Of course, once the vegetation has been contaminated, it cannot be used for subsequent determination unless the trace chemical in the tank mix is changed. If multiple studies of the same site are planned, areas of vegetation must be tented and protected until required. Alternatively, fresh, uncontaminated foliage can be brought in from unsprayed areas or foliage simulators may be used (Riley 1992).

Plant canopies contain vertical and horizontal elements. These elements collect a range of droplet sizes, from large ones that fall almost vertically under the influence of gravity and deposit by sedimentation, to small ones that are carried predominantly horizontally in air currents to be deposited by turbulent impaction. Consequently, foliage simulators must be calibrated and their use validated with natural foliage (Uk 1977). Foliage simulators may be deployed at various levels throughout the canopy. Total canopy deposit can be calculated if the correlation between natural and artificial surface deposit is known and if parameters such as plant density and leaf area index are documented (Picot et al. 1993).

An alternative method of assessing the potential off-target deposit on the canopy and the ground combined is to deploy a series of flat-plate collectors horizontally on a platform at the top of the plant canopy. The deposit subsequently observed on these collectors provides an estimate of the total spray flux through a horizontal plane corresponding to the top of the canopy. Except in very low wind-speed conditions, this technique provides an underestimate of spray flux in that flat-plate collectors have a poor collection efficiency for small droplets (Parkin and Merritt 1988). The technique assumes that spray droplets, having descended below the top of the plant canopy, will deposit and not subsequently reemerge. In some cases, researchers have used collectors with raised edges, such as petri dishes or foil pie tins, as an alternative to flat-plate collectors (Maybank et al. 1978). Such collectors lead to problems of interpretation in that the raised edges are efficient collectors of droplets moving horizontally, and may exert a sheltering effect on the interior flat portion of the collector.

If a dyed spray formulation is used, qualitative deposit information can be obtained directly from elements of the canopy itself. Canopy elements can be spray painted to improve deposit visibility, cards can be attached to canopy elements, or foliage simulators can be deployed.

Bioassays using potted plants also can be used to assess spray deposit within the plant canopy (Byass and Lake 1977). For herbicidal sprays, susceptible indicator plants can be used as the test organisms and, for insecticidal sprays, plants can be exposed and then returned to the laboratory and used as host plants in insect exposure assays (Sinha, Lakhani, and Davis 1990).

2-3-4 Sampling Techniques for Airborne Spray Drift

Perhaps the most widely used or accepted technique for obtaining an absolute measure of airborne spray is the Rotorod (Fig. 2-7). The Rotorod is an active sampler with a sampling surface that rotates at 2,400 rpm and sweeps the air, thereby collecting airborne particles. Various sampling surfaces or collector rods are available, depending upon sampling requirements. The combination of a high rotational speed and a narrow impaction surface (0.159 or 0.0483 cm) gives the collector rods a high collection efficiency for small, drift-prone particles (Edmonds 1972). Such sampling devices are supported at various levels above the ground from towers, masts, platforms, or tethered balloons. Spray flux is calculated from the spray deposit, corrected for wind-speed effects, and used to describe a vertical profile of time-integrated spray flux. By having such sampling arrays at different distances downwind of the target area, decreases in drift-cloud concentration due to spray deposition and vertical dilution of the spray cloud can be observed (McCooeye et al.

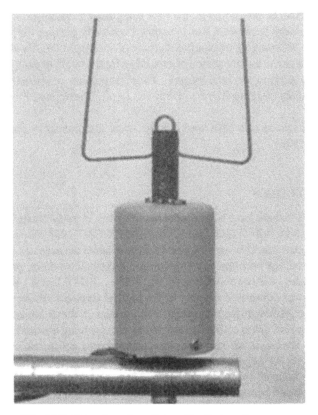

Figure 2-7 Static Rotorod sampler with type-U collector rod.

1993). Static wires (Riley and Wiesner 1990) and strings (Andrews et al. 1983) with high collection efficiencies for small droplets also have been used to obtain similar data.

Low- and high-volume air samplers or aspirators also have been used, mounted either at specific heights above the ground (Grover et al. 1985; Glotfelty et al. 1990) or in an aircraft (Riley, Wiesner, and Ecobichon 1989). Aspiration devices with filters can collect pesticide vapors as well as pesticide droplets but there are specific problems associated with nonisokinetic sampling and intake orientation which can complicate data interpretation.

Arrays of sampling devices can be used to take sequential measurements over time and hence document changes in airborne pesticide concentrations with the passing of the spray cloud (Crabbe, Elias et al. 1980).

Quantitative assessments of the drifting spray cloud can be obtained from vertically oriented, static samplers such as cards, ribbons, sensitive papers, or magnesium oxide-coated slides. Static surfaces have a low collection efficiency for small droplets at low wind speeds. Cards attached to Rotorod surfaces, magnesium oxide-covered Rotorod surfaces (May 1950), or cascade impactors (Crabbe and McCooeye 1989) have a higher collection efficiency and produce more accurate results.

The spray-drift cloud can be monitored in real time using laser technology. Rapid-acquisition lidar emits a pulsed laser beam that is scanned vertically through the spray cloud. The intensity of the returning light pulse is a measure of the droplet density within the reflecting volume of the cloud, and the period of time taken for the return of each pulse gives the distance of the cloud from the lidar location. Thus it is possible to visualize the behavior and fate of the spray cloud as it drifts downwind (Hoff, Mickle, and Fropude 1989).

Bioassays with caged insects also have been used to assess airborne spray concentrations (Pankiw and Jay 1992).

2-4 SPRAY-DRIFT STUDIES

Beginning in the early 1980s there have been concentrated efforts in many countries to review and/or document spray drift (Elliott and Wilson 1983; CCAC 1993; Nicholson 1994). Over the past five years, the U.S. agrochemical industry, under the auspices of the U.S. Spray Drift Task Force, has been undertaking a major, multimillion-dollar project on spray drift which includes over twenty extensive drift studies (SDTF 1995). A similar series of studies also was completed recently in the Federal Republic of Germany (Ganzelmeier et al. 1995). Although there are many drift studies in the published literature, there are very few that are specifically targeted at documenting bystander exposure. Studies typically document off-target deposits and airborne concentrations as a function of distance from the edge of the target area but the majority do not discuss or even mention the issue of bystander exposure. Many spray-drift studies have been carried out not to evaluate bystander exposure from spray operations but with the objective of investigating the effects of various parameters such as formulation type, meteorological conditions, droplet size, or equipment modifications on off-target deposit.

As discussed earlier, controlling droplet size and reducing the production of small droplets during application is one of the easiest ways to minimize drift. This can be achieved through the type of nozzle, the operating pressure of the nozzle, the size of the orifice, and the carrier volume. Spray nozzles that produce a coarse spray, such as raindrop, low-pressure flat fan, and flooding nozzle tips, have been demonstrated by Permin (1979) and Bode, Butler, and Goering (1976) to produce less drift than regular flat-fan nozzles (Table 2-3). Reduced nozzle pressure typically increases droplet size and has been demonstrated to reduce off-target deposit (Table 2-4) (Bode, Butler, and Goering 1976).

Sanderson et al. (1986) evaluated the drift of water- and oil-based formulations under low-humidity (17%), high-temperature (22°C) conditions and showed approximately twice as much on-target deposit and reduced airborne concentrations when ULV sprays were applied from the air using oil instead of water as a carrier. McDaniel et al. (1983) compared the drift from low-volume aerially applied aqueous sprays with that from aerially applied ULV oil-based sprays. Deposit decay profiles produced from a series of replicated trails showed greater amounts of airborne material from the oil-based spray in the first 200 m downwind of the target area. However, from 200 to 800 m, the greatest drift was observed from the water-based spray.

Table 2-3 Summary of off-target deposit comparisons by type of nozzle

	Nozzle type			
	Regular flat fan (8002) @ 207 kPa[a]	Flooding flat fan (8002LP) @ 103 kPa[b]	Low pressure (TK-2) @ 69 kPa[c]	Raindrop (33974-2) @ 276 kPa[d]
Wind speed (m/s) (range/average)	2.6–6.8/4.2	1.4–5.6/3.3	1.4–5.1/3.0	3.4–7.0/5.3
Air temperature (°C) (range/average)	11–34/21	13–34/29	18–33/26	19–31/23
Spray volume unaccounted for (% of applied)	16.5	15.5	15.6	1.9
Off-swath deposit, i.e., drift (% of total recovered)	16.5	4.0	2.6	4.4
Deposit beyond 2.4 m (% of total recovered)	1.50	0.47	0.18	0.43
Calculated deposit beyond 2.5 m when spraying 400-m width (% of total applied)	2.99	0.66	0.23	0.67

Source: Bode, Butler, and Goering (1976).
[a] Average of 8 tests.
[b] Average of 6 tests.
[c] Average of 5 tests.
[d] Average of 6 tests.

Drift-control additives can be used to increase the viscosity of aqueous-based spray mixes and increase the average droplet size produced. The drift-prone droplets cannot be completely eliminated.

Tests carried out by Bode, Butler, and Goering (1976) have shown, under a range of wind-speed conditions, an approximate fourfold reduction in drift from ground-based

Table 2-4 Summary of drift tests with flat-fan 8002 nozzles at 276, 207, and 172 kPa

	Average of 9 tests @ 276 kPa	Average of 5 tests at 207 kPa	Average of 3 tests at 172 kPa
Wind speed (m/s) (range/average)	2.2–6.3/4.1	2.6–5.1/4.0	3.2–6.4/5.0
Air temperature (°C) (range/average)	12–30/21	17–34/40[a]	21–29/26
Spray volume unaccounted for (% of applied)	17.7	19.6	20.1
Off-swath deposit, i.e., drift (% of total recovered)	14.9	7.5	20.1
Deposit beyond 2.4 m (% of total recovered)	2.79	1.43	1.81
Calculated deposit beyond 2.4 m when spraying 400-m width (% of total applied)	4.49	2.59	2.23

Source: Bode, Butler, and Goering (1976).
[a] This appears incorrect.

Table 2-5 Summary of drift tests with flat-fan nozzles at 276 and 207 kPa with and without spray thickener

	276 kPa without thickener[a]	207 kPa without thickener[b]	276 kPa with thickener[c]	207 kPa with thickener[d]
Wind speed (m/s) (range/average)	2.2–6.3/4.1	2.6–5.1/4.0	2.2–8.0/4.8	2.1–4.9/4.1
Air temperature (°C) (range/average)	12–30/21	17–34/40[e]	12–30/18	17–33/23
Spray volume unaccounted for (% of applied)	17.7	19.6	15.8	10.7
Off-swath deposit (% of total recovered)	14.9	7.5	5.9	4.2
Deposit beyond 2.4 m (% of total recovered)	2.79	1.43	0.79	0.46
Calculated deposit beyond 2.4 m when spraying 400-m width (% of total applied)	4.49	2.59	1.13	0.67

Source: Bode, Butler, and Goering (1976).
[a] Average of 9 tests.
[b] Average of 5 tests.
[c] Average of 7 tests.
[d] Average of 4 tests.
[e] This appears incorrect; a value of 23 to 24°C would seem more appropriate.

boom sprays when a thickener was used (Table 2-5). Huitink, Walker, and Lavy (1990) compared the drift of an aerially applied, aqueous spray of 2,4-D containing Nalcotrol (Nalco Chemical Co., Naperville, Illinois) drift retardant with a Visko-Rhap invert emulsion spray of 2,4-D (Agrolinz, Inc., Memphis, Tennessee). At a distance of 15 m downwind from the edge of the nominal swath, off-target deposits from the invert emulsion were less than half the magnitude of those from the thickened aqueous spray. The relative difference between the off-target deposits increased with increasing distance such that, at distances greater than 500 m, the off-target deposits for the invert formulations were only approximately 5% of those from the thickened spray.

Clark, Tessier, and Marion (1994) evaluated the drift of ethyl parathion from cranberry bogs in the northeastern United States. Cranberry bogs in this highly populated area typically are located adjacent to residential and business districts and, consequently, there is potential for adverse acute and chronic health effects from chemical trespass. The study comprised a total of four aerial applications by helicopter and two applications by chemigation (application of agrochemicals with irrigation water) each with and without spray adjuvant. High-volume air samplers and horizontal targets were used to determine airborne pesticide concentrations and off-target deposits up to 200 m downwind of the 4-ha target area. Drift was evaluated over a number of sequential sampling intervals from 10 hours prior to, and 48 hours following, the application.

The maximum concentrations measured by the active samplers were observed during the 5-min period corresponding to aerial application and then during the next 2-h sampling period between 1 and 3 h after spray application. In each case, airborne pesticide

Table 2-6 Summary of sprayer operational parameters for drift trials conducted with shielded and unshielded sprayers from 1986 to 1991

Sprayer types	Tip type[a]	Boom pressure (kPa)	Output volume (l/ha)	Boom height (cm)	Nozzle orientation	Boom width (m)
Unshielded						
Standard I	8001	275	50	45	vertical	13
Standard II	8002	275	100	45	vertical	12
Shielded						
Solid shield I[b]	8001	275	50	40	30° forward	12
Solid shield II[c]	8001	275	50	40	30° forward	10
Solid shield II	800017	550	15	40	30° forward	10
Perforated shield[d]	8001	275	50	40	30° forward	13.5
Perforated shield	11001	275	50	40	30° forward	13.5
Cones[e]	8001	275	50	45	vertical	13

Source: Wolf and Grover (1992).
[a] Spraying Systems, Wheaton, Illinois.
[b] Ag Shield Manufacturing, Benito, Manitoba.
[c] Rogers Engineering, Saskatoon, Saskatchewan.
[d] Flexi-Coil Ltd., Saskatoon, Saskatchewan.
[e] Brandt Industries Ltd., Regina, Saskatchewan.

concentrations were less than 1 $\mu g/m^3$. Off-target deposits at 23 m from the downwind edge of the target area were equivalent to approximately 9% of the nominal application rate of 841 g a.i./ha but typically decreased by an approximate order of magnitude at the 100-m sampling location. The levels of off-target pesticide residues associated with the aerial applications were approximately two to four times greater than those with chemigation, and the use of spray adjuvant reduced overall drift residues by approximately 80% with both application methods.

In recent years, there has been a lot of attention given to reducing spray drift through the design of new or modified spray equipment. Various commercially available shields and shrouds have been shown to reduce the drift from boom sprayers. Maybank, Shewchuk, and Wallace (1990) demonstrated that cone shields placed around flat-fan nozzles on boom sprayers could reduce the drift from approximately 4% to 1.5% with 110-degree nozzles and from approximately 2% down to 1% with 80-degree nozzles.

The equipment was evaluated over a range of wind speeds from 22 to 36 km/h and the authors calculated that a boom sprayer equipped with cone shields and TeeJet 8002 nozzles (Spraying Systems Co., Wheaton, Illinois) could be used to spray in winds of 40 km/h and produce no more droplet drift than unshielded sprayers at 20 km/h, that is, approximately 1%.

Wolf and Grover (1992) reported the results of almost fifty experiments in which drift from tractor-drawn sprayers equipped with four different types of shielding devices were evaluated from 1986 to 1991 (Table 2-6). Two different types of solid shields were used, one made of plastic and the other of sheet metal. Both covered the outside of the entire boom down to the crop canopy. The perforated shield sprayer was equipped with a perforated plastic shield at the front and the rear of the boom and was open across the top to allow access to the nozzles.

Table 2-7 Summary of regression line statistics and estimated airborne drift amounts at four wind speeds for various sprayers

Sprayer	Nozzle tips[a]	Number of trails	Calculated % drift at various wind speeds				% Reduction in drift compared to Standard 1
			10 km/h	20 km/h	30 km/h	40 km/h	
Standard 1	8001	10	3.0	7.7	12.4	17.1	0
Standard 2	8002	5	1.9	2.7	3.4	4.2	65
Cones[b]	8001	11	2.5	5.7	8.9	12.2	26
Solid shield[c]	8001	4	0.3	1.2	2.1	3.0	84
Solid shield[d]	8001	4	0.6	2.9	5.0	7.3	63
Perforated shield[e]	8001	5	1.2	2.8	4.3	5.8	64
Perforated shield[e]	11001	5	2.0	3.9	5.8	7.6	49
Solid shield[e]	800017	5	7.5	9.7	12.0	14.2	−26

Source: Wolf and Grover (1992).
Note: Drift is measured 5 m downwind of the swath.
[a] Spraying Systems, Wheaton, Illinois.
[b] Brandt Industries Ltd., Regina, Saskatchewan.
[c] Rogers Engineering, Saskatoon, Saskatchewan.
[d] Ag Shield Manufacturing, Benito, Manitoba.
[e] Flexi-Coil Ltd., Saskatoon, Saskatchewan.

The cone shields used were the same type previously evaluated by Maybank (1990) and served as a collar around the first 10 cm of the spray pattern from each nozzle tip. Drift from the various shielded sprayers over a wind-speed range of 10 to 35 km/h was compared to two standard types of unshielded sprayers equipped with TeeJet 8001 or 8002 nozzles.

The results presented in Table 2-7 show that, in all trials, the addition of a shielding device over TeeJet 8001 tips, operated at 275 kPa and applying spray solution at the rate of 50 l/ha, reduced airborne drift amounts compared to those from an unshielded boom. A 65 to 85% drift reduction was accomplished with the combination of solid or perforated shielding and lowering the sprayer boom. The use of protective cones with TeeJet 8001 tips without lowering the boom reduced airborne drift by 33% compared to that from an unshielded standard sprayer. Drift from very fine sprays produced by TeeJet 800017 tips operated at 615 kPa and applying 15 l/ha, protected by a solid shield, was greater than drift from an unshielded TeeJet 8001 standard sprayer. The use of TeeJet 8002 tips without the use of a shield resulted in amounts of drift similar to that from booms covered by perforated or solid shields using TeeJet 8001 tips. Off-target drift increased with increasing wind speeds for all sprayers, but the magnitude of this effect was less for shielded sprayers and coarser sprays.

Air-assisted boom sprayers have been introduce to apply broadcast sprays. These sprayers typically use an air curtain that envelops the spray in a downward-moving air mass that carries the spray to the target and reduces the effect of side winds. These systems typically have all of the basic characteristics of a conventional hydraulic nozzle boom sprayer but are equipped with an axial fan that blows air through ducting to an air slit located above the entire width of the conventional spray boom.

Table 2-8 Summary of scenarios for recent studies on spray drift of plant protection products in the Federal Republic of Germany

Target	No. of trials	Application rate (l/ha)	Driving speed (km/h)	Pressure (bar)	Wind speed (m/s)	Temp-erature (°C)	Relative humidity (%)	Sprayer type
Field crops	16[a]	300	6	2.4–2.5	0.8–3.6	10–17	57–83	Hardi 361 HardiTwin (without air assistance)
Grapevines	21[b]	400–600	6–61	8–9	1.1–3.4	14–25	43–83	Holder AS 310/TU 50 Krobath Axial
Fruit crops	61[c]	250–1000	5.5–6.6	3–17	0.1–6.4	2–22	36–90	Platz AX360 Holder TU 60 Holder TU 61 Sorarui Axial Holder Querstr Myers Axial Hardi Radial
Hops	21[d]	100–4700	0.8–2.7	16–29	0–5	16–28	34–82	Myers 20/105/Z Myers Axial

Source: From Ganzelmeier et al. (1995).
[a] 8 on bare soil, 8 on late-stage cereals. No air assistance.
[b] 10 on early-stage, 11 on late-stage vines. No air assistance on early-stage canopies.
[c] 31 on early-stage, 30 on late-stage crops.
[d] 9 on mid-stage, 12 on late-stage canopy.

Taylor, Anderson, and Cooper (1989) have reported that the Hardi Twin System (Hardi Inc., Davenport, Indiana), an air-curtain sprayer, can reduce spray drift over cereal stubble by 60% and over growing cereals by up to 90%. Increasing the sprayer speed over the range of 4 to 10 km/h increased spray drift when air assistance was not used but, with air assistance, spray drift was relatively constant. Air assistance also reduced the drift from small nozzles applying 100 l/ha to that of larger nozzles applying 200 l/ha and reduced drift in winds of 8.5 m/s to the level of 2% observed without air assistance in winds of 3 m/s. Rutherford et al. (1989) observed lower drift from air-assisted Airtec nozzles (Cleanacres Machinery Ltd., Cheltenham, UK) than from conventional hydraulic nozzles.

One of the most recent and extensive experimental studies of agricultural spray drift is described in a report by Ganzelmeier et al. (1995). The report describes 119 spray-drift trials carried out between 1989 and 1992 and documents the off-target deposit of tracer materials applied using conventional application practices to several types of crop canopy. A description of the various types of treatments and the results thereof are summarized in Tables 2-8 and 2-9.

The lowest levels of off-target deposit were measured in the trials employing hy-draulic boom and nozzle spray equipment and sprayers without air-assisted spray trans-port, that is, in applications to field crops and grapevines in the early stages of growth. The highest level of off-target deposit was observed in spraying fruit crops (mostly tree fruit) using an airblast sprayer to obtain the necessary thorough coverage of the upper

Table 2-9 Summary of off-target deposit results observed by Ganzelmeier et al. (1995)

Canopy type and development	Distance downwind of target area (m)			
	5	10	20	50
Field crops, early growth stage (bare ground)	0.35	0.20	0.06	—[a]
Field crops, late growth stage	0.75	0.33	0.12	—
Grapevines, early growth stage	1.20	0.33	0.05	0.06
Grapevines, late growth stage	3.80	1.28	0.39	—
Fruit crops, early growth stage	17.81	9.60	3.98	0.15
Fruit crops, late growth stage	10.46	4.46	1.51	—
Hops, early growth stage	14.12	4.15	0.73	0.10
Hops, late growth stage	10.86	7.37	3.49	0.31

Note: Results are expressed as percentages of the application rate and represent upper 95% percentile values based on mean values from each trial.
[a] Not measured.

canopy in the early stages of foliar development. In all scenarios, the predicted upper 95% confidence limit for the spray deposit at 50 m downwind of the treated area was less than 0.5% of the nominal application rate.

Air-blast sprayers typically have a high-speed axial fan capable of delivering 150 to 700 m^3 of air per minute. The pesticide is pumped through hydraulic or twin fluid (shear) nozzles and is introduced into the airstream produced by the high-speed fan. The air blast shatters the spray liquid into droplets and entrains them. The air blast is directed to one or both sides as the sprayer moves forward, or it may be delivered through a movable head. The air blast displaces air in the tree or space being sprayed and the droplets being carried are deposited on the target, usually trees or row crops, according to plant size, shape, and distance from the sprayer. The probability of spray drift with air-blast sprayers is much higher than with boom and nozzle spray equipment. Small droplets, high-velocity airstreams, and upward or sideward spray emissions combine to make the spray very susceptible to drift.

Several other studies document drift from air-blast sprayers used in orchards. Riley and Wiesner (1990) found that off-target spray deposition decreased rapidly with increased distance downwind of the orchard. In each of the five applications (Fig. 2-8), deposition on ground samplers decreased by approximately three orders of magnitude over a distance of 0 to 50 m (as similarly observed by Fox et al. [1990]). Regression equations of deposit relative to downwind distance were used to predict cumulative, worst-case deposits that realistically might be expected from multiple-row applications. The results indicated that spray applications made to the first five rows, closest to the downwind edge of the orchard, cumulatively make up more than 99% of off-target deposits on the ground (Fig. 2-9). Predicted deposits from multiple-row applications at downwind distances of 10 m, 50 m, and 100 m were equivalent to approximately 11.9%, 0.5%, and 0.05%, respectively, of the nominal dosage rate. These values were somewhat lower than those reported by MacCollom, Currier, and Baumann (1986).

The airborne spray cloud was sampled in each of the above experiments. However, none of the experiments was designed to address the mass balance. Fox et al. (1990)

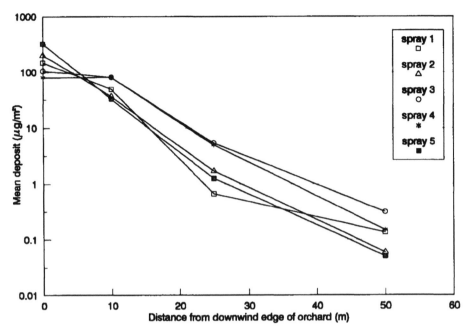

Figure 2-8 Spray deposit on flat-plate collectors.
Source: Adapted from Riley and Wiesner (1990).

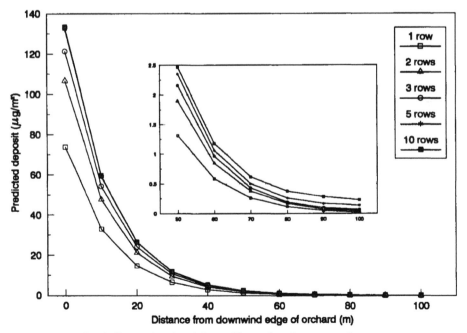

Figure 2-9 Predicted off-target deposit as influenced by number of rows sprayed.
Source: Adapted from Riley and Wiesner (1990).

concluded that the total airborne spray passing 122 m downwind of the orchard was about 3.5% of the total applied. This compares with approximately 0.3% past 50 m in the study by Riley and Wiesner (1990).

In addition to air-blast sprayers, mist blowers, concentrate sprayers, and air-carrier and air-curtain sprayers are all examples of air-assisted sprayers. In all cases, a generated airstream is used to direct, transport, and assist the impaction of the pesticide spray. Air-assisted sprayers typically are used in agriculture to apply pesticides to tree and bush fruit and to some row crops. In forestry, air-assisted sprayers are used to apply pesticides to individual trees or small areas of high-value trees such as Christmas tree plantations. Air-assisted sprayers also are used to apply larvicides and residual insecticides for control of biting flies.

As the name implies, mist blowers are used to apply very fine sprays or mists of droplets, typically 50- to 100-μm diameter, that are susceptible to drift. Nordby (1982) investigated the drift of herbicide sprays applied using a tractor-mounted mist blower. Spray emitted into a wind of 3.4 m/s resulted in spray deposit to a distance of only 5 m upwind from the point of emission but to the maximum sampling distance of 100 m downwind of the point emission.

Motorized, backpack mist blowers with tank capacities of ca. 20 l are used in small-scale treatments or in areas that are inaccessible by larger machines. They typically are used in agriculture, forestry, and public health to apply liquid sprays of fungicides and insecticides.

These sprayers have the advantage of access to places that are hard to reach. They use concentrated sprays and can cover a much greater area per tankful of concentrate than is possible with the dilute spray used in the knapsack or compressed-air sprayers. There is a greater chance for drift with mist blowers than with the portable sprayers used to apply dilute sprays. Sundaram, de Groot, and Sundaram (1987) investigated spray drift from a backpack mist blower in a crosswind of 9.3 km/h. Off-target deposit decreased exponentially and the greatest deposits were found with the first 30 m from the sprayer. Off-target deposit and airborne drift were detected beyond 100 m.

Pesticide applications made with aerosol generators and foggers are particularly prone to drift. These sprayers, sometimes referred to as ULV sprayers and used almost exclusively for insect control, emit concentrated liquid insecticide formulations in clouds of droplets that typically are less than 50 μm in diameter. Some aerosol generators, known as thermal foggers, use heat to vaporize a special oil formulation of a pesticide. As the pesticide vapor is released into the cooler air, it condenses to produce a fog of very fine droplets less than 15 μm in diameter. Other aerosol generators (cold foggers) break the pesticide into aerosols by using mechanical methods such as rapidly spinning discs, high-pressure nozzles, or strong blasts of air. This specialized equipment often is used in greenhouses, barns, and warehouses as well as in the control of biting flies and mosquitos.

Outdoor fogging operations to control adult biting flies are based on drift-spray strategies. Insecticide clouds are carried by the wind, permeating vegetation and ventilated structures, and contacting any flying insects. For best results, fogging operations should coincide with peak flight activity of the target insect. Typically, this is in the early morning, early to mid evening, or on heavily overcast days when the atmosphere is stable, winds are light, and humidity is high. Fogging in the absence of wind is undesirable because the cloud just lingers around the area of emission; fogging in high winds or unstable conditions is undesirable because the cloud is diluted rapidly by vertical and horizontal transport. Obviously, with these types of application, the potential for exposure is very great. Operators, bystanders, and animals must be kept out of the fog mist and enclosed spaces must be well ventilated before being reentered.

Aircraft often are used to apply pesticides where there are large areas to be treated, when time is very limited, or when ground-based application equipment cannot operate effectively.

The drift of aerially applied pesticide sprays has received a great deal of attention both in agriculture and forestry. In Canada, where insecticides historically have been applied over vast areas of forest, there have been extensive efforts by the Spray Efficacy Research Group (SERG) to evaluate spray drift and deposit under various meteorological conditions (Picot and Kristmanson 1997).

In the 1986 Dunphy Study, maximum spray deposits in the forest canopy were found between 100 and 200 m downwind of a fixed spray line with the largest deposits at 850 m typically being equivalent to approximately 10% of maximum deposit (Van Vliet 1987). Mean integrated spray deposits, on the ground and on the foliage, at distances of 400 m and 1200 m downwind of the spray line were approximately equivalent to 29% and 34% of the material applied from a Grumman Avenger and 23% and 29%, respectively, from a Cessna Ag Truck aircraft. The range of integrated deposit values over the eleven sprays was 14–50% at 400 m and 16–57% at 1200 m for the Grumman Avenger, and 7–45% at 400 m and 10–53% at 1200 m for the Cessna Ag Truck.

Field studies over a range of meteorological conditions produced maximum deposits at a distance of 200 to 400 m downwind of the fixed spray line with deposits at 850 m being equivalent to almost 5% of maximum deposit (Riley 1991a). Experiments in 1991 over the same site produced maximum deposits within 100 m downwind of the spray line with deposits at 400 m approximately equivalent to 10% of maximum deposit (Riley 1991b).

Mass balance experiments conducted by Crabbe and McCooeye (1989) on an immature forest site indicated that drift from aerially applied ULV sprays could be as high as 77% of the material applied at a distance of 400 m downwind of the spray line and up to 50% at a distance of 2,200 m downwind of the spray line.

Three mass balance studies to investigate the effects of various operating parameters including aircraft height, wind speed, and droplet size on the drift and deposition of herbicide sprays applied aerially over clear-cut sites have been carried out by SERG since 1988. In the latest of these studies, three Cessna 188 Ag Truck aircraft were equipped with either Thru Valve Boom (TVB), D8-46 hydraulic nozzles or Micronair AU4000 rotary atomizers to deliver sprays with approximate $D_{v0.5}$ values of 600 μm, 250 μm, and 150 μm, respectively (Riley 1992; Crabbe, McCooeye, and Stimson 1992a). Twelve experiments were completed with each aircraft under a range of atmospheric stability conditions in which wind speeds ranged from 1.7 to 5.2 m/s at aircraft height. For nine cases in which drift results were obtained, the lowest measured drift always was associated with the TVB. Drift from this spray system varied from 2% to 30% past 300 m and from 2% to 19% past 600 m over the range of conditions studied. Drift from the D8-46 spray system ranged from 19% to 67% at 300 m and from 4% to 50% at 600 m, whereas the range for the AU4000 configuration was 13% to 56% at 300 m and 14% to 68% at 600 m.

Four mass balance experiments completed by Shewchuk, Wallace, and Maybank (1991) evaluated herbicide spray drift and deposit from a helicopter equipped with D8-46 and D6-46 nozzles. Spray deposit within the designated swath ranged from 80 to 94%. Integrated drift flux beyond 50 m downwind of the spray line was calculated at less than 1% of the material emitted. Wind speeds in all cases were less than 1 m/s.

Studies by Dostie et al. (1988), Dostie and Lécuyer (1991), Payne, Feng, and Reynolds (1990), and Riley, Wiesner, and Sexsmith (1991) present off-target spray-deposit profiles for various herbicide sprays applied aerially using a range of atomizer types and wind speeds. A report by Riley (1993) describes an analysis of 200 spray-deposit profile data sets from three simulated aerial herbicide spray applications and three simulated aerial herbicide spray application studies. Deposit decay profiles typically can be described using an equation of the form $y = ax^b$, where y = deposit and x = distance.

A number of studies have evaluated the drift and deposit of aerially applied agricultural sprays. Maybank et al. (1976) examined spray drift from four herbicide sprays applied at a rate of 5.5 l/ha under high-temperature, low-humidity conditions with a Cessna Ag Wagon operating with TeeJet D4-25 nozzle tips at a pressure of 210 kPa. Initial drift, that is, drift beyond 5 m downwind from the edge of the designated swath, is stated as being around 25 to 30% of the material emitted. This compares with 5 to 10% observed from the Spracoupe (Melroe Ag Products, Bismark, North Dakota) self-propelled sprayer equipped with TeeJet 730077 nozzles at 280 kPa. Wind speeds ranged from 1.8 to 2.6 m/s in the aircraft trials and from 2.6 to 6.0 m/s in the Spracoupe trials.

In subsequent tests, the effects of wind speed (1.8 to 4.2 m/s), application volume (5 or 25 l/ha), and aircraft type on spray deposit and drift were investigated (Maybank et al. 1978). Over the 16 trials, calculated total drift flux at a distance of 5 m downwind of the nominal swath ranged from 6.3 to 44% of the total amount of material recovered. Airborne drift at 30 m ranged from 2.8 to 26.4% and at 200 m from 0.2 to 13.8%. However, the authors pointed out that, even at a distance of 5 m, the spray could have extended above the height of the sampling devices.

Crabbe and McCooeye (1985b) found that the initial deposit from aerially applied agricultural sprays was approximately 80% in wind speeds up to 2 m/s. The figure decreased to approximately 60% in wind speeds from 2 to 4 m/s. (Initial deposit is the recovered deposit to within 5 m downwind of the nominal swath of the aircraft.) Nine experiments were completed using application volumes from 23 to 38 l/ha and a range of nozzle types and orientations. Wind speeds ranged from 0.7 to 4.2 m/s. Spray flux at 200 m was measured in only three of the nine experiments and the maximum value observed was 1.4% of the emitted material.

Semmes, Cromwell, and Shoup (1990) investigated the downwind transport of aerially applied, simulated herbicide sprays using three different atomizers under high-temperature and high-humidity conditions. Airborne drift and off-target deposit was detected to a distance of 3.2 km downwind of the treated area and was found to be correlated with the emitted droplet spectrum.

Wilson, Harper, and Baker (1986) examined the drift of aerially applied insecticides in seven experiments in which they obtained mean off-target deposits of 6.8%, 1.0%, 0.13%, and 0.002% of the on-target deposit at distances of 15 m, 90 m, 165 m, and 315 m, respectively. Insecticides were applied as low-volume aqueous sprays or oil-based ULV sprays. Temperature ranged from 16 to 28°C and wind speed from 1.0 to 7.4 m/s. Off-target deposits at 100 m downwind of an aerial insecticide application on potatoes observed by Riley, Wiesner, and Ernst (1989) were $\leq 0.5\%$ of the nominal target dose.

The drift of aerial application of an ULV malathion formulation over a cotton crop was evaluated in an extensive study by Mierzejewski et al. (1993) using passive and active samplers. Deposit, dosage, and total flux values were presented at distances up to 1 km downwind of a single line spray. Spray flux at 1,000 m ranged from 0.003 to 0.512% of the amount applied and off-target deposit at 400 m ranged from 0.04 to 0.47% of the nominal application rate.

The aerial application of chemical pesticides for the control of adult biting flies is common in many countries. These applications typically rely on small droplets and drift for efficient and effective control (Mount, Biery, and Haile 1996). Dukes et al. (1985) observed 92 to 100% mortality of stable flies at distances up to 900 m downward of aerial applications of naled at a rate of 37 g a.i/ha. Even more emphatic were the results of Davies (1981) who observed over 98% mortality of tsetse flies at distances of 6 to 12 km and 64% mortality at distances of 10 to 20 km downwind of large areas treated with ULV applications of endosulfan at 6 to 12 g a.i./ha. Unfortunately, these studies present little, if any, information on quantitative spray drift at these distances.

The effects of aerial ULV applications of malathion on caged mosquitoes and honey-bees were investigated by Pankiw and Jay (1992). Highly significant correlations were obtained between insect mortality and spray deposit at distances up to 1 km downwind of the target area but no information was given on absolute deposit as a function of downwind distance. Davies and Williams (1990) present and use off-target drift results from several ground and aerial application studies to determine buffer zone widths to protect honeybees.

Absolute measures of long-range off-target drift have been obtained by several workers. Riley, Wiesner, and Ecobichon (1989) measured airborne concentrations of fenitrothion up to 4 μg/m^3 at distances of 2 to 3 km downwind of sprayed forest areas, whereas Wood and Stewart (1976) found fenitrothion residues equivalent to 0.02 to 0.15 g/ha in blueberry fields at distances of 3 to 8 km. Airborne concentrations of fenitrothion equivalent to 6 to 16% of the amount applied to forested areas were found by Crabbe, Elias et al. (1980) at a distance of 7.5 km from the target area.

Active samplers also have been used in some urban areas to monitor airborne pesticide concentrations. Results over a five-week period correlated well with the use of pesticides in the local area (Seiber, McChesney, and Woodrow 1989). Recently, growing concerns over volatile pesticide concentrations in the air have led to the development of correlation techniques to estimate pesticide volatilization, flux, and downwind concentrations (Woodrow et al. 1997).

2-5 BYSTANDER EXPOSURE

Human exposure to agricultural and forestry pesticides can occur as a consequence of occupational or non-occupational contact. Occupational exposure may result from manufacturing, formulating, packaging, and shipping operations as well as from mixing/

loading, application, reentry of treated areas, overspray, and spray drift from the operational application of pesticides in the field. Apart from the possible ingestion of pesticide residues in food, non-occupational exposure is most likely to occur through contact with pesticide particles or droplets that have drifted away from the intended target area during the application process. This latter form of contact is commonly referred to as *bystander exposure* even though the bystander may be located at a significant distance from the spray operation and may be unaware of its occurrence.

Occupational exposure to pesticides has been investigated intensively and continues to be the primary focus of concern because of the potential for very high levels of dermal and respiratory contact. In Australia it has been reported that 50 to 80% of workers mixing or loading pesticides for aerial application had severely depressed cholinesterase levels (Simpson 1973).

Numerous studies have been conducted to quantify the exposure of workers to specific pesticides in a wide range of application scenarios (e.g., Lavy, Shepard, and Bouchard 1980; Wolfe, Armstrong, and Durham 1966; Richter et al. 1992; Batchelor and Walker 1954). Half a century of such studies has led to the implementation of significant changes in the way pesticides are handled and applied and has prompted the development of new, safer pesticides and pesticide formulations.

Although the risks of pesticide exposure, that is, both the probability and the concentration, generally are manyfold higher for workers than for the bystander. The former exposes himself or herself willingly and, it is hoped, with adequate awareness of both risks and safety precautions. In contrast, the bystander is an innocent, and generally unwilling, recipient of these chemicals. Consequently and understandably, the social, political, and legal repercussions of bystander exposure tend to be far greater than might be anticipated from the actual, relatively low levels of exposure.

As the preceding sections of this review have indicated, in the past twenty years, considerable effort has been devoted to the elucidation of the mechanisms of spray drift and to the development of a detailed understanding of the factors that influence it. The ultimate distillation of our knowledge of aerial spray transport processes has been the development of computer models that, given the input of appropriate parameters such as droplet spectrum, swath width, aircraft characteristics, meteorological conditions, altitude, and/or crop or forest stand structure, will predict the pattern of spray deposit on and downwind of the target area.

Ground applications are regarded as being less likely to generate significant amounts of drift (Crabbe and McCooeye 1985b; Riley and Wiesner 1990), although, as is the case with aerial applications, inappropriate practices (e.g., spraying in high winds and during the heat of the day) can result in large off-target losses.

2-5-1 Exposure Pathways

The two main pathways by which pesticides can enter the system of the bystander are through *dermal absorption* of spray deposits on the skin and *inhalation* of airborne particles or droplets. It generally is accepted that inhalation provides a highly efficient route of entry for most chemicals. Droplets smaller than ca. 7 μm are retained by the trachea, bronchi, and alveoli; larger particles are deposited on the mucous membranes of the upper respiratory tract. These larger particles may be either absorbed by the mucous membrane or swallowed to become available for absorption in the gastrointestinal tract (Durham and Wolfe 1962; Chester and Ward 1984). Consequently, as a first approximation, it can be assumed that 100% of inspired pesticide is absorbed.

Dermal deposition has been determined directly using collection patches attached to various potentially exposed parts of the body (Durham and Wolfe 1962). Following exposure, these patches are extracted and analyzed for the pesticide using GC or LC and the results are extrapolated on the basis of the area of skin usually exposed. As much as a threefold difference in deposit level has been observed among different parts of the body (Batchelor and Walker 1954). Alternatively, dermal deposition has been estimated from spray deposits on the ground in the vicinity of the bystander by making the assumption that all exposed skin received the same deposit per square meter as the adjacent ground (Draper and Street 1981).

Numerous studies have shown that ordinary clothing is an effective barrier to dermal contact, provided that it remains unsaturated (Batchelor and Walker 1954; Wolfe 1976). Thus, only the unclothed skin need be considered as the site of potential exposure and absorption. Because the skin presents a relatively effective barrier to systemic absorption, with most pesticides only 5 to 30% of the deposited material may enter the bloodstream (Ecobichon 1995).

Respiratory exposure has been estimated either by the analysis of pesticide trapped in special respirator mask filters or, alternatively, by measurement of pesticide concentrations in air coupled with breathing-rate calculations. Exposure values determined by these two methods have been shown to agree quite well (Durham and Wolfe 1962; Durham 1976).

In the extreme cases of workers using air-blast orchard sprayers, it has been demonstrated that dermal deposit vastly exceeds inhalation. Laborers spraying parathion in orchards using air blast equipment were shown to have received dermal exposures of 8.9 mg/kg of body weight whereas their respiratory exposure totaled only 0.018 mg/kg (Batchelor and Walker 1954). A later study (Wolfe, Armstrong, and Durham 1966) found similar absolute levels, with dermal exposures ca. 1,000 times higher than respiratory exposures.

Both the total amount of exposure and its significance must, of course, be assessed in the light of specific application or drift conditions and with respect to the chemical and toxicological properties of the pesticide being applied. For example, paraquat, a specific pulmonary toxicant, is ca. 1,000 times more toxic by inhalation of respirable-size particles than by the oral route. Consequently, under certain conditions, paraquat may present a greater risk to bystanders than to workers who are exposed to greater concentrations of larger particles (Ames, Howd, and Doherty 1993).

2-5-2 Off-Target Drift

There are relatively few published studies specifically examining bystander exposure to off-target drift. This paucity of information is probably due to the focusing of regulatory attention on the higher-risk occupational exposures in which acute poisoning is quite a frequent occurrence (Simpson 1973). Indeed, because off-target drift from a wide range of typical pesticide applications has been shown to result in exponential decline of downwind deposits, it is clear that, in the majority of individual drift incidents, exposure of bystanders can be expected to fall far below occupational exposure levels (Crabbe and McCooeye 1985b; Riley, Wiesner, and Ecobichon 1989). Nevertheless, in residential areas surrounded by intensive agricultural activity, the potential for significant cumulative exposure exists. For example, nonfarming residents of some communal agricultural settlements in Israel were shown to have pesticide exposures, as indicated by urinary pesticide metabolites, only marginally lower than those of field-workers (Richter et al. 1992).

2-5-3 Studies

Bystander exposure studies fall into three general categories:

1. Studies documenting that exposure has occurred but without any attempt being made to quantify the relationship between the degree of exposure and the specific conditions of the causal pesticide application (e.g., Richter et al. 1992; Ratner and Eshel 1986).
2. Studies that measure exposure resulting from an actual, operational field application. Interpretation of this type of study is inevitably complicated by inadequate control or definition of specific application parameters and variables. Thus, the results may accurately define potential exposure in the particular conditions of the study but are likely to have limited value as a predictive aid (Draper and Street 1981; Draper, Gibson, and Street 1981). An excellent study of drift and exposure from aerial application of paraquat to cotton in California generated 1,000-fold differences in airborne concentrations of paraquat 800 m downwind of ostensibly replicate applications (Chester and Ward 1984).
3. Studies that essentially are dedicated to the gathering of drift and exposure data in which application parameters, meteorological conditions, sampling layout, and methodology are strictly controlled or defined.

As shown by the earlier drift sections of the current review, many field experiments of the latter type have been carried out in the process of developing our understanding of spray-drift mechanisms. Unfortunately, few of them have incorporated bystander exposure measurements into their experimental designs. A representative example of one that did is a National Research Council of Canada (NRCC) study comprising eleven trials measuring drift, deposit, and bystander exposure to 500 m downwind of a single spray line (Crabbe and McCooeye 1985b).

This controlled field experiment collected data from four aerial spray trials with a Grumman/Schweizer Ag-cat aircraft, five aerial trials with a Piper Pawnee, and two with a ground rig, all fitted with conventional boom and nozzle systems.

To generate a line-source spray and ensemble-averaged data for each trial, the pilots flew a single line three times over a 15-min interval. Only one pass per trial was made with the ground rig. Spray deposit was measured on stainless steel plates deployed at ground vegetation level. Airborne drift concentrations were determined using both gimbal-mounted air samplers and Rotorods at 1.5 m above ground level as well as at five levels on an 18-m sampling mast at 200 m downwind of the spray line. Bystander dermal exposure was estimated from deposit on stainless steel badges fitted on clothed dressmaker forms. Meteorological instrumentation on an 18-m mast measured mean vertical profiles of wind speed, temperature gradient, turbulent momentum, and heat flux. During the trials, meteorological conditions varied from light wind speed with thermally unstable flow to moderate wind speed flow with slight lapse and inversion. Wind speed at 2 m above ground level ranged from 0.7 to 4.2 m/s.

The authors concluded that, within experimental error, aircraft differences between the double-wing Ag-cat and the single-wing Pawnee had no influence on drift. Cumulative deposit to 75 m ranged from 64.1 to 94.0% for the aerial applications and 85.5 to 110.9% for the ground applications. Light winds on the order of 1 m/s resulted in less than 1% drift at 200 m.

Bystander dermal exposure increased with wind speed but fell rapidly with distance. The authors calculated that dermal deposit at 75 m would total ca. 600 μg on a calculated 3,500-cm^2 exposed frontal area from a multiswath field spray in a wind speed of 2 m/s. These values were based upon an active ingredient application rate of 50 g/l and 30 l/ha, and an effective skin exposure area of 3,500 cm^2. Respiratory exposure, estimated from time-integrated atmospheric concentrations and a breathing rate of 20 l/min, was determined to be negligible in comparison.

Aircraft height is a critical operational parameter that has been shown repeatedly to be one of the most important factors influencing drift (Picot et al. 1993; Teske and Barry 1993). Because this parameter was not varied as part of the NRCC study and the spray aircraft flew consistently with their wheels within approximately half a meter off the ground, it is clear that higher spray heights would result in significantly greater drift and exposure.

2-6 MODELS FOR PREDICTING SPRAY BEHAVIOR

Over the past 20 years, there has been much progress in the development of mathematical models to simulate the behavior of pesticide sprays. These models are used as decision support systems by spray applicators, resource managers, researchers, and regulatory agencies to predict the dispersion and deposition of pesticide sprays. Much of the effort has been concentrated on the development of models to predict the behavior and fate of aerial spray applications. A prime example of such a model is the Forest Service–Cramer–Barry–Grim (FSCBG) model which has been heavily supported by the USDA Forest Service and the U.S. Army (Teske 1996). The FSCBG model incorporates the agricultural dispersal (AGDISP) model (Bilanin et al. 1989) and is the basis of the AGDRIFT model which is being developed and validated as a regulatory model to be used in North America. Other aerial application models include the Picot–Kristmanson–Basak-Brown–Wallace (PKBW) model (Picot and Kristmanson 1997) and the model of Atias and Weihs (1985).

Each of the above models requires the entry of detailed droplet spectrum data, meteorological information, and various application and equipment parameters. The models differ in their capabilities to differentiate between total surface deposit versus ground or canopy deposit and to predict airborne concentrations. Data typically can be

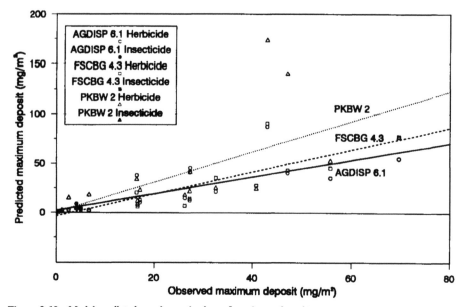

Figure 2-10 Model-predicted vs. observed values of maximum deposit.
Source: Reproduced with kind permission of Environment Canada from Riley (1995).

presented either numerically or graphically. The PKBW model has a graphical interface that enables the user to view spray deposition and airborne concentrations as functions of time and distance.

Various aerial spray models have been described and compared in several publications (de Brossia, M. 1986; Bilanin and Teske 1987; Atias and Weihs 1985; Mickle 1987). A recent report by Riley (1995) for Environment Canada compared predicted off-target deposit values from the AGDISP, FSCBG, and PKBW models with deposit profile data from eighteen field experiments. The accuracy of the above models was calculated using four characteristics of the spray deposit profile, that is, the size of the maximum deposit, the location of the maximum deposit, the total accumulated or integrated deposit, and the medium integrated deposit location. Regression analysis of the data indicated that all of these models typically predicted values for each of the four characteristic parameters that differed from observed results by a factor of two or less (Figs. 2-10 to 2-13).

Unlike FSCBG 4.3 and PKBW 2, the AGDISP model also can be used to predict the behavior of sprays from ground applications in terms of total surface deposit. Other computer simulation models for ground applications include those discussed by Holterman and Van de Zande (1996), Miller and Hadfield (1989), and Walklate (1987). Huddleston et al. (1996) describe a model being developed for the U.S. Environmental Protection Agency to predict the deposit and drift from orchard air-blast sprayers.

The authors were unable to identify any models or routine regulatory procedures for predicting bystander exposure on the basis of spray-drift data. In view of the abundance of spray-drift studies carried out over the past two decades and the excellent spray-drift models that have been developed from drift study data, it would seem logical to extend

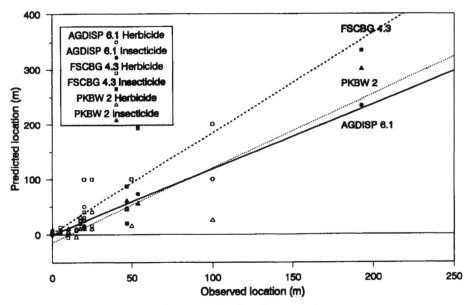

Figure 2-11   Model-predicted vs. observed location of maximum deposit.
Source: Reproduced with kind permission of Environment Canada from Riley (1995).

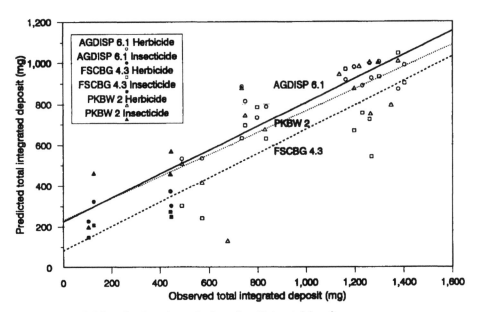

Figure 2-12   Model-predicted vs. observed values of total integrated deposit.
Source: Reproduced with kind permission of Environment Canada from Riley (1995).

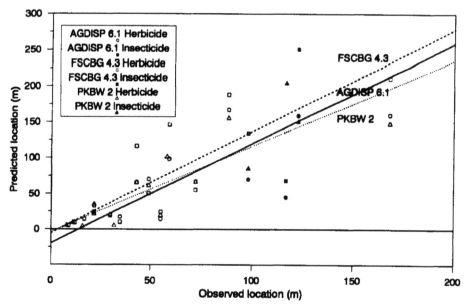

Figure 2-13 Model-predicted vs. observed location for 50% integrated deposit.
Source: Reproduced with kind permission of Environment Canada from Riley (1995).

the utility of those models by incorporating into them some form of bystander exposure module. Because dermal deposit has been shown, with very few exceptions, to be the overwhelming contributor to human exposure, it is likely that relatively simple relationships could be developed to convert off-target deposit calculations to some standard dermal exposure. Although these relationships would require experimental validation, the process would provide a permanent linkage to the credible and scientifically rigorous drift models.

REFERENCES

Ames, R. G., R. A. Howd, and L. Doherty. 1993. Community exposure to a paraquat drift. *Arch. Environ. Health* 48:47–52.

Andrews, M., L. S. Flower, D. R. Johnston, and C. R. Turner. 1983. Droplet assessment and insecticide drift studies during the large scale aerial application of endosulfan to control *Glossina morsitans* in Botswana. *Trop. Pest Manage.* 29:239–48.

Atias, M., and D. Weihs. 1985. On the motion of spray drops in the wake of an agricultural aircraft. *Atomization Spray Technol.* 1:21–36.

Barry, J. W., M. E. Teske, H. W. Thistle, P. J. Skyler, and A. Z. MacNichol. 1996. Spray deposition and drift in orchards. Paper no. AA96-006 presented at National Agricultural Aviation Association/American Society of Agricultural Engineers Joint Technical Session, Reno, Nevada, December 9, 1996.

Batchelor, G. S., and K. C. Walker. 1954. Health hazards involved in use of parathion in fruit orchards of North Central Washington. *AMA Arch. Ind. Hyg. Occup. Med.* 10:522-29.

Bierly, T. A., and Hewson, E. W. (1962). Some restrictive meteorological conditions to be considered in the design of stacks. *J. Appl. Meteorol.* 1:383–390.

Bilanin, A. J., and M. E. Teske. 1987. Comparisons of AGDISP code predications with program WIND deposition results. In *Proceedings of the symposium on the aerial application of pesticides in forestry*, Report no. NRC 29197 (AFA-TN-18) Ottawa, Ontario: National Research Council of Canada, 139–43.

Bilanin, A. J., M. E. Teske, J. W. Barry, and R. B. Ekblad. 1989. AGDISP: The aircraft spray dispersion model, code development and experimental validation. *Trans. Am. Soc. Agric. Eng.* 32:327–34.

Bode, L. E., B. J. Butler, and C. E. Goering. 1976. Spray drift and recovery as affected by spray thickener, nozzle type and nozzle pressure. *Trans. Am. Soc. Agric. Eng.* 19 (2):213–18.

Bode, L. E., and S. B. M. Zain. 1987. *Spray drift deposits from low volume application using oil and water carriers. Vol. 7 of Pesticide formations and application systems*, ASTM STP 968, edited by G. B. Beestman and D. I. B. Vander Hooven, 93–107. Philadelphia: American Society for Testing and Materials.

Byass, J. B., and J. R. Lake. 1977. Spray drift from a tractor-powered field sprayer. *Pestic. Sci.* 8:117–26.

Chester, G., and R. J. Ward. 1984. Occupational exposure and drift hazard during aerial application of paraquat to cotton. *Arch. Environ. Contam. Toxicol.* 13:551-63.

Clark, J. M., D. M. Tessier, and J. R. Marion. 1994. Mitigation of airborne parathion residues from treated cranberry bog environments bordering suburban areas by a spray adjuvant. *J. Environ. Sci. Health* A29 (1):215–36.

Cooke, B. K., E. C. Hislop, P. J. Herrington, N. M. Western, and F. Humpherson-Jones. 1990. Air-assisted spraying of arable crops, in relation to deposition, drift and pesticide performance. *Crop Prot.* 9: 303–11.

Coordinating Committee on Agricultural Chemicals (CCAC). 1993. Review of agricultural chemical spray drift. Commonwealth Scientific and Industrial Research Organization Report, East Melbourne, Australia.

Courshee, R. J., and M. J. Ireson. 1961. Experiments on the subjective assessment of spray deposits. *J. Agric. Eng. Res.* 6:175–82.

Crabbe, R., L. Elias, M. Krzymien, and S. Davie. 1980. New Brunswick forestry spray operations: Field study of the effect of atmospheric stability as long range pesticide drift. Laboratory Technical Report LTR-UA-52, National Research Council, Ottawa, Ontario.

Crabbe, R., M. Krzymien, L. Elias, and S. Davie. 1980. New Brunswick spray operations: Measurement of atmospheric fenitrothion concentrations near the spray area. Laboratory Technical Report no. LTR-UA-56, National Aeronautical Establishment, National Research Council of Canada, Ottawa, Ontario.

Crabbe, R. S., and M. McCooeye. 1985a. Effect of atmospheric stability and windspeed on wind drift in aerial forest spray trials, neutral to unstable conditions. Report no. LTR-UA-82, National Research Council, Ottawa, Ontario.

Crabbe, R. S., and M. McCooeye. 1985b. A field study of ground deposition, wind drift and bystander exposure from agricultural aircraft spray emissions. Report no. 24745, National Research Council, Ottawa, Ontario.

Crabbe, R. S., and M. McCooeye. 1989. Kapuskasing field study relating atmospheric stability to wind drift from aerial forest spray operations. Laboratory Technical Report no. LTR-UA-99, National Research Council, Ottawa, Ontario.

Crabbe, R. S., M. McCooeye, and E. B. Stimson. 1992. Measurement of drift in the 1992 Sevogle Field Study: A preliminary report. Report no. L.O. 5520, National Research Council, Ottawa, Ontario.

Craymer, H. E., and D. G. Boyle. 1973. The micrometeorology and physics of spray particle behaviour. Pesticide Spray Technology Workshop, Emeryville, Calif.

Crossland, N. O., S. W. Shires, and D. Bennett. 1982. Aquatic toxicology of cypermethrin. III. Fate and biological effects of spray drift deposits in fresh water adjacent to agricultural land. *Aquat. Toxicol.* 2:253–70.

Davies, J. E. 1981. Insecticide drift and reinvasion of spray blocks in aerial spraying experiments against *Glossina morsitans centralis* Machado (Diptera: Glossinidae). *Bull. Entomol. Res.* 71:499–508.

Davis, B. N. K., and C. T. Williams. 1990. Buffer zone widths for honeybees from ground and aerial spraying of insecticides. *Environ. Pollut.* 63:247–59.

de Batista, G. C., J. H. Stamper, H. N. Nigg, and J. L. Knapp. 1985. Dislodgeable residues of carbophenothion in Florida citrus: Implications for safe worker reentry. *Bull. Environ. Contam. Toxicol.* 35:213–21.

de Brossia, M. 1986. Selected mathematical models in environmental impact assessment in Canada. Canadian Environmental Assessment Research Council Report, Ottawa, Ontario.

Dostie, R., and H. Lécuyer. 1991. Dépôt du glyphosate à l'intérieur et à l'extérieur des aires traitées par voie aérienne en 1988. Gouvernement du Québec, Direction de l'Environment, Ministére des Forêts. 29 p.

Dostie, R., H. Lécuyer, L. Major, and G. Mamarbachi. 1988. Évaluation du dépôt et de la dérive du glyphosate lors de pulérisations àeriennes en 1987. Publ. no. 3324, Gouvernement du Québec, Ministére des Forêts, Service du Suivi Environnemental.

Draper, W. M., R. D. Gibson, and J. C. Street. 1981. Drift from and transport subsequent to a commercial, aerial application of carbofuran: An estimation of potential human exposure. *Bull. Environ. Contam. Toxicol.* 26:537–43.

Draper, W. M., and J. C. Street. 1981. Drift from a commercial aerial application of methyl and ethyl parathion: An estimation of potential human exposure. *Bull. Environ. Contam. Toxicol.* 26:530–36.

Dukes, J. C., C. F. Hallmon, J. P. Ruff, and J. C. Moore. 1985. Downwind drift and droplet distribution of Naled aerial sprays applied for stable fly control over Gulf beaches. *J. Fla. Anti-Mosquito Assoc.* 56:86–90.

Durham, W. F. 1976. Human health hazards of respiratory exposure to pesticides. In *Air pollution from pesticides and agricultural processes*, ed. R. E. Lee, Jr. Cleveland, Ohio: CRC Press.

Durham, W. F., and H. R. Wolfe. 1962. Measurement of the exposure of workers to pesticides. *Bull. WHO* 26:75–91.

Ecobichon, D. J. 1995. Aerial application of agricultural chemicals and off-target drift. In *Aerial application of pesticides in agriculture—set backs from habitation*. Report, New Brunswick Department of Agriculture and Rural Development Committee Fredericton, New Brunswick, Canada.

Edmonds, R. L. 1972. Collection efficiency of Rotorod samplers for sampling *Figures fugus* spores in the atmosphere. *Plant Dis. Rep.* 52:704–08.

Elliott, J. G., and B. J. Wilson, eds. 1983. *The influence of weather on the efficiency and safety of pesticide application: The drift of herbicides.* Occasional Publ. no. 3, British Crop Protection Council, Croydon, UK.

Ernst, W. R., P. Jonah, K. Doe, G. Julian, and P. Hennigar. 1991. Toxicity to equatic organisms of off-target deposition of endosulfan applied by air. *Environ. Toxicol. Chem.* 10:103–14.

Ford, R. J. 1986. Field trials of a method for reducing drift from agricultural sprayers. *Can. Agric. Eng.* 28:81–83.

Fox, R. D., R. D. Brazee, D. L. Reichard, and F. R. Hall. 1990. Downwind residue from air spraying of a dwarf apple orchard. *Trans. Am. Soc. Agric. Eng.* 33 (4):1104–08.

Ganzelmeier, H., D. Rautmann, R. Spangenburg, M. Streloke, M. Herrmann, H. J. Wenzelburger, and H. F. Walter. 1995. Studies on the spray drift of plant protection products. Blackwell Wissenshafts—Verlag GmbH Berlin/Wien.

Glotfelty, D. E., C. J. Schomburg, M. M. McChesny, J. C. Sagebiel, and J. N. Siebor. 1990. Studies of the distribution, drift and volatilization of diazinon resulting from spray application to a dormant peach orchard. *Chemosphere* 21:1303–14.

Goering, C. E., L. E. Bode, and M. R. Gebhardt. 1972. Mathematical modelling of spray droplet and evaporation. *Trans. Am. Soc. Agric. Eng.* 15 (2):220–25.

Grover, R. 1991. *Nature, transport, and fate of airbrone residues.* In *Environmental chemistry of herbicides*. Vol. 2. ed. R. Grover and A. J. Cessna, 89–117. Boca Raton, Fla.: CRC Press.

Grover, R., S. R. Shewchuck, A. J. Cessna, A. E. Smith, and J. H. Hunter. 1985. Fate of 2,4-D iso-octyl ester after application to a wheat field. *J. Environ. Qual.* 14:203–10.

Gunn, D. L., J. F. Graham, E. C. Jacques, F. C. Perry, W. G. Seymour, T. M. Telford, J. Ward, E. N. Wright, and D. Yeo. 1948. Aircraft spraying against desert locust in Kenya 1945. *Anti-Locust Bull.* 4. Tropical Pesticides Research Unit, Porton Down, Salisbury, UK.

Hoff, R. M., R. E. Mickle, and F. A. Fropude. 1989. A rapid acquisition lidar for aerial spray diagnostics. *Trans. Am. Soc. Agric. Eng.* 32:1523–28.

Holterman, H. J., and J. C. Van de Zande. 1996. Drift reduction in crop protection: Evaluation of technical measures using a drift model. In *Proceedings of the British Crop Protection Council*, Croydon. Conference proceedings (Brighton Crop Protection Conference), Brighton, UK. 111–16.

Huddleston, E. W., T. M. Ledson, J. B. Ross, R. Sanderson, D. R. Miller, W. E. Steinke, D. E. Aylor, S. G. Perry. 1996. Development of a pesticide drift model for orchard airblast spraying. In *Proceedings of the British Crop Protection Council.* Croydon, UK. 1187–92.

Huitink, G., J. T. Walker, and T. L. Lavy. 1990. Downwind deposition of 2,4-dichlorophenoxyacetic acid herbicide (2,4-D) in invert emulsion. *Trans. Am. Soc. Agric. Eng.* 33 (4):1051–56.

Kim, C. H., D. R. Mackenzie, L. J. Jordon, and K. C. Collin. 1988. Tests of controlled droplet applicators for the control of potato late blight with less fungicide drift. *Appl. Agric. Res.* 7:24–28.

Lagerfelt, P. 1988. Spray drift deposits during application with field crop sprayers. *Weeds and Weed Contr.* 1:183–93.

Lavy, T. L., J. S. Shepard, and D. C. Bouchard. 1980. Field worker exposure and helicopter spray pattern of 2,4,5-T. *Bull. Environ. Contam. Toxicol.* 24:90–96.

Law, S. E. 1980. Droplet charging and electrostatic deposition of pesticide sprays—research and development in the U.S.A. In *Spraying systems for the 1980's.* BCPC Monograph Series, no. 24, ed. J. O. Walker, 85–94. Croydon, UK: British Crop Protection Council.

Lawson, T., and S. Uk 1980. The influence of the aircraft wake on the downwind dispersal of ULV sprays. *Agric. Aviat.* 21 (2):59–67.

MacCollom, G. B., W. E. Currier, and G. L. Baumann. 1986. Drift comparisons between aerial and ground orchard application. *J. Econ. Entomol.* 79 (2):459–64.

MacNeil, J. D., and M. Hikichi. 1986. Phosmet residues in an orchard and adjacent recreational area. *J. Environ. Sci. Health* B21:375–85.

May, K. R. 1950. The measurement of airborne droplets by the magnesium oxide method. *J. Sci. Instrum.* 2:128–30.

Maybank, J. 1992. The physics of spray drift. In *Proceedings of Appli-Tech '92. Agricultural Chemical Application Technology for the 90's*, ed. F. A. Holm. University of Saskatchewan Extension Division, Regina.

Maybank, J., S. R. Shewchuk, and K. Wallace. 1990. The use of shielded nozzles to reduce off-target herbicide spray drift. *Can. Agric. Eng.* 32:231–35.

Maybank, J., K. Yoshida, and R. Grover. 1978. Spray drift from agricultural pesticide applications. *J. Air Pollut. Contr. Assoc.* 28:1009–14.

Maybank, J., K. Yoshida, S. R. Shewchuk, and R. Grover. 1976. Comparison of swath deposit and drift characteristics of groundrig and aircraft herbicide spray systems. Report of the 1975 field trials. Report no. P76-1, Saskatchewan Research Council, Saskatoon.

Maybank, J., K. Yoshida, S. R. Shewchuk, and R. Grover. 1978. Spray drift behaviour of aerially-applied pesticide: Report of the 1977 field trials. Report no. P. 78-2, Saskatchewan Research Council, Saskatoon.

McCooeye, M. A., R. S. Crabbe, R. E. Mickle, A. Robinson, E. B. Stimsen, J. A. Arnold, and D. G. Alword. 1993. Strategy for reducing drift of aerially applied pesticides. *Report*. National Research Council, Ottawa, Ontario.

McDaniel, S. G., B. M. McKay, R. K. Houston, and L. D. Hatfield, 1983. Aerial drift profile of oil and water sprays. *Agric. Aviat.* 10 (2):25–29.

Mickle, R. E. 1987. A review of models for ULV spraying scenarios. In *Proceedings of the symposium on the aerial application of pesticides in forestry*. Report no. 29179 (AFA-TN-18), 179–88. G. W. Green National Research Council of Canada, Ottawa, Ontario.

Mickle, R. E. 1996. Influence of aircraft vortices on spray cloud behaviour. *J. Am. Mosquito Contr. Assoc.* 12 (2):372–79.

Mierzejewski, K., W. G. Yendol, W. McLane, M. Legendre, J. Ford, B. Tanner, T. Roland, and K. Ducharme. 1993. Study of off-site deposition of malathion using operational procedures for the Southeastern Cotton Boll Weevil Eradication Program. Aerial Application Technology Laboratory Report no. AATL 93-3. Pennsylvania State University. Stare College, PA.

Miller, P. C. H., and D. J. Hadfield. 1989. A simulation model of the spray drift from hydraulic nozzles. *J. Agric. Eng. Res.* 42:135–47.

Mount, G. A., T. L. Biery, and D. G. Haile. 1996. A review of ultralow-volume aerial sprays of insecticide for mosquito control. *J. Am. Mosquito Contr. Assoc.* 12 (4):601–18.

Nicholson, I. 1994. The Canadian Interdepartmental Task Force on Pesticide Drift. Paper presented at symposium, *Overview of spray drift and deposit*. Annual meeting of the Canadian Association of Pesticide Control Official (CAPCO), Ottawa, 25 October. Pub.: Natural Resources Canada, Canadian Forest Service, Hull, Quebec.

Nordby, A. 1982. Investigations using a tractor-mounted mist-blower in control of brushwood. Deposition of spray, reach, drift. In *Weeds and weed control: proceedings of the 23rd Swedish Weed Conference*, Uppsala, Sweden.

Pankiw, T., and S. C. Jay. 1992. Aerially applied ultra-low-volume malathion effects on caged honey bees

(Hymenoptera: Apidae), caged mosquitoes (Diptera: Culicidae), and malathion residues. *J. Econ. Entomol.* 85 (3):687–91.

Parkin, C. S., and C. R. Merritt. 1988. The measurement and prediction of spray drift. *Aspects Appl. Biol.* 17:351–60.

Parkin, C. S., and J. C. Wyatt. 1981. The measurement of drop spectra in agricultural sprays using a particle measuring systems optical array spectrometer. In *Spraying systems for the 1980's*, ed. J. O. Walker. Monograph Series no. 24, 241–250. Croydon, UK: British Crop Protection Council.

Payne, N. J., J. C. Feng, and P. E. Reynolds. 1990. Off target deposits and buffer zones required around water for aerial glyphosate applications. *Pestic. Sci.* 30:183–98.

Payne, N. J., B. V. Helson, K. M. S. Sundaram, and R. A. Fleming. 1988. Estimating buffer zone widths for pesticide applications. *Pestic. Sci.* 24:147–61.

Permin, O. 1979. The extent of spray drift and the effect on weeds when spraying herbicides with foam nozzles or raindrop nozzles. In *Weeds and weed control: Proceedings of the 20th Swedish Weed. Conference*, Uppsala, Sweden.

Picot, J. J. C., and D. D. Kristmanson. 1997. *Forestry pesticide aerial spraying*. Environmental Science and Technology Library, vol. 12. Dordrecht, The Netherlands: Kluwer Academic.

Picot, J. J. C., D. D. Kristmanson, R. E. Mickle, R. B. B. Dickison, C. M. Riley, and C. J. Wiesner. 1993. Measurements of foliar and ground deposits in forestry aerial spraying. *Trans. Am. Soc. Agric. Eng.* 36:1013–24.

Quantick, H. R. 1985. *Aviation in crop protection, pollution and insect control*. London: Collins.

Ratner, D., and E. Eshel. 1986. Aerial pesticide spraying: An environmental hazard. *J. Am. Med. Soc.* 256:2516–17.

Richter, E. D., P. Chuwers, Y. Levy, M. Gordon, F. Grauer, J. Marzouk, S. Levy, S. Barron, and N. Greuner. 1992. Health effects from exposure to organophosphate pesticides in workers and residents in Israel. *Isr. J. Med. Sci.* 28:584–97.

Riley, C. M. 1991a. The effects of aircraft vortices on spray deposition and drift. Report no. C/89/632. New Brunswick Research and Productivity Council, Fredericton, N.B., Canada.

Riley, C. M. 1991b. A quantitative and qualitative field evaluation and comparison of three aerial insecticide application systems used in forest management. Report no. C/91/1195. New Brunswick Research and Productivity Council, Fredericton, N.B., Canada.

Riley, C. M. 1992a. Continuation of studies to examine the effects of aircraft application parameters on deposit and drift of forestry herbicides. Report no. C/92/443 (a). New Brunswick Research and Productivity Council, Fredericton, N.B., Canada.

Riley, C. M. 1992b. Development and compilation of the Spray Drift Database for the Canadian Interdepartmental Task Force on Spray Drift. RPC contract report for Environment Canada.

Riley, C. M. 1993. Continued development of the Environment Canada Spray Drift Database. Report no. C/93/301, Research and Productivity Council, Fredericton, N.B., Canada.

Riley, C. M. 1995. A sensitivity analysis and validation of the AGDISP 6.1, FSCBG 4.3 and PKBW2 spray drift and deposit models. Report no. B/95/0008, New Brunswick Research and Productivity Council, Fredericton, N.B., Canada.

Riley, C. M., and C. J. Wiesner. 1990. *Off-target pesticide losses resulting from the use of an air-assisted orchard sprayer*. Vol. 10 of *Pesticide formulations and application systems*. ASTM STP 1078, edited by L. E. Bode, J. L. Hazen, and D. G. Chasin. Philadelphia: American Society for Testing and Materials.

Riley, C. M., C. J. Wiesner, and D. J. Ecobichon. 1989. Measurement of aminocarb in long distance drift following aerial application to forests. *Bull. Environ. Contam. Toxicol.* 42:891–98.

Riley, C. M., C. J. Wiesner, and W. R. Ernst. 1989. Off-target deposition and drift of aerially applied agricultural sprays. *Pestic. Sci.* 26:159–66.

Riley, C. M., C. J. Wiesner, W. A. Sexsmith. 1991. Estimating off-target spray deposition on the ground following the aerial application of glyphosate for conifer release in New Brunswick. *J. Environ. Sci. Health* B26 (2):185–208.

Rutherford, I., G. J. Bell, J. B. S. Freer, P. J. Herrington, and P. C. H. Miller. 1989. An evaluation of chemical application systems. *Proceeding of the British Crop Protection Council*, Brighton Crop Protection Conference, 601–13. Brighton, UK.

Sanderson, R., E. W. Huddleston, J. B. Ross, J. A. Henderson, and E. W. Ferguson. 1986. Deposition and drift of Pydrin® in cottoneed oil and water under and conditions applied with a dual spray system aircraft. *Trans. Am. Soc. Argic. Eng.* 29 (2):378–81.

Seiber, J. N., M. M. McChesney, and J. E. Woodrow. 1989. Airborne residues resulting from use of methyl parathion, molinate and thiobencarb on rice in the Sacramento Valley, California. *Environ. Toxicol. Chem.* 8:577–88.

Semmes, R. H., R. P. Cromwell, and W. D. Shoup. 1990. Downwind transport of aerially applied sprays under high temperature, high humidity conditions. *Trans. Am. Soc. Agric. Eng.* 36 (3):257–61.

Shewchuk, S. R., K. Wallace, and J. Maybank. 1991. Spray drift and deposit pattern from a forest herbicide application. Report no. E-2310-4-E-90. Saskatchewan Research Council, Saskatoon.

Simpson, G. R. 1973. Aerial spraying of organic phosphate pesticides lowered blood cholinesterase levels of aerial spray operators at Wee Waa. *Med. J. Aust.* 1:735–36.

Sinha, S. N., K. H. Lakhani, and B. N. K. Davis. 1990. Studies on the toxicity of insecticidal drift to the first instar larvae of the large white butterfly *Pieris brassicae* (Lepidoptera: Pieridae). *Ann. Appl. Biol.* 116:27–41.

Spencer, W. F., W. J. Farmer, and M. M. Clath. 1973. *Pesticide Volatilization Residue Rev.* 49:1.

Southcombe, E. S. E., ed. 1985. *Application and biology.* Monograph Series, no. 28. Croydon, UK: British Crop Protection Council.

Spillman, J. J. 1977. Air velocities induced by aircraft. Cranfield Institute of Technology Short course on the aerial application of pesticides. Cranfield, Bedford, UK.

Spray Drift Task Force (SDTF). 1995. Bulletin. Stewart Agricultural Research Services, Inc., Macon, Mo.

Sundaram, K. M. S., P. de Groot, and A. Sundaram. 1987. Permethrin deposits and airborne concentrations downwind from a single swath application using a back pack mist blower. *J. Environ. Sci. Health*, B22 (2):171–93.

Taylor, A. W., and D. E. Glotfelty. 1988. *Evaporation from soils and crops.* In *Environmental chemistry of herbicides*, Vol. I. ed. R. Grover, 89–129. Boca Raton, Fla: CRC Press.

Taylor, W. A., P. G. Anderson, and S. Cooper 1989. The use of air assistance in a field crop sprayer to reduce drift and modify drop trajectories. *British Crop Protection Council—Weeds.* Brighton Crop Protection Conference, 631–39. Brighton, UK.

Teske, M. E. 1996. An introduction to aerial spray modelling with FSCBG. *J. Am. Mosquito Contr. Assoc.* 12 (2):353–58.

Teske, M. E., and J. W. Barry. 1993. Parametric sensitivity in aerial application. *Trans. Am. Soc. Agric. Eng.* 36:27–33.

Thom, A. S. 1975. Momentum, mass and heat exchange of plant communities. In *Vegetation and the atmosphere*, vol. 1, Principles, ed. J. L. Monteith, 57–109. London: Academic Press.

Trayford, R. S., and L. W. Welsh. 1977. Aerial spraying: A simulation of factors influencing the distribution and recovery of liquid drops. *J. Agric. Eng. Res.* 22:183–96.

Uk, S. 1977. Spray droplet capture. The aerial application of pesticides. Short course notes, Cranfield Institute of Technology, Cranfield, Bedford, UK.

Van Vliet, S. J. 1987. Aerial Sprays of Forests: Measurements of ground and canopy deposits. B.Sc. Thesis, Department of Chemical Engineering, University of New Brunswick, Canada.

Waite, R. 1978. Spread factor of pesticide spray formulation on cards. In *Methods for sampling and assessing deposits of insecticide sprays released over forests.* Tech. Bull. 1596. U.S. Department of Agriculture, Washington, D.C.

Walklate, P. J. 1987. A random-walk model for dispersion of heavy particles in turbulent air flow. *Boundary Layer Meteorol.* 39:157–90.

Weihs, D., and M. Atias. 1987. Calculation of aerosol distributions in aerial spraying of forested terrain. Proceedings of symposium on the aerial application of pesticides in forestry. Report no. NRC 29197 (AFA-TN-18), 171–74. National Research Council of Canada, Ottawa, Ontario.

White, A. W., L. A. Harper, R. A. Leonard, and J. W. Turnbull. 1977. Trifluralin losses from a soybean field. *J. Environ. Qual.* 6:105.

Wickens, R. H. 1977. Aircraft wake vortices and their effects on aerial spray application. Report no. LTR-LA-219. National Research Council of Canada, Ottawa, Ontario.

Williamson, R. E., and E. D. Threadgill. 1974. A simulation of the dynamics of evaporation spray droplets in agricultural spraying. *Trans. Am. Soc. Agric. Eng.* 17 (2):254–61.

Wilson, A. G. L., L. A. Harper, and H. Baker. 1986. Evaluation of insecticide residues and droplet drift following aerial application to cotton in New South Wales. *Aust. J. Exp. Agric.* 26:237–43.

Wolfe, H. R. 1976. Field exposure to airborne pesticides. In *Air pollution from pesticides and agricultural processes*, ed. R. E. Lee, Jr. Cleveland, Ohio: CRC Press.

Wolf, T. M., and R. Grover. 1992. The role of application factors in the effectiveness and drift of herbicides. Report. Agriculture Canada Research Station, Regina, Saskatchewan.

Wolfe, H. R., J. F. Armstrong, and W. F. Durham. 1966. Pesticide exposure from concentrate spraying. *Arch. Environ. Health* 13:340–44.

Wood, G. W., and D. R. K. Stewart. 1976. Fenitrothion residues in blueberry fields after aerial forest-spraying. *Bull. Environ. Contam. Toxicol.* 15:623

Woodrow, J. E., J. N. Seiber, and L. W. Baker. 1997. Correlation techniques for estimating pesticide volatilization flux and downwind concentrations. *Environ. Sci. Technol.* 31:523–29.

Wooley, D. H. 1962. A note on helicopter spray distribution. *Agric. Aviat.* 4:43–47.

Yates, W., R. E. Cowden, and N. B. Akesson. 1985. Drop size spectra from nozzles in high speed airstreams. *Trans. Am. Soc. Agric. Eng.* 28 (2):405–14.

THREE

DISLODGEABLE FOLIAR RESIDUES OF PESTICIDES IN AGRICULTURAL, LANDSCAPE, AND GREENHOUSE ENVIRONMENTS

Gerald R. Stephenson and Gwen M. Ritcey

Department of Environmental Biology, University of Guelph, Guelph, Ontario, Canada N1G ZW1

3-1 INTRODUCTION

In the simplest of terms, human hazard with respect to pesticides is dependent on the toxicities of the pesticides involved and the chance for exposure to those pesticides. Thus, to assess the hazards of pesticides in the workplace or recreational areas, it is essential to know the toxicities of the chemicals that are introduced as well as to measure the concentrations in the air or on surfaces to which humans could be exposed. Because most pesticides are applied to plant or soil surfaces, we must understand the potential for human exposure to pesticides in outdoor or greenhouse environments. The following review is focused on the factors influencing the dissipation and the toxicity of dislodgeable pesticide residues from plant foliage with respect to human exposure. Plant-related factors, climatic factors, the properties of the pesticides, and differing sampling methods are examined for agricultural, greenhouse, and landscape environments.

3-2 FACTORS INFLUENCING DISSIPATION OF DISLODGEABLE FOLIAR RESIDUES (DFRs)

3-2-1 Pesticide Compartmentalization on or in Leaf or Fruit Cuticles

When chemical pesticides or growth regulators are applied intentionally to the above-ground portions of plants, they are deposited initially on the cuticle of a leaf, a stem, or a fruit surface. The ultimate fate for all of these chemicals is very important for efficacy as well as for health and environmental considerations. For protectant fungicides and insecticides, dissipation rates on the leaf surfaces must be known to estimate reapplication times for effective pest control as well as safe reentry times for workers.

Many insecticides or fungicides act as surface active protectants that do not actually have to enter plant cells. However, most of the research in this area has focused on the foliar penetration of postemergence herbicides and growth regulators and to a lesser extent on other systemic insecticides or fungicides. All of these latter chemicals actually must penetrate the cuticles of leaves, stems, or fruits to move within living plant cells and reach their appropriate sites for protective, phytotoxic, or hormonal action to occur. In general terms, there are various possible fates for a chemical that is deposited on the plant cuticles.

1. Removal of the chemical by volatility, irrigation, or rainfall.
2. Decomposition of the chemical by photolysis in sunlight.
3. Penetration of the chemical through the cuticle to enter adjacent epidermal cells.

Our major concerns in this review are with the pesticide residues that remain on the surface of cuticles as potentially DFRs (dislodgeable foliar residues).

Plant cuticles are nonliving, lipoidal membranes that cover all of the aboveground portions of plants including the leaves, stems, and fruits (Martin and Juniper 1970). The cuticle prevents excessive losses of water and desiccation of the plant. Cuticles are also important barriers to the penetration of chemicals deposited on them. However, they are somewhat permeable to organic chemicals, gases, and water and this fact has been exploited in the development of foliage-applied plant nutrients, pesticides, and growth regulators. The outermost layer of a cuticle is composed of irregular deposits of epicuticular waxes imbedded in a biopolymer called cutin (Fig. 3-1). There is tremendous diversity in the deposition of epicuticular waxes. The deposits can be amorphous or crystalline and can be found as platelets, granules, ribbons, rodlets, aggregates, or crusts, depending on the plant species and the weather conditions (Chamel 1986). The cutin layer, the main structural framework of the cuticle, is made up of polyesters of long chain (C-16 or C-18) hydroxylated fatty acids. Because of its numerous $-COOH$ and $-OH$ groups and its numerous $-CH_2 -$ and $-CH_3$ groups, it has both hydrophilic and lipophilic properties (Chamel 1986) for the penetration by diffusion of both polar and nonpolar compounds (Fig. 3-1). On its lower surface, the cutin layer is bound to epidermal cells by a layer of pectin (Bukovac and Petrocek 1993). Incubating portions of leaves or fruits in pectinase enzyme preparations can yield isolated, intact cuticles which have been used extensively in penetration studies with herbicides or growth regulators (Bukovac and Petrocek 1993).

Despite the fact that a very large and increasing number of herbicides and plant-growth regulators require foliar application to be active, the foliar penetration of these chemicals often is very limited and can vary from nearly 0 to more than 90% of the chemical that is deposited on the cuticular surface. It is for this reason that the fate of these chemicals within cuticles has been studied so extensively. For our purposes here, a better understanding of the factors that influence the foliar penetration and dissipation of pesticides should be very helpful in understanding the probability that unpenetrated pesticides could become DFRs that could adversely affect human health.

Cuticular penetration can be described as a three-phase process (Bukovac and Petrocek 1993) involving sorption (deposit) on the cuticular surface, diffusion through

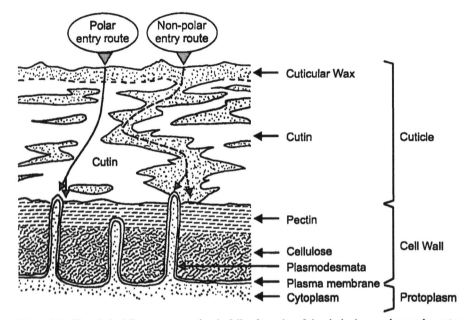

Figure 3-1 Hypothetical diagram representing the foliar absorption of chemicals via nonpolar or polar routes through the cuticle.
Source: Modified from Ashton and Crafts (1981), *Mode of Action of Herbicides.* Adapted by permission of John Wiley and Sons, Inc., New York.

the cuticle, and desorption from the cuticle into the epidermal cells or cell walls. This penetration process can vary tremendously from plant to plant and often can be related to differences in leaf morphology, including length and density of surface hairs (trichomes), quantity and deposition patterns for epicuticular waxes, as well as overall thickness and chemical properties of the cuticle.

Hess, Bayer, and Falk (1981) utilized scanning electron microscopy (SEM) to examine the influence of formulation type on the deposits of various pesticides on leaf cuticles. Pesticides applied as aqueous solutions tend to accumulate as net-like deposits above the anticlinal (perpendicular to leaf surface) walls of epidermal cells as the spray droplets dry on the surface of the cuticle. Pesticides applied as wettable powder or flowable suspensions in water appear as crystalline films above both the periclinal (parallel with leaf surface) and anticlinal cell walls of epidermal cells. The crystalline deposits are non-uniform and appear as concentric rings as the droplets of the aqueous suspensions dry on the surface of the cuticle. In some types of emulsifiable concentrate formulations, where the active ingredient, carrier solvent, and or adjuvant is a slow-drying liquid, the pesticide deposits appear as a fairly uniform film on the cuticular surface. In contrast, fast-drying emulsifiable concentrate applications leave very nonuniform crystalline deposits of the pesticides on the surface of the cuticle (Hess, Bayer, and Falk 1981).

The relative affinity of pesticides for plant cuticles (adsorption or binding) as compared to water within or on the surface of the cuticle is directly related to the octanol/water partition coefficient of the chemical (Kerler and Schönherr 1988). Some ester formulations of hormone herbicides (clopyralid or 2,4-D) readily penetrate into the cuticular waxes but then are much less likely to penetrate into the living epidermal cells than are their acidic analogues (Kloppenburg and Hall 1990). Some pesticides containing free carboxyl groups can become covalently bound to epoxides in the cuticles of some plant species (Riederer and Schönherr 1987). The SEM investigations of Hess et al. (1981), showed that the addition of surfactants greatly increased the surface area of the cuticle that was contacted by aqueous pesticide applications to plant foliage. Thus it was not surprising that, in his review of the effects of spray adjuvants, Kirkwood (1993) cited numerous studies that showed that surfactants could dramatically increase the cuticular penetration of water-based pesticide sprays.

Nigg et al. (1981), Thompson, Stephenson, and Sears (1984), and Sears et al. (1987) have studied DFRs in relation to the compartmentalization of pesticides within leaf cuticles and leaf cells. In studies of parathion applied to citrus leaves, Nigg et al. (1981) showed that the leaves were not structurally injured following a 20-min extraction with the standard dislodgeable solution (0.08 ml of 70% sodium dioctyl sulfosuccinate/1,000 ml water [Iwata et al. 1977]) or a 1-min extraction in methanol. The 1-min methanol rinse provided the best estimate of surface residues because the longer time (20 min) required for the dislodgeable solution permitted some of the desorbed residues of parathion to enter living cells or bind to the epicuticular wax fraction. Electron microscopy confirmed that a subsequent extraction with chloroform removed the waxes and parathion bound to the waxes without causing cell damage. Penetrated parathion was estimated via a methylene chloride–acetone extraction of the remaining cells. In an indoor study with 2,4-D on pots of turfgrass, Thompson, Stephenson, and Sears (1984) showed that DFRs on a cheesecloth wipe averaged 10% of the amount applied but there was always a large fraction of the surface residues removed with a methanol rinse of the grass blades (usually >50% of the amount applied) (Fig. 3-2). Wax-bound (toluene-extractable) residues of 2,4-D increased with time during the 15-day study, whereas the amount of 2,4-D that penetrated into leaf cells reached a peak after 2 days and then declined as a result of degradation or translocation to other parts of the plant.

A very similar study with diazinon on turfgrass (Sears et al. 1987) revealed very different results. All fractions associated with the grass blades (cheesecloth wipe, surface residues, wax residues, or tissue residues) dissipated very rapidly, whereas, at all times after treatment, most of the diazinon was recovered in the thatch. These studies (Nigg et al. 1981; Thompson, Stephenson, and Sears 1984) utilized a similar approach for examining the compartmentalization of DFRs on leaf tissue. However, the studies by Thompson, Stephenson, and Sears (1984) showed that leaf rinses with detergent solutions or solvents were likely to overestimate the residues that are actually dislodgeable by physical contact.

3-2-2 Nature of the Pesticide and the Formulation

Photolysis. Obviously, the greater the molecular stability of the pesticide while on leaf tissue, the greater its potential for protecting the plant and, unfortunately, the greater the opportunity for DFRs to be a problem. However, degradation products as well as the original active ingredient can be a concern. Although the photolysis of pesticides has been studied extensively, most studies have been conducted on pesticides exposed to

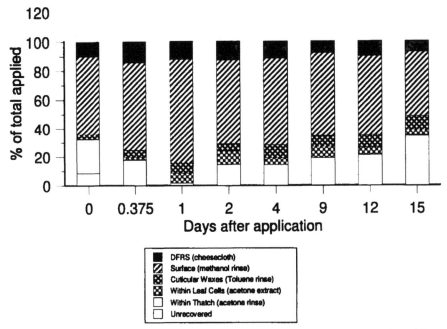

Figure 3-2 Dislodgeability and compartmentalization of 2,4-D following application to turfgrass foliage. *Source:* Thompson, Stephenson, and Sears (1984).

ultraviolet (UV) light in water or on glass or soil surfaces. Few researchers have studied the photolysis of pesticides on leaf surfaces and fewer still have conducted studies in which dissipation by photolysis could be separated from dissipation by volatility or actual penetration into the plant (Bentson 1990). Dissipation by photolysis can be the major fate for some pesticides. This is particularly true for the organophosphorus (OP) insecticides such as parathion (Joiner and Baetcke 1973) which can be oxidized to their more toxic oxon analogues while on leaf or fruit tissue (Stamper, Nigg, and Winterlin 1981). One particularly interesting observation was that worker illnesses due to DFRs of oxons was a more serious problem in California citrus groves than when the same OP insecticides were used in Florida citrus groves (Nigg et al. 1977). Dry, dusty conditions with high sunlight and high levels of oxidizing air pollutants eventually were related to the faster production of the toxic oxons under California conditions (Popendorf and Leffingwell 1978; Stamper, Nigg, and Winterlin 1981). Most organic molecules that have activity as pesticides or growth regulators can absorb light and undergo photochemical reactions that will result in the degradation and eventual deactivation of the molecule (Crosby 1976). Although more research is needed, differences in rates of photolysis on plant cuticles is potentially a very major factor in dissipation rates for DFRs.

Pesticide volatility. The loss of pesticides as vapors from leaf surfaces is dependent on a number of factors. In most situations, the atmosphere can be regarded as an infinite reservoir, with the wind as a dispersing agent and equilibrium between the air and the

Table 3-1 Comparative volatility of selected pesticides

Pesticide[a]	VP[b] (mPa)	Pesticide	VP (mPa)
	Nonvolatile		
2,4-D amines (H)	negligible	Imazethapyr (H)	0.013
Glyphosate (H)	negligible	Azinphos-methyl (I)	0.018
Rimsulfuron (H)	0.002	Fenvalerate (I)	0.019
Cypermethrin (I)	0.002	Amidosulfuron (H)	0.022
Fenoxaprop-ethyl (H)	0.005	Carbofuran (I)	0.033
Benomyl (F)	0.005	Carbaryl(I)	0.040
Primisulfuron (H)	0.005	Permethrin (I)	0.045
Nicosulfuron (H)	0.008	Chlorothalonil (F)	0.076
	Low volatile		
Bensulide (H)	0.133	Diclofop-methyl(H)	0.250
Ethion (I)	0.200	Mecoprop acid (H)	0.310
Chlorthal-dimethyl (H)	0.210	Dithiopyr (H)	0.530
Isofenphos (I)	0.220	Parathion (I)	0.890
	Moderately volatile		
Captan (F)	1.300	Dicamba acid (H)	4.500
Clopyralid (H)	1.330	Bendiocarb (I)	4.600
Chlorpyrifos (I)	2.700	Malathion (I)	5.300
Pendimethalin (H)	4.000		
	Volatile		
Tefluthrin (I)	8.000	2,4-D isooctyl ester (H)	13.300
Trifluralin (H)	9.500	Fenitrothion (I)	18.000
2,4-D acid (H)	11.000	Terbufos (I)	36.400
Diazinon (I)	12.000		
	Highly volatile		
2,4-D n-butyl ester (H)	519.000	Dichlorvos (I)	2,100.000
2,4-D methyl ester (H)	1,333.000	Nicotine (I)	5,650.000

Sources: Que Hee and Sutherland (1981); Tomlin (1994).
[a]F = fungicide, H = herbicide, I = insecticide.
[b]VP = vapor pressure at 20–25°C.

leaf surface a very rare occurrence. Volatility, of course, is related directly to the vapor pressure of the chemical. Although the vapor pressures of common pesticides vary over several orders of magnitude (Table 3-1), volatility is also a function of wind speed, contact volume, surface area, type of compound and temperature (Que Hee and Sutherland 1981). OP insecticides vary from the slightly volatile azinphos-methyl (0.018 mPa) to the highly volatile dichlorvos (2,100 mPa) developed to act as a vapor (Table 3- 1). Two soil-active, dinitroaniline herbicides, pendimethalin (4 mPa) and trifluralin (9.5 mPa) are interesting in that trifluralin must be incorporated into the soil to be effective whereas for pendimethalin, incorporation is much less important (Ahrens 1994). Some of the newer synthetic pyrethroid insecticides and sulfonylurea herbicides are very, very nonvolatile whereas the older, natural insecticide, nicotine (5,650 mPa) is one of the most volatile pesticides ever used (Table 3-1).

Although some insecticides and fungicides (captan, 1.3 mPa) can be moderately to highly volatile, their movement and environmental impact are less visible than is the case for herbicides such as 2,4-D. As a synthetic plant hormone, 2,4-D can readily injure nontarget, broadleaf plants when it moves by spray-droplet drift or volatility/vapor drift to nontarget areas. Some short-chain (C-1 to C-4) esters are much more volatile than the pure 2,4-D acid, whereas the longer-chain esters (C-8) are much less volatile and amine salts are essentially nonvolatile (Table 3-1) (Que Hee and Sutherland 1981). If added surfactants increase the cuticular penetration of the pesticide, they may decrease losses by volatility. Conversely, if a surfactant competes with the pesticide (i.e., 2,4-D ester) for binding sites in the cuticle, it may increase volatility losses and perhaps DFRs as well. Pesticides applied as small droplet sprays can volatilize off leaf surfaces faster than when applied as large droplets (Que Hee and Sutherland 1981). As the droplet dries, the water carrier evaporates first, followed by some of the inert ingredients until the active compound and possibly the surfactant are all that are left in or on the cuticle. At this stage, the penetration rate for 2,4-D esters increases and volatility decreases. Under these drying conditions, penetration of amine salts of 2,4-D would be reduced and would be left deposits on the cuticle. Thus they may be more susceptible to removal by rainfall (Que Hee and Sutherland 1981) or as potential DFRs. In some areas (e.g., Ontario, Canada), only amine formulations of 2,4-D are allowed for use in landscape situations and only amine or low-volatile ester formulations are permitted for use in agriculture. Regulations of this type certainly have minimized the risks of 2,4-D damage to sensitive nontarget plants. However, the relative dislodgeability of amine versus ester formulations of 2,4-D from plant foliage has not been examined.

It is unfortunate that most studies of volatilization have focused on the volatility of materials from soil or water or from surfaces that are dissimilar to foliar surfaces (Bentson 1990). Rapid volatilization from leaf surfaces obviously would decrease the efficacy of foliarly applied pesticides and it also would decrease the persistance of DFRs. However, it obviously increases the unwanted movement of pesticides into nontarget environments and, during the first days after application, it could increase the risks of human exposure via inhalation of vapors. Clearly, more research is needed to assess the impact of volatility on the probability of human exposure to pesticides in outdoor as well as in indoor environments.

Rain-fast properties of pesticides. Rain (or irrigation) is a major weather factor to consider in many aspects of pesticide use. Too little or too much rainfall can influence the efficacy as well as the environmental fate of pesticides. For pesticides applied to plant foliage, it is perhaps one of the major factors influencing the persistence of DFRs.

Most of the insecticides or fungicides used on crops or landscapes are surface acting, protectant pesticides. Their protective effects on plants are related directly to their persistence on plant foliage. However, if the pesticide is highly toxic, even moderate persistence becomes a disadvantage because it increases concerns about both preharvest spray intervals and DFRs in relation to reentry times for field-workers. Generally speaking, pesticides that act on or in plant leaves should not be applied when rain is imminent because, if the rainfall is significant, a reapplication may be required to ensure effective pest control. In a series of studies with the insecticides malathion and permethrin, sulprofos, azinphosmethyl and fenvalerate, and carbaryl, Willis et al. (1992, 1994a, 1996) established that the time interval between application and the first rainfall had a major influence on the amount of protectant insecticide that stayed on the plant. Soluble concentrate or emulsifiable concentrate formulations have been shown to be more rainfast than wettable powder formulations of some pesticides such as carbaryl (Willis et al. 1996) or maneb and mancozeb (Kudsk and Mathiassen 1991). Oil-based, sticker adjuvants have improved the rainfast properties of some pesticides, for example, maneb and mancozeb on peas (Kudsk and Mathiassen 1991) and dimilin on balsam fir (Sundaram and Sundaram 1994), but their effects are not predictable for different formulations or different crops (Kudsk and Mathiassen 1991).

For pesticides that act systemically in plants (e.g., postemergence herbicides, systemic insecticides, and fungicides), the goal is to get the active pesticide through the plant cuticle rather than to keep it on the cuticular surface. Obviously, the best systemic pesticides, in all respects, would be those that rapidly and efficiently penetrate into plants before photolysis, volatility, washoff, or problems as DFRs could occur. Ester formulations of 2,4-D and related herbicides, although more volatile (Table 3-1), penetrate into plants faster and thus are more rainfast than when formulated as the more water-soluble, amine salts (Behrens and Elakkad 1981). It is a reasonable hypothesis that, in the short term or until the first rainfall, the amine formulations might be more dislodgeable than the ester formulations which also have a greater affinity for the waxy cuticle than for water. Glyphosate is perhaps the most widely used systemic herbicide. It is highly water soluble and not very rainfast (Ahrens 1994). However, using organosilicone surfactants, Roggenbuck et al. (1993) and Roggenbuck, Burow, and Penner (1994) have shown that the rainfast properties of glyphosate and a number of other herbicides can be increased significantly by increasing both the rate and amount of active ingredient that penetrates the cuticle. Although more research is needed, methods that improve the rainfast properties of systemic pesticides by improving their foliar penetration also should decrease their problems with respect to DFRs.

3-3 DFRs OF PESTICIDES IN AGRICULTURAL CROPS

During the late 1960s and early 1970s, the use of DDT and other organochlorine insecticides was decreasing in North American agriculture because of pest insect resistance, environmental persistence, and problems associated with deleterious effects on nontarget organisms, including fish, birds, and beneficial insects (McEwen 1977). During this period, there was a shift in insecticide use to the far more acutely toxic, OP and carbamate insecticides. As the use of these more toxic pesticides expanded, incidents of human poisonings and even deaths alerted direct users of the products to carefully follow handling instructions. Government regulators also imposed low maximum residue limits for these pesticides in fruits and vegetables, and minimum days-to-harvest restrictions for pesticide applications appeared on product labels to ensure that these residues were not exceeded. Although there were a few reports of field-worker illnesses following reentry of fields recently treated with parathion in the late 1940s and early 1950s, California farmworkers and researchers became much more aware of the "worker reentry problem" in the 1960s and 1970s. Many incidents were reported for harvesters of citrus, grapes, and cotton in California but there were also problems in other areas where cotton and tobacco crops were grown (Gunther et al. 1977). Field-workers involved in hand harvesting, fruit thinning, or summer pruning operations who had extensive contact with

DFRs of pesticides on foliage or other plant parts were seen to be the people at greatest risk. However, pest management scouts and agricultural researchers were also at risk because they were often required to reenter fields for extended periods of time and at intervals soon after the pesticides had been applied. An extensive research effort by a number of different laboratories was focused on the persistence of DFRs and worker reentry concerns, particularly in citrus, grapes, and cotton during the 1970s and there are several excellent reviews of their findings (McEwen 1977; Gunther et al. 1977; Knaak 1980; Popendorf and Leffingwell 1982). Although this timely research helped to develop a process for reentry decisions for important pesticides in those crops at that time, among the conclusions were the following concerns: (1) that it is difficult to predict the persistence of DFRs for a particular pesticide on different crops and under different conditions, (2) different pesticides act differently, and (3) risks to workers will vary with different activities that are required in different crops.

In the following, our goals are to summarize the excellent definitive research on DFRs conducted primarily on the OP insecticides in citrus, grapes, and cotton and to compare that research with more recent research involving new pesticide chemistry and different crops. More importantly, the effectiveness of methodologies for preventing acute health problems among agricultural workers is evaluated.

3-3-1 Citrus

In their review of the citrus reentry problem, Gunther et al. (1977) pointed out that it was primarily a problem associated with the use of the insecticide parathion in citrus production. Although worker illnesses associated with parathion use in California citrus groves were increasingly common during the 1970s (Popendorf and Leffingwell 1982), such reports were very rare in Florida. An early and important discovery was that of Spear et al. (1978) who showed that parathion degraded to its more toxic and more persistent paraoxon analogue on citrus foliage. In Florida, DFRs of both parathion and paraoxon were dissipated quite rapidly by rainfall (Nigg, Allen, and King 1979). In contrast, the dry, rainfree, dusty conditions in California were related to higher rates of paraoxon production and greater paraoxon persistence (Popendorf and Leffingwell 1978) and thus the need for longer reentry times. Adams, Iwata, and Gunther (1976) showed that, in California, even soil type and the type of dust landing on citrus foliage could influence the persistence of DFRs for both parathion and paraoxon. In studies conducted in Florida citrus groves on the nonpersistent malathion, Nigg and Stamper (1981) observed significant residues of the malaoxon metabolite under cool dry conditions, whereas it was rarely detectable under hot, wet, humid conditions. DFRs for the carbamate insecticide carbosulfan and its metabolite carbofuran (Nigg, Stamper, and Knaak 1984) and the growth regulator diflubenzuron (Nigg, Cannizzaro, and Stamper 1986) also were observed to be less persistent in Florida citrus groves under hot, wet conditions as compared to cool, dry conditions. It was clear from these studies and a number of others that dissipation curves for pesticide DFRs in citrus were very dependent on a number of postapplication environmental and climatic factors and that realistic reentry intervals would have to vary with different conditions. In response to this, systematic leaf disk sampling and extraction methods were developed to estimate DFRs in citrus, which were in turn related to the toxicity of the particular pesticide, in order to calculate safe reentry intervals (Gunther et al. 1973; Iwata et al. 1977; Knaak 1980). Nigg, Stamper, and Queen (1984) developed a simple regression model for predicting transfer of insecticide DFRs from foliage to the bodies of Florida citrus workers and found close agreement with an earlier California model developed by Popendorf and Leffingwell (1982). They felt that differences in application rates, initial residues, and climate were more than adequate to explain the different dissipation rates of DFRs between the two regions and that perhaps the role of foliar dust in California had been overemphasized.

3-3-2 Peaches and Apples

Because of the importance of summer pruning and fruit thinning activities, workers in apple and peach orchards may have a greater risk for pesticide exposure at times closer to pesticide application. In a study involving both peaches and apples, Staiff, Comer, and Foster (1975) did not observe any net accumulations of either parathion or paraoxon on foliage as a result of weekly applications during the fruit thinning period. McEwen et al. (1980) compared the dissipation rates for total versus glove DFRs on apple foliage for both parathion and phosalone. Although DFRs and total residues were higher for phosalone than for parathion for the first two days, by day 5, total residues and glove DFRs were similarly low for both insecticides. Davis et al. (1982) and Davis, Stevens, and Staiff (1983) studied the exposures of experienced apple thinners to glove DFRs of phosalone and azinphos-methyl. In their study, DFRs were similar for nylon and cotton gloves and were not higher at 24 h than at 48 h after phosalone application (Table 3-2). Their results were comparable to the 24-h results of Wolfe et al. (1975) and confirmed that, at the low rates of parathion applied to apples (1 kg/ha), reentry risks at 24 h after treatment were minimal. In their study of the potential exposure of apple thinners to azinphosmethyl Davis, Stevens, and Staiff (1983) compared hand exposures by using ethanol handwash and two types of absorbent gloves, cotton and nylon (Table 3-2). There were no differences between exposures obtained by using the different types of gloves, although significantly lower hand exposure results were obtained by using ethanol hand rinses. Fenske et al. (1989) assessed DFRs on cotton gloves versus DFRs in a isoproponal–water handwash following captan application prior to the harvesting of peaches. The 1.5- to

Table 3-2 Comparison of DFRs of two OP insecticides as measured on nylon and cloth gloves and ethanol-rinsed bare hands of apple thinners

| | DFRs | |
| | Phosalone | Azinphosmethyl |
Surface	mg/h	mg/h
Cotton glove	5.5/24	8.5/48
Nylon glove	5.5/48	7.0/48
Ethanol-rinsed bare hands		1.8/48

Source: Davis, Stevens, and Staiff (1983); Davis et al. (1982).

2.5-fold difference between gloves and handwash in exposure rates was lower than the five-fold difference found by Davis, Stevens, and Staiff (1983). According to Fenske et al. (1989) the differences between the two studies could have been dependent on the weight or type of cotton glove. The isoproponal–water handwash could be more efficient than the ethanol handwash, and exposure rates could be partially dependent on the properties and formulation of the particular chemical. Regardless of the active ingredient, formulation differences can affect both pest control efficacy and worker exposure to DFRs. Although microencapsulated (ME) formulations of parathion are only 1/20 as dermally toxic as emulsifiable concentrate (EC) formulations, Davis et al. (1981) concluded that studies of DFRs were still essential in calculating reentry intervals because the ME formulation was more dislodgeable than the EC formulation from apple foliage. The incidence of worker poisonings in apple orchards have decreased with the decreased use of parathion during the fruit thinning periods in these crops. Another factor that has helped to reduce the problem is the increased use of chemicals such as naphthalene acetic acid and carbaryl for fruit thinning as alternatives to hand thinning.

3-3-3 Grapes

California grape harvesters were also among the first groups of workers to suffer illnesses as a result of reentering vineyards recently treated with OP insecticides. In their review, Gunther et al. (1977) summarized numerous incidents in California during the 1970s that involved exposure to DFRs of OP insecticides such as ethion, phosalone, phosmet, dialifor, carbophenothion, naled, dimethoate, and the carbamate insecticide methomyl. Parallel to the efforts to resolve the reentry problem for citrus harvesters, there was also extensive research to develop appropriate methodology for sampling, extracting, and estimating DFRs to establish acceptable reentry times for toxic pesticides (primarily OP insecticides) used in grape vineyards. Popendorf and Spear (1974) studied the work activities of grape harvesters and showed that, in comparison to harvesters of tree fruits, workers spent a lot more time kneeling on the ground. The average harvester would pick three to four bunches per minute and would contact up to ten leaves per bunch. They would also have a lot of head and shoulder contact with DFRs on overhanging foliage. In their 1975 study of DFRs for ethion, phosalone, and azinophos-methyl, Leffingwell, Spear, and Jenkins (1975) confirmed that it was far better to base reentry-interval decisions on DRFs than on total grape-leaf residues. Furthermore, for ethion, from one week after application onward, DFRs for the oxon metabolites were higher than for unaltered ethion. They concluded that photodecomposition or oxidation were the important factors in production of oxon metabolites because there was a much faster conversion of ethion to its oxon metabolite in the DFRs fraction (recovered in the leaf-disk rinses) than in the penetrated fraction (recovered in the extracts of leaf disks after rinsing). In spite of 25% higher application rates of active ingredients for ethion or zolone in high versus low spray volume treatments, there were no significant differences in the initial DFRs for these two insecticides. There was a high correlation between DFRs and soil surface residues of ethion, and dissipation rates for both were correlated directly with increases and decreases in temperature (Leffingwell, Spear, and Jenkins 1975). In later studies, Winterlin et al. (1980) observed that dissipation rates for DFRs of dialifor were slower than for other OP insecticides. However, DFRs of the oxon metabolites never exceeded those for the parent dialifor. Furthermore, neither the dusting of the vineyards before or after spraying nor the adding of nutrients to the spray had a significant effect on the dissipation rates for DFRs of dialifor or its conversion to dialifor oxon. Most of the research on DFRs in grape vineyards has focused on OP insecticides in California vineyards; thus regional comparisons (California vs. Florida) are not yet possible as they are for citrus. However, it is clear that dissipation rates for DFRs of OP insecticides on grape foliage in California are far slower than for the same insecticides in Florida citrus groves.

3-3-4 Strawberries and Blueberries

In row crops such as strawberries, the fruit is handpicked while the worker squats, bends, kneels, or sits between the rows. In contrast, workers in bush crops such as blueberries, which are 4–6 ft high, are usually standing and reaching into the crop. However, in spite of these differences, Zweig, Leffingwell, and Popendorf (1985), showed that exposure to DFRs of pesticides on various parts of the body were similar for the two groups of workers, with highest exposures being on the hands (60–80%), lower arms (7–21%), and lower legs (1–10%). Studies with the fungicides captan and benomyl (Zweig, Gao, and Popendorf 1983) showed that the ratio of the two fungicides in the DFRs rinsed off strawberry leaves was the same as the ratio in the DFRs on gloves and forearm patch monitors worn by the strawberry harvesters. Furthermore, the dermal exposure rate (DER) to the fungicides could be correlated with the amount of fruit picked. Following studies of DFRs with carbaryl in strawberries, and vinclozolin, captan, and methiocarb on blueberries, Zweig, Leffingwell, and Popendorf (1985) calculated the ratio of DER to DFRs for carbaryl and compared them to the ratios for captan, benomyl, chlorobenzilate, or OP insecticides that could be calculated from the literature (Zweig et al. 1984; Zweig, Leffingwell, and Popendorf 1985). On the basis of these examples, Zweig et al. (1984) and Zweig, Leffingwell, and Popendorf (1985) proposed that a first approximation of the DER for field harvesters or workers might be obtained by a simple multiplication, once the DFRs were known, as estimated by the water (surten), leaf-disk rinsing method of Iwata et al. (1977); that is,

$$\text{Dermal exposure rate (mg/h)} = \text{DFRs}(\mu g/cm^2) \times 5{,}000$$

(Zweig, Leffingwell, and Popendorf 1985). The research of Ripley and Ritcey (1997), as shown in Table 3-3 illustrates that the DFRs on hands or gloves are many times higher than the DFRs on leggings of strawberry harvesters even though the residues on the mulch are usually much higher than the residues on leaves or berries. Their research also provides a comparison of DFRs for some of the newer and older insecticides and fungicides. Although the DFRs for the newer synthetic pyrethroid cypermethrin are quite persistent, cypermethrin is far less toxic than any of the other insecticides except dimethoate. Although carbofuran and parathion are the most toxic, their rapid dissipation could permit acceptable reentry times. This might be more difficult with endosulfan, which is quite toxic and quite persistent. The DFRs for the fungicides captan, iprodione, and metalaxyl are all quite persistent but their acute toxicities are quite low (acute oral lethal dose > 600 mg/kg) (Table 3-3).

Table 3-3 Total pesticide residues (TRS) on strawberries, leaves, and mulch, 2 and 7 days after treatment and DFRs on gloves or leggings of harvesters after 10 min of picking

Pesticides[a]	AOLD$_{50}^{b}$ (mg/kg)	Days after Treatment	TRS(μg/g)			DFRs(μg/cm^2)	
			Berries	Leaves	Mulch	Gloves	Leggings
Cypermethrin (I)	250	2	0.36	5.48	6.15	2.96	0.130
		7	0.13	1.85	3.10	1.77	0.046
Captan (F)	850	2	8.75	47.8	84.8	69.0	24.9
		7	4.53	25.0	24.0	60.1	2.92
Dimethoate (I)	290	2	0.110	1.28	12.7	0.728	0.067
		7	0.028	0.490	3.1	0.126	0.012
Parathion (I)	2-6	2	0.320	1.50	11.6	0.853	0.034
		7	0.090	0.130	4.7	0.080	0.011
Carbofuran (I)	8	2	1.63	13.4	46.3	5.65	0.826
		7	0.670	2.00	7.4	0.489	0.035
Metalaxyl (F)	633	2	3.65	33.3	223	6.53	0.621
		7	1.830	17.5	26.3	1.63	0.053
Iprodione (F)	2000	2	3.63	112	145	30.5	10.4
		7	3.03	313	40.3	5.20	0.251
Endosulfan (I)[c]	70	2	2.87	29.5	88.4	11.5	0.634
		7	0.92	8.46	40.4	1.78	0.179

Source: Unpublished data, Ripley and Ritcey Laboratory, University of Guelph, 1997.

[a] I = insecticide, F = fungicide.

[b] AOLD = acute oral lethal dose (Tomlin, 1994).

[c] Endosulfan = endosulfan 1 + endosulfan 2 + endosulfan sulfate.

Although not acutely toxic like the OP insecticides, captan has been the subject of considerable toxicological concern because of its effects as a mutagen or teratogen in test animals. In response to these concerns, Winterlin et al. (1984) monitored DFRs of captan and its metabolites on leaf disks as well as on gloves or patches worn by strawberry harvesting crews. In addition, they monitored the air for vapors. Because captan is at least moderately volatile (Table 3-1), recovery ratios for captan vapors were more dependent on time after treatment than on the dissipation rates for residues on the foliage (Table 3-4). DFRs on various parts of the body were much more variable than DFRs on gloves (Table 3-5). These results indicate that these exposures may be very dependent on picking styles such as bending at the waist versus crawling along the

Table 3-4 Captan residue on strawberry foliage and in air at low elevations

Days after application	Foliage (μg/cm^2)	air (μg/m^2)
0	4.961	1.024
3	4.787	0.445
14	1.437	0.380

Source: Winterlin et al. (1984).

Table 3-5 Captan residue on patches and gloves worn by strawberry havesters

Sample area	Patches ($\mu g/cm^2$)	SD[a]
Chest	0.837	0.612
Sleeve	5.909	3.814
Back	0.624	0.199
Thigh	3.708	3.733
Shin	1.936	0.820
Glove	2.198	0.598

Source: Winterlin et al. (1984)

[a] SD = standard deviation.

rows, whereas hand exposure could be quite consistent. However, more detailed studies would be needed to confirm this. No members of the harvesting crew were provided with a respirator; thus it was not possible to compare exposure via DFRs versus those by inhalation of captan vapors. However, captan residues were detected on the pads of the respirator worn by the mixer-loader. Although there were no detectable residues in urine samples from the mixer-loader, there were detectable residues of captan metabolites in the urine of the harvesters.

3-3-5 Cotton

During the past 30 or 40 years, it is possible that more pesticides have been applied to cotton than to any other single crop. According to some estimates, more than one third of the insecticides used agriculturally in the United States are applied to cotton. During the 40 or 50 years that cotton pests have been managed chemically with pesticides, most cotton has been harvested mechanically. Thus, the reentry problem following pesticide treatment has not involved a large number of workers actually picking cotton. The concern is more focused on pest management scouts who enter fields and monitor for pests on a regular basis so that pesticide applications can be timed to those periods when pest levels are above the economically damaging thresholds. A worker may scout more than 400 ha (1,000 acres) per week and may be exposed to pesticides in treated fields for 8 to 10 h day (Wicker et al. 1979). Although exposure via the hands is significant, later in the season when the cotton is 3 to 5 ft tall and the inter-row spaces are narrowed, exposure of the hips and legs is also significant, particularly when the foliage is wet. Parathion and other OP insecticides have been the chemicals of greatest concern with respect to DFRs in cotton. However, thanks to an excellent series of studies at the University of Arizona (Estesen and Buck 1990; Ware, Buck, and Estesen 1983; Estesen, Buck, and Ware 1982; Ware, Estesen, and Buck 1980; Buck, Estesen, and Ware 1980) cotton provides an excellent opportunity to compare dissipation rates for DFRs of OP, carbamate, and pyrethroid insecticides (Table 3-6). From these data, it is clear that DFRs of OP insecticides on cotton foliage dissipate to 30% or less of initial residues within 24 h and to 20% or less within 48 h. The newer pyrethroid insecticides are more persistent. However, their application rates, initial DFRs, and acute mammalian toxicities are so low that illnesses of scouts would not be expected with immediate reentry of treated

Table 3-6 Comparative dissipation rates for DFRs of OP, carbamate, and pyrethoid insecticides on cotton foliage

Insecticide (type)	AOLD$_{50}$ in rats (mg/kg)[a]	Initial DFRs (mg/cm^2)	Percentage of initial DFRs remaining at various times after treatment (h)				
			24	48	72	96	144
Methalparathion (OP)	6	4.5[b]	17	7	2	1	
Methamidiphos (OP)	20	0.57[c]			19		6
Chlorpyrifos (OP)	>130	3.64[d]	3	2	2	1	
Sulprofos (OP)	200	4.00[d]	14	4	3	2	
Profenofos (OP)	358	3.50[d]	31	21	14	10	
EPN (OP)	36	2.60[b]	27	18	15		
Carbaryl (CB)[e]	850	7.20[c]			82		8
Methomyl (CB)	17	1.10[b]	14	15	9	3	
Thiodicarb (CB)	120	4.70[b]	120	102	100		
Permethrim (PY)[f]	>400	0.39[b]	77	61	49	41	
Fenvalerate (PY)	>400	0.85[d]	42	45	33	33	
Cypermethrin (PY)	250	0.19[c]				84	50

[a] Tomlin (1994).
[b] Ware, Estesen, and Buck (1980).
[c] Estesen, Buck, and Ware (1982).
[d] Buck, Estesen, and Ware (1980).
[e] CB = carbamate.
[f] PY = pyrethroid.

fields. Generalizations about reentry times for carbamate insecticides would be difficult. In spite of its persistence on cotton foliage, reentry problems with carbaryl would not be expected because of its low toxicity. In contrast, thiodicarb is moderately toxic and its DFRs are highly persistent on cotton foliage. DFRs for methomyl are less persistent, but it is highly toxic to mammals and initial DFRs are quite high (Table 3-6). An additional problem for the OP insecticides is the fact that the first degradation products on plant foliage are their respective oxon analogues which are usually severalfold more toxic to humans. In other studies, Ware, Buck, and Estesen (1983), showed that DFRs for most insecticides are more persistent on cotton foliage following ultralow-volume (ULV) applications in cottonseed oil than when applied as water-based sprays (Table 3-7).

Ware and Morgan (1976) conducted ten exposure studies over a three-year period for scouts at prescribed reentry times in cotton fields in Arizona. In addition, Wicker et al. (1979) reviewed an eight-year monitoring program for pest management scouts exposed to OP insecticides in cotton fields in North Carolina. Both studies would support the following conclusions with respect to reentry of cotton fields treated with OP insecticides:

1. Unlike the experiences with citrus or grapes, there were few if any confirmed cases of worker illnesses associated with reentry and exposure to DFRs of OP insecticides on cotton foliage.
2. When recommended application rates and worker procedures are followed (i.e., 48-h reentry intervals, daily clothing changes, avoidance of wet fields) depressions of cholinesterase (ChE) levels in field scouts by more than 25% would be rare and adverse effects on worker health would not be expected.

Table 3-7 DFRs from cotton for various
insecticides applied as water-based sprays of
ECS or as ULV applications in cottonseed oil

Insecticide	(Type)[a]	DFRs at 48 h (% of initial)	
		EC	ULV
Permethrin	PY	60	100
Flucythrinate	PY	68	80
Sulprofos	OP	46	77
Fenvalerate	OP	59	73
Cypermethrin	PY	62	64
Methidathion	OP	7	29
Azinphosmethyl	OP	26	27
Methomyl	CB	4	13
Chlorpyrifos	OP	7	12
EPN	OP	17	10
Methyl parathion	OP	2	7

Source: Ware, Buck, and Estesen (1983).
[a] PY = synthetic pyrethroid, CB = carbamate, OP = organophosphorous.

3. Nevertheless, the margin of safety is not great. Routine monitoring of ChE levels in exposed workers and information on the dissipation rates for highly toxic insecticides used in cotton fields are the minimal precautions needed for early detection of possible problems associated with unusual conditions.

3-3-6 Mechanically Harvested Tomatoes

With more and more mechanical harvesting of fruit and vegetable crops in North America, the number of hand-harvested crops will decrease. Field scouts may become the workers most likely to be exposed to DFRs on crop foliage, and workers in processing or cleaning operations may become the first to be exposed to DFRs on the harvested products. It may be some time before such trends are evident in less developed countries.

Spencer et al. (1991) investigated exposure to the fungicide chlorothalonil for workers involved in sorting on mechanical tomato harvesters. DFRs of chlorothanonil on the fruit averaged 1.13 μg/cm^2. Gloves and normal work clothing were acceptably effective for reducing worker exposure. These investigators calculated a very low transfer factor for residues from fruit to worker and suggested that this was due to a much lower exposure of workers to the fruit than would be expected in hand-harvesting operations.

3-4 DIFFERENCES IN SAMPLING METHODOLOGY

Any attempt to compare risks of worker exposure to agricultural pesticides is hindered by the different efficiencies of the methods to estimate DFRs of pesticides on crop plants that are reported in the literature. Although the Iwata et al. (1977) aqueous surfactant leaf-disk rinsing method (20 min of shaking in 0.08 ml of 70% sodium dioctyl sulfosuccinate in 1,000 ml of water) is used most frequently as a standard, it often is criticized for overestimating the physically dislodgeable residues. Bissell et al.(1991) showed that sampling variation was significantly greater with leaf punches or disks than with whole leaves and that both tissue rinse methods overestimated the DFRs compared to DFRs obtained by wiping the leaves with moistened gauze pads. Glove dislodgeable residues often have been employed but, with these methods, DFRs will vary with type of glove (i.e., cotton vs. nylon), picking time, and glove saturation time (Davis, Stevens, and Staiff 1983). Furthermore, DFRs as estimated by gloves also may overestimate the actual DFRs by physical contact, as estimated with analyses of residues in hand rinses (Davis, Stevens, and Staiff 1983; Fenske et al. 1989). There are also obvious problems with solvent rinses of the hands. Aqueous detergent rinses and organic solvent rinses are all likely to be irritating to the skin and they will vary in efficiency for chemically different pesticides. Furthermore, because there will be some exposure for the human volunteers, it would be difficult to use such methods for the most important, most toxic pesticides.

For data to be used for regulatory purposes, there is a clear need for regulatory agencies to specify a more standardized, physical method in addition to a more standardized leaf tissue rinsing method to estimate DFRs in agricultural situations.

3-5 GREENHOUSE CROPS

The total crop area and the total use of pesticides in greenhouse or glass house production is a very low percentage of the totals for agriculture in general. However, greenhouse production of flowers, ornamentals, or vegetables represents an important area of concern with respect to worker exposure to pesticides. The very high value of the crops supports a very intensive management system. The enclosed nature of the environment can lead to frequent use of insecticides and fungicides to prevent epidemics when pests are introduced. Because of the lack of irrigation directly on the foliage and the filtering out of UV light by the glass, it is not uncommon for DFRs of pesticides to be more persistent in greenhouses than on outdoor crops exposed to full sunlight, wind, rainfall, and dew (Brouwer, Brouwer, De Mik et al. 1992; Archibald, Solomon, and Stephenson 1994b; Frank et al., 1989). Furthermore, there is also a concern for exposure of greenhouse workers to vapors of the pesticides soon after application and via volatilization of residues from the plant foliage. Greenhouse horticulture is very international and importation regulations for pest-free products result in higher pesticide use in crops destined for the international market (Brouwer, Brouwer, Tijssen, and van Hemmen 1992).

3-5-1 Greenhouse Workers

Most applications of pesticides in greenhouses are made by trained, full-time, male employees who are often members of the families that own the businesses (Archibald, Solomon, and Stephenson 1994a). In contrast, most of the workers involved in manipulating the plants or harvesting the crops are female employees. Many of these women are of child-bearing age. However, because of the year-around nature of the job and the pleasant working conditions, long-term employment for a woman in a particular greenhouse operation is quite common (Archibald, Solomon, and Stephenson 1994a).

In The Netherlands, approximately 25,000 people are potentially exposed to pesticides in commercial greenhouse operations (Brouwer, Brouwer, De Mik et al. 1992). In carnation production for the cut-flower market, where contact with the crop is high, typical activities include disbudding, cutting of the flowers, collecting them as a bundle on the arm and sorting them for quality into bunches. During the 13-week production period for potted chrysanthemums, disbudding activities are also important and individual workers spend around 80 h in the crop, of which 50–60 h is spent handling the plants (Frank et al. 1989).

3-5-2 DFRs for Common Greenhouse Pesticides

A large number of different pesticides are used in greenhouses. In their survey of chrysanthemum growers, Archibald, Solomon, and Stephenson (1994a) found that the nine most commonly used insecticides included four organochlorines, two OP compounds, two carbamates, and one synthetic pyrethroid. Benomyl was the most commonly used fungicide. In a study of DFRs for four different fungicides in greenhouse carnations, Brouwer, Brouwer, Tijssen, and van Hemmen (1992) obtained transfer factors from foliage to the gloves of workers that were within the same order of magnitude as those observed by other investigators for field-grown fruit crops (Zweig, Leffingwell, and Popendorf 1985; Nigg, Stamper, and Queen 1984). They observed that 66% of the differences in glove dislodgeable residues of chlorothalonil, thiophanate-methyl, thiram, and zineb could be accounted for by differences in DFRs as estimated by the leaf-disk rinsing method with water and surfactant (Iwata et al. 1977). In a study with pyrazophos applied to greenhouse-grown chrysanthemums, Frank et al. (1989) observed that DFRs removed by swabbing the stems, leaves, and flowers for 5 min with muslin cloth ranged from 2 to 7% of the total residues extracted from the plant tissue. In another study with chrysanthemums, Archibald, Solomon, and Stephenson (1994b) observed that DFRs for pirimicarb as estimated by the muslin swabbing technique averaged 5.8% at 1 h after treatment and decreased to 2.1% or less after 12 h. Nigg, Brady, and Kelly (1992) utilized the water/surfactant method (Iwata et al. 1977) to estimate the DFRs following treatment of greenhouse-grown azaleas with bendiocarb. DFRs ranged from 10.9 μg/cm^2 on the day of treatment to 4% of that or less after 21 days. Because the calculated safe reentry level is 13.0 μg/cm^2, Nigg, Brady, and Kelly (1992) concluded that there would be minimal risks associated with reentering a greenhouse after bendiocarb sprays had dried on the foliage.

3-5-3 Application Methods

In their study with the fungicide chlorothalonil on carnations, Brouwer, Brouwer, Tijssen, and van Hemmen (1992) observed that DFRs were higher immediately after spray applications than after dust applications. However, they felt that these differences could be due to the higher application rates for the active ingredient for sprays as compared to dusts. In other studies, Nilsson et al. (1996) showed that DFRs for vinclozolin fungicide applied to tomatoes were 0.25 μg/cm^2 following application by a fan-assisted, low-volume mist applicator. In contrast, when vinclozolin was applied as a high-volume spray, the DFRs were 0.68 μg/cm^2. As would be expected, air sampling revealed higher concentrations of vinclozolin in the first several hours after the low-volume mist application. In contrast to these results, Giles et al. (1992) showed that an electrostatic, air-assisted, low-volume application of permethrin to greenhouse-grown chrysanthemums resulted in fourfold higher foliage deposits than was observed with the conventional high-volume spray application. They suggested that the electrostatic applicator was so effective at directing the pesticide to the chrysanthemum foliage instead of to inert surfaces in the greenhouse that it might facilitate a three- to fourfold reduction in application rates.

3-5-4 Differences in Sampling Methodology

Giles et al. (1992) also suggested that the leaf-disk rinsing/extracting method of Iwata et al. (1977) may overestimate the DFRs that would be possible with physical contact by human hands. Using permethrin-treated chrysanthemum leaves, they compared the DFRs obtained by the Iwata method to those obtained by a small brushing machine to physically dislodge the residues. With the Iwata method, they observed much greater differences between electrostatic applications and conventional sprays than was apparent with the mechanical brushing method of removing DFRs.

In other studies, Archibald, Solomon, and Stephenson (1994b) utilized a fluorescence Video Imaging Technique for Assessing Exposure (VITAE), originally developed by Fenske, Leffingwell, and Spear (1986) and Fenske et al. (1986), of greenhouse workers to pirimicarb as used in chrysanthemum production. (See Chapter 6.) A fluorescent dye, MDAC (4-methyl-7-diethylaminocoumarin) was combined with the pirimicarb and was quantified by video imaging techniques following 1 to 4 h of disbudding at 36 or 48 h after spraying. The fluorescent tracer gave the workers immediate feedback on the parts of their bodies that were most exposed to the plant foliage. As might be expected, the hands were the most exposed, followed by the arms, legs, head and neck, and torso. They concluded that as long as the dissipation rates for DFRs for both the fluorescent dye and a particular pesticide were known, accurate exposure studies could be conducted with the nontoxic dye instead of the pesticide. Their results also support recommendations that greenhouse workers should wear gloves and wash their hands regularly when they have frequent contact with plant foliage.

3-6 TURFGRASS AND LANDSCAPES

One of the leading controversies involving pesticide use during the 1980s and 1990s has been the use of pesticides on turfgrass, particularly in schoolyards, parks, playing fields, and other public areas. Obviously, the underlying concern pertains to possible risks for children because of their high use of these areas and because it is possible that children may be more sensitive to pesticides than is true for adults. Concern has also been raised for the impact of turfgrass pesticides on the health of domestic pets (Hayes et al. 1991).

Pest management in landscapes, particularly in turfgrass, is a rapidly growing industry in the United States and southern Canada. Herbicides to control broadleaf weeds in lawns are the predominant pesticides used. However, use of herbicides to control annual grasses and use of insecticides to control pests such as chinch bugs and white grubs also are rapidly growing segments of the business. Fungicides represent a large proportion of the pesticides used on golf courses, but fungicides rarely are used on nonirrigated turf-grass in domestic lawns, parks, and schoolyards. In the United States, approximately $2.2 billion is spent annually on the application of pesticides in the home and land-scape market, with about one-half being spent on home-use products and one-half on professional applications (Hodge 1993). Lawn care companies treat at least 10% of the home lawns each year and, in total, nearly 70 million pounds of pesticides are applied annually to lawns in the United States (GAO 1990). The province of Ontario represents a large proportion of the lawn care industry in Canada. In 1993, licensed landscape pes-ticide applicators applied more than 2 million lb (922, 113 kg) of pesticides in Ontario, with more than 70% being for broadleaf weed control in lawns (OMAFRA 1995). The herbicide 2,4-D and mixtures of 2,4-D with closely related herbicides are the predomi-nant treatments for control of broadleaf weeds whereas chlorpyrifos and diazinon are two of the major treatments for insects in turfgrass. Chlorthal-dimethyl and bensulide are the major spring or fall treatments for annual grasses. However, pendimethalin, dithiopyr, and fenoxaprop-ethyl are newly registered treatments, particularly for crab-grass. Chlorothalonil and benomyl are the major fungicides used for disease control on golf courses.

This extensive use of pesticides in urban landscapes has led to considerable concern in numerous cities across North America. Health and environmental activist groups have questioned the use of landscape pesticides by their municipal governments and by their neighbors. In some communities, opposing activist groups have fought to retain the use of pesticides to maintain a healthy turf to prevent physical injuries on playing fields during unfavorable weather conditions. Governments have responded in a variety of ways. In some municipalities (e.g., near Montreal), pesticide use in landscapes is banned completely, while higher courts debate the legalities of the decision. A more common approach has been for municipal governments to reduce pesticide use as much as possible in public areas through the implementation of integrated turfgrass management programs. State and provincial governments have imposed notification requirements to ensure that pesticide-treated areas, public and private, are clearly posted with signs so that potential users of these areas can minimize their exposure. In most cases, these notices are required only for areas treated by licensed professional applicators. In Ontario, it was the proposed pesticide notification legislation that prompted much of the research on factors influencing the persistence of DFRs in turfgrass environments. In 1993, the American Chemical Society sponsored a symposium on the Fate and Significance of Pesticides in Urban Environments. A portion of the research summarized here comes from relevant papers presented and published in the proceedings of that symposium (Racke and Leslie 1993).

3-6-1 Herbicides for Broadleaf Weed Control

In their initial studies on DFRs in turfgrass, Thompson, Stephenson, and Sears (1984) conducted a growth-room study in which they monitored the total dissipation of 2,4-D amine applied to pots of turfgrass. DFRs of 2,4-D in a cheesecloth wipe as well as residues in methanol or toluene rinses or acetone extracts of grass blades, roots, and soil were monitored over a 15-day period. Most of the chemical intercepted by each pot was associated with the surface residue fraction that was removed with the methanol rinse. This fraction decreased from 77 to 44% of that applied over the 15 days. DFRs in the cheesecloth wipe ranged from 10% of applied initially to 7% at 15 days (Fig. 3-2), thus only a small fraction of the surface residues have the potential to become DFRs. Extractable residues that had penetrated into the grass blades increased initially and then decreased via degradation and/or translocation to the roots. SEM revealed that the toluene rinse removed the epicuticular waxes. Gas chromatographic (GC) analysis of the toluene revealed that approximately 10% of the applied 2,4-D was bound to the waxes and most likely was not dislodgeable or available for penetration into the leaf cells.

In outdoor studies, Thompson, Stephenson, and Sears (1984) and Bowhey, McLeod, and Stephenson (1987) have shown repeatedly that initial DFRs of 2,4-D and related herbicides are lower and dissipate more rapidly than initially was observed in their growthroom studies. Possible explanations for this could include more cuticular waxes on outdoor plants, absence of moisture on the foliage (rainfall, irrigation, or dew) of indoor plants, absence of UV irradiation for photolysis of the herbicide on indoor plants, and differences in the sampling process (pinching and rubbing the grass blades between folds of cheesecloth for the pots of turfgrass grown indoors versus two-way scuffling of the turfgrass with cheesecloth pads on the feet in the field plots of treated turfgrass). In outdoor studies, DFRs for sprays of 2,4-D from a healthy Kentucky bluegrass turf ranged from 4 to 8% at 0 time to 1% or less of that applied after 1 to 3 days (Thompson, Stephenson, and Sears 1984; Solomon, Harris, and Stephenson 1993). DFRs for mecoprop, a close analogue of 2,4-D, were very similar; however, DFRs for dicamba, a common component in three-way mixtures of herbicides with 2,4-D, were usually much lower (Bowhey, McLeod, and Stephenson 1987). With sprays of 2,4-D, as application rate was increased, there was a more than proportional increase in DFRs. On the day of treatment, DFRs were 2.6, 6.9, and 18.6 mg/m^2 from plots treated with 1, 2, and 4 kg/ha, respectively. From day 1 onward, DFRs continued to be higher for the higher application rates, but when expressed as a percentage of the 2,4-D applied, they were quite similar. DFRs for 2,4-D (Thompson, Stephenson, and Sears 1984) or 2,4-D–mecoprop–dicamba mixtures (Bowhey, McLeod, and Stephenson 1987) were much lower for granular formulations than with the same amount of active ingredient applied as a spray, particularly on the day of treatment. However, DFRs for the granular formulations usually increased sharply during the first 24 h as the granules absorbed moisture and released the herbicide. Furthermore, when sprays were compared at their respective application rates for the active ingredient, which is 3 to 5 times higher for the granular, DFRs were not significantly different, particularly from 1 day after treatment. Although mowing treated turfgrass areas resulted in a significant reduction in DFRs, the amount of reduction was too small to be of any practical significance (Thompson, Stephenson, and Sears 1984). One of the most significant factors influencing the persistence of DFRs of broadleaf herbicides on turfgrass was rainfall or irrigation. In their initial studies, Thompson, Stephenson, and Sears (1984) showed that an 18-mm rainfall on the day of treatment reduced DFRs to barely detectable levels when sampled after 24 h. More recent studies (Schwan, Gatahi, and Stephenson 1997) have established that irrigation is similarly effective for reducing DFRs for either granular or liquid applications of 2,4-D on turfgrass foliage (Table 3-8).

Table 3-8 Impact of formulation and irrigation on dislodgeable 2,4-D residues from treated turfgrass foliage

Days after treatment	Dislodgeable residues (mg/m^2)[a]	
	Liquid 2,4-D (1.1 kg/ha)	Granular 2,4-D (5.0 kg/ha)
0	8.73a	7.02b
1 (before irrigation)	2.77c	9.36a
1 (after irrigation)	0.11d	0.34d
2	0.03e	0.09e

Source: Unpublished data from the G. Stephenson Laboratory, University of Guelph.

[a] Means followed by unlike letters are significantly different (P < 0.05, ANOV), as measured by the cheesecloth bootie scuffling method.

3-6-2 Herbicides for Control of Annual Grasses

Chlorthal-dimethyl is a soil-active, preemergence herbicide that can be applied in the spring or fall to control germinating annual grasses in established turfgrass. Pendimethalin and dithiopyr are new herbicides for annual grasses that have preemergence activity or preemergence and early postemergence activity, respectively. In studies with a wettable powder formulation of chlorthal-dimethyl and a dispersable granule formulation of pendimethalin, Hurto and Prinster (1993) used the detergent extracting technique (Iwata et al. 1977) to estimate the surface dislodgeable residues in irrigated and nonirrigated turfgrass. They observed that, with these formulations, surface residues for either chlorthal-dimethyl or pendimethalin were significantly lower from irrigated plots than from nonirrigated plots. However, regardless of the irrigation treatment, surface residues were usually 8 times higher for chlorthal-dimethyl (1.7 and 3.3 μg/cm^2) than for pendimethalin (0.2 and 0.4 μg/cm^2), most likely because of the fivefold difference in application rate. Cowell et al. (1993) used polyurethane foam pads and the scuffling technique and observed that DFRs were very low for an emulsifiable concentrate formulation of dithiopyr at time of treatment (3.14 mg/m^2 or <2% of applied) and decreased more than 80% within 24 h.

3-6-3 Insecticides for Turfgrass

Using methods very similar to those employed earlier by Thompson, Stephenson, and Sears (1984), Sears et al. (1987) found that DFRs for diazinon spray (125 g/L EC) were much lower than those observed for 2,4-D. Most of the residues were recovered as surface residues in the methanol or in the acetone extracts of the thatch. Very little was bound to or entered the leaf, and DFRs in the cheesecloth wipe decreased from 6% to less than 1% of applied within 24 h. Using the cheesecloth scuffling technique of sampling, DFRs for diazinon applied as a spray (125 g/L EC) were more dislodgeable than DFRs for diazinon applied as a granular (50 g a.i./kg) but only on the day of treatment (1.26% vs. 0.06% of that applied). Thereafter, DFRs for both formulations were well below 1% of that applied. Similar to observations for 2,4-D, DFRs were reduced quickly by irrigation for either the EC or the granular formulation. This is particularly significant for turfgrass insecticides, because a thorough irrigation usually is recommended immediately after application to aid the movement of the insecticide down into the thatch or soil for better control of insects such as white grubs. In further field studies, Sears et al. (1987) observed that DFRs for chlorpyrifos, isofenphos, and diazinon ranged from 1.5 to 2.5% of that applied at time of treatment and that there was a very rapid dissipation by movement into the thatch even without irrigation. In contrast to the enhanced dissipation of DFRs observed for the EC formulation (125 g/l) of diazinon as a result of irrigation (Sears et al. 1987), Hurto and Prinster (1993) saw little effect of irrigation on the dissipation of EC formulations of diazinon (Diazinon AG-500) or chlorpyrifos (Dursban 4EC). However, in their studies, they did observe up to a 50% reduction in DFRs for non-EC formulations of other pesticides such as pendimethalin (Pre-M, 60DG), chlorthal-dimethyl (Dachthal 75WP), and isophenfos (Oftanol 2F). The fact that their sprays were always mixtures of the pesticides with solubilized fertilizers is a possible explanation for their differing results from those of other investigators with some pesticides. However, it is clear that the effects of irrigation on DFRs are more dependent on formulation type and other factors in addition to differences in the active ingredient.

3-6-4 Differences in Sampling Methodology

Researchers at the University of Guelph (Thompson, Stephenson, and Sears 1984; Bowhey, McLeod, and Stephenson 1987; Soloman, Harris, and Stephenson 1993) have carried out the field sampling in all of their studies with a two-directional scuffling of 1 m^2 plots in which DFRs were collected onto water-moistened cheesecloth booties fastened over plastic bags placed over rubber boots. The exposed portions of the cheesecloth were quickly placed in bottles containing the appropriate solvent and returned to the laboratory for GC analysis. They have expressed their data on DFRs as a percentage of the chemical that was theoretically applied to the 1-m^2 plot.

In their study of DFRs with dithiopyr, Cowell et al. (1993) made an interesting comparison of sampling techniques. Their rinsing of grass blades with acetonitrile and water was found to be equivalent in efficiency to the detergent/water rinsing method developed by Iwata et al. (1977) but with a reduced problem of interfering compounds in the GC analysis. They also compared the cheesecloth scuffling technique (Thompson, Stephenson, and Sears 1984) with scuffling of the DFRs onto polyurethane foam (PUF) pads attached to the soles of boots. With dithiopyr, the PUF pads were found to be superior and they generally recovered higher DFRs than did cheesecloth pads, possibly because of a greater affinity for organic compounds. However, the aqueous acetonitrile rinse overestimated the quantity and persistence of DFRs at all times except immediately after treatment. The apparent half-lives for DFRs were 6.0 days and 0.9 days with the acetonitrile rinse method and the PUF scuffling method, respectively.

Harris and Solomon (1992) used the cheesecloth scuffling method and observed an excellent relationship between the dissipation of DFRs and actual human exposure to 2,4-D as monitored by urinalysis. At 1 h after treatment when DFRs were 8% of that applied, human volunteers in shorts with bare feet "who used the park for one hour" had detectable residues of 2,4-D in their urine that were well below the acceptable daily intake (ADI) for 2,4-D (0.3 mg/kg body weight per day). At 1 day after treatment, when DFRs were 1% of that applied, barefoot, bare-legged volunteers "who used the park for one hour" had no detectable residues of 2,4-D in their urine (limit of detection [LOD] was 0.4 μg/L). Thus, for 2,4-D, the Ontario requirement to keep sign notices of pesticide treatment up for 48 h after treatment seems to be more than adequate to prevent exposure.

For most pesticides used in turfgrass, there is a wide margin of safety for people to use the area even immediately after treatment. Requirements for notification rarely are needed to prevent acute toxic effects or illness. They simply are intended to help people to minimize their exposure or the exposure of their children if they wish to. Of course, the margin of safety could be much narrower for some of the more toxic insecticidal inhibitors of acetylcholinesterase than for the less toxic herbicides. In fact, Goh et al. (1986) observed that DFRs for dichlorvos exceeded the estimated safe levels for plant foliage (0.1 μg/cm^2 vs. 0.06 μg/cm^2) immediately after treatment. In the same study, DFRs for chlorpyrifos were well below the estimated safe level (0.14 μg/cm^2 vs. 0.5 μg/cm^2) at the time of treatment. However, Goh et al. (1986) employed the surfactant/water rinse method (Iwata et al. 1977), which may overestimate the DFRs that could have occurred by actual human contact.

As stated earlier, the greatest concern about DFRs of pesticides in turfgrass relates to the possibility for exposure and adverse health effects in children. However, studies of actual exposure can be conducted only with thoroughly informed adult volunteers (Harris and Solomon 1992). Black and Fenske (1996) have attempted to overcome this problem by correlating DFRs for chlorpyrifos from turfgrass with the DFRs of a nontoxic fluorescent dye applied simultaneously. By this approach, exposure of children to just the dye in a turfgrass environment could be employed to estimate the theoretical exposure to chlorpyrifos.

During the past decade in particular, the extensive amount of research that has been conducted by several groups of investigators indicates that DFRs for turfgrass pesticides are very low and drop quickly within 1 to 3 days after application. For the major turfgrass pesticides that have been studied, there is a wide margin of safety for immediate use of treated turfgrass areas. However, more research is needed to fully understand why rainfall or irrigation can quickly reduce DFRs for some but not all formulations of particular pesticides. It is unfortunate that no studies of DFRs have been conducted with some of the major fungicides that are used extensively on golf courses. In the likely event that studies of DFRs may be required for the registration of new turfgrass pesticides, more research to standardize sampling techniques, perhaps with the PUF scuffling method, certainly would be worthwhile. Finally, these authors are not aware of any studies of DFRs for pesticides used in other segments of the landscape environment, such as trees, shrubs, and flowers.

3-7 SUMMARY, CONCLUSIONS, AND OUTLINE OF RESEARCH NEEDS

In agricultural crops and in greenhouses, agricultural workers can be exposed to moderately toxic and highly toxic pesticides as DFRs or as vapors from the recently treated crops. With the initial use of OP insecticides during the the 1960s and 1970s, numerous cases of worker illness were encountered, particularly in citrus crops as grown and treated under California conditions. In response to these problems, worker exposure studies initially were required to determine reentry times that would be long enough to minimize health effects such as ChE inhibition or other physiological effects (Knaak 1980). Because of the logistics and cost and possible health effects of human exposure studies, Knaak, in 1980, was among the first to recommend the use of measurements of DFRs (Iwata et al. 1977) as an important criterion for establishing safe reentry intervals. Popendorf and Leffingwell (1982) went on to develop a unified field model for determining safe reentry intervals for workers in pesticide-treated crops. With their model, the first step was to determine the maximum acceptable change in some measurable health-related response such as the maximum inhibition of blood ChE that would not have an adverse effect on health. Using information on dermal dose response, dissipation rates for DFRs, initial DFRs, and application rates, acceptable reentry intervals then could be calculated with their model. Using a 4% daily inhibition and a one-time 50% inhibition of ChE as acceptable, Popendorf (1992) compared U.S. Environmental Protection Agency (EPA) and California reentry regulations for 20 different OP and carbamate pesticides. Reentry times as established in California were adequate according to their model for thirteen of the pesticides, excessive for five and inadequte for two. In contrast, the EPA reentry regulations were adequate for ten, inadequate for nine, and excessive for one. Popendorf (1992) concluded that the system is working well in California which, because of its dry, sunny climate, may be a relative worst-case scenario. However, he also concluded that there is a tremendous need for data on DFRs of pesticides on other crops and in other geographical areas.

It is apparent from this review that, although many of the newer agricultural pesticides are more persistent than some of the highly toxic OP or carbamate insecticides, it is fortunate that most are far less toxic. Furthermore, application rates and initial DFRs are often far lower. However, because of slower dissipation rates, greenhouse environments should be viewed as a special area of concern with respect to DFRs of pesticides. In the commercial environment of field, orchard, vineyard, or greenhouse agriculture, it is fortunate that we can use environmental measurements, laboratory analyses, toxicology, and mathematical models (i.e., good science) to determine reentry intervals with acceptable margins of safety for agricultural workers. In contrast, despite extensive efforts to educate and notify people in our cities that a potentially dislodgeable (but nontoxic) residue of a pesticide may be present in a public area or on a neighbor's lawn, there are many people who continue to find the use of any pesticides in such areas to be unacceptable.

Clearly, if we are to continue to provide good regulations to protect agricultural workers in either field or greenhouse environments and if we are to continue to inform and to provide good regulatory systems for safe pesticide use in urban environments, more research is needed.

We need a full understanding of the toxicity, chemical properties (volatility, photodegradation, activation, inactivation, water leachability, and cuticular penetration), and dislodgeability of toxic pesticides while on plant foliage wherever there is significant opportunity for human contact. Furthermore, we need to understand the effects of formulation components and technology on the dissipation of DFRs. At present, there appear to be too many contradictions to permit useful generalizations.

Early methods for estimating DFRs that involved shaking or rinsing leaf tissue in solvents (Iwata et al. 1977) were consistent and useful as long as they were followed carefully. More recent researchers suggest that these methods overestimate the residues that are actually dislodgeable by human contact (hands, gloves, clothing). An extra margin of safety is understandable in efforts to prevent illnesses among agricultural workers who will be reentering crops treated with "toxic" pesticides. However, with increased public awareness, there is an increasing concern about any chance for exposure to pesticides, no matter how nontoxic they are, particularly in urban areas, public areas, or areas used by children. For these purposes, glove DFRs or DFRs estimated by the cheesecloth–bootie–scuffling technique may be more accurate estimates of the DFRs to which people actually could be exposed. However, these methods may not be sufficiently consistent for regulatory use. Mechanized methods, such as brushes (Giles et al. 1992) or synthetic foam pads (Cowell et al. 1993), deserve more evaluation as possibly more consistent and more accurate methods for estimating the foliar residues of pesticides that are dislodgeable by human contact.

As scientists, it is too easy, and perhaps too self-serving simply to point out the need for more research. Few governments and few companies can afford the needed research indicated in just this review for all pesticides. Perhaps our first step would be for toxicologists to develop a list of low-risk, nonpersistent pesticides of low mammalian and human toxicity, for which information on DFRs would not be needed. We could then concentrate our research dollars and efforts on the study of DFRs where we have the greatest concerns for human health.

REFERENCES

Adams, J. D., Y. Iwata, and F. A. Gunther. 1977. Worker environment research, part 4: The effect of dust derived from several soil types on the dissipation of parathion and paraoxon dislodgable residues on citrus foliage. *Bull. Environ. Contam. Toxicol.* 15:547–54.

Ahrens, W. H., ed. 1994. *Herbicide handbook* 7th ed. Ahrens. Champaign, Ill.: Weed Science Society of America.

Archibald, B. A., K. R. Solomon, and G. R. Stephenson. 1994a. Survey of pesticide use by Ontario greenhouse chrysanthmum producers. *Bull. Environ. Contam. Toxicol.* 53:486–92.

Archibald, B. A., K. R. Solomon, and G. R. Stephenson. 1994b. Estimating pirimicarb exposure to greenhouse workers using video imaging. *Arch. Environ. Contam. Toxicol.* 27:126–29.

Ashton, F. M., and A. S. Crafts. 1981. *Mode of action of herbicides.* 2d ed. New York: Wiley.

Behrens, R., and M. A. Elakkad. 1981. Influence of rainfall on the phytotoxicity of foliarly applied 2,4-D. *Weed Sci.* 29:349–55.

Bentson, K. P. 1990. Fate of xenobiotics in foliar pesticide deposits. *Rev. Environ. Contam. Toxicol.* 114:125–61.

Bissell, S., J. Troiano, J. Marade, E. Graham, and M. del Valle. 1991. Comparison of sampling methods for determination of pesticide residue on leaf surfaces. *Bull. Environ. Contam. Toxicol.* 46:397–403.

Black, K. G., and R. A. Fenske. 1996. Dislodgeablility of chlorpyrifos and fluorescent tracer residues on turf: Comparison of wipe and foliar wash sampling techniques. *Arch. Environ. Contam. Toxicol.* 31:563–70.

Bowhey, C., H. McLeod, and G. R. Stephenson. 1987. Dislodgeable residues of 2,4-D on turf. *Proc. Brit. Crop. Protec. Conf.* 8A–10:799–805.

Brouwer, D. H., R. Brouwer, G. De Mik, C. L. Maas, and J. J. van Hemmen. 1992. Pesticides in the cultivation of carnations in greenhouses; part 1: Exposure and concomitant health risk. *Am. Ind. Hyg. Assoc. J.* 53:575–81.

Brouwer, R., D. H. Brouwer, S. C. H. A. Tijssen, and J. J. van Hemmen. 1992. Pesticides in the cultivation of carnations in greenhouses; part 2: Relationship between foliar residues and exposures. *Am. Ind. Hyg. Assoc. J.* 53:582–87.

Buck, N. A., B. J. Estesen, and G. W. Ware. 1980. Dislodgable insecticide residues on cotton foliage: Fenvalarate, permethrin, sulprofos, chlorpyrifos, methyl parathion, EPN, oxamyl, and profenofos. *Bull. Environ. Contam. Toxicol.* 24:283–88.

Bukovac, M. J., and P. D. Petrocek. 1993. Characterizing pesticide and surfactant penetration with isolated plant cuticles. *Pestic. Sci.* 37:179–94.

Chamel, A. 1986. Foliar absorption of herbicides: Study of the cuticular penetration using isolated cuticles. *Physiol. Veg.* 24 (4):491–508.

Cowell, J. E., S. A. Adams, J. L. Kunstman, M. G. Mueth. 1993. Comparison of foliar dissipation and turf dislodgeable residue sampling techniques. In *Pesticides in urban environments*, ed. K. D. Racke and A. R. Leslie, 100–112. Washington, D.C.: American Chemical Society.

Crosby, D. G. 1976. Herbicide photodecomposition. In *Herbicides: Chemistry, degradation and mode of action*, ed. P. C. Kearney and D. D. Kaufman, 835–90, New York: Marcel Dekker.

Davis, J. E., D. C. Staiff, L. C. Butler, and E. R. Stevens. 1981. Potential exposure to dislodgable residues after application of two formulations of methyl parathion to apple trees. *Bull. Environ. Contam. Toxicol.* 27:95–100.

Davis, J. E., E. R. Stevens, and D. C. Staiff. 1983. Potential exposure of apple thinners to azinphosmethyl and comparison of two methods for assessment of hand exposure. *Bull. Environ. Contam. Toxicol.* 31:631–38.

Davis, J. E., E. R. Stevens, D. C. Staiff, L. C. Butler. 1982. Potential exposure of apple thinners to phosalone. *Bull. Environ. Contam. Toxicol.* 29:592–98.

Estesen, B. J., and N. A. Buck. 1990. Comparison of dislodgable and total residues of three pyrethroids applied to cotton in Arizona. *Bull. Environ. Contam. Toxicol.* 44:240–45.

Estesen, B. J., N. A. Buck, and G. W. Ware. 1982. Dislodgable insecticide residues on cotton foliage: Carbaryl, cypermethrin, and methamidophos. *Bull. Environ. Contam. Toxicol.* 28:490–93.

Fenske, R. A., S. G. Birnbaum, M. M. Methner, and R. Soto. 1989. Methods for assessing fieldworker hand exposure to pesticides during peach harvesting. *Bull. Environ. Contam. Toxicol.* 43:805–13.

Fenske, R. A., J. T. Leffingwell, and R. C. Spear. 1986. A video imaging technique for assessing dermal exposure, part 1: Instrument design and testing. *Am. Ind. Hyg. Assoc. J.* 47:764–70.

Fenske, R. A., S. M. Wong, J. T. Leffingwell, and R. C. Spear. 1986. A video imaging technique for assessing dermal exposure, part 2: Fluorescent tracer testing. *Am. Ind. Hyg. Assoc. J.* 47:771–75.

Frank, R., H. E. Braun, G. Ritcey, and J. Stanek. 1989. Pyrazophos residues from treated greenhouse and growth chamber grown chrysanthemums. *Can. J. Plant Sci.* 69:961–66.

General Accounting Office (GAO). 1990. Lawn care pesticides. Pul. no. RCED-90-134, GAO, Washington, D.C.

Giles, D. K., T. C. Blewett, S. G. Saiz, A. M. Welsh, and R. L. Krieger. 1992. Foliar and nontarget deposition from conventional and reduced-volume pesticide application in greenhouses. *J. Agric. Food Chem.* 40: 2510–16.

Goh, K. S., S. Edmiston, K. T. Maddy, D. D. Meinders, and S. Margetich. 1986. Dissipation of dislodgeable foliar residues of chlorpyrifos and dichlorvos on turf. *Bull. Environ. Contam. Toxicol.* 37:27–32.

Gunther, F. A., Y. Iwata, G. E. Carman, and C. A. Smith. 1977. The citrus problem: Research on its causes and effects, and approaches to its minimization. *Residue Rev.* 67:1–133.

Gunther, F. A., W. E. Westlake, J. H. Barkley, W. Winterlin, and L. Langbehn. 1973. Establishing dislodgeable pesticide residues on leaf surface. *Bull. Environ. Contam. Toxicol.* 9:243–49.

Harris, S. A., and K. R. Solomon. 1992. Human exposure to 2,4-D following controlled activities on recently sprayed turf. *J. Environ. Sci. Health* B27 (1):9–22.

Hayes, H. M., R. E. Turone, K. R. Cantor, C. R. Jessen, D. M. McCurnin, and R. C. Richardson. 1991. Case-control study of canine malignant lymphoma: Positive association with dog owner's use of 2,4-dichlorophenoxyacetic acid herbicides. *J. Nat. Cancer Inst.* 83:1226–31.

Hess, F. D., D. E. Bayer, and R. H. Falk. 1981. Herbicide dispersal patterns, part 3: As a function of formulation. *Weed Sci.* 29:224–29.

Hodge, J. E. 1993. Pesticide trends in the professional and consumer markets. In *Pesticides in urban environments*, ed. K. D. Racke and A. R. Leslie, 10–17. Washington, D.C.: American Chemical Society.

Hurto, K. A., and M. G. Prinster. 1993. Dissipation of turfgrass foliar dislodgeable residues of chlorpyrifos, DCPA, diazinon, isofenphos and pendimethalin. In *Pesticides in urban environments*, ed. K. D. Racke and A. R. Leslie, 86–99. Washington, D.C.: American Chemical Society.

Iwata, Y., J. B. Knaak, R. C. Spear, and R. J. Foster. 1977. Worker reentry into pesticide-treated-crops, part 1: Procedure for the determination of dislodgeable pesticide residues on foliage. *Bull. Environ. Contam. Toxicol.* 18:649–55.

Joiner, R. L., and K. P. Baetcke. 1973. Parathion: Persistence on cotton and identification of its photoalteration products. *J. Agric. Food Chem.* 21:391–96.

Kerler, F., and J. Schönherr. 1988. Accumulation of lipophilic chemicals in plant cuticles: Prediction from octanol/water partition coefficients. *Arch. Environ. Contam. Toxicol.* 17:1–6.

Kirkwood, R. C. 1993. Use and mode of action of adjuvants for herbicides: A review of some current work. *Pestic. Sci.* 38:93–102.

Kloppenburg, D. J., and J. C. Hall. 1990. Efficacy of five different formulations of clopyralid on *Cirsium arvense* (L.) Scop. and *Polygonum convolvulus* L. *Weed Res.* 30:227–34.

Knaak, J. B. 1980. Minimizing occupational exposure to pesticides: Techniques for establishing safe levels of foliar residues. *Residue Rev.* 75:81–96.

Kudsk, P., and S. K. Mathiassen. 1991. Influence of formulations and adjuvants on the rainfastness of maneb and mancozeb on pea and potato. *Pestic. Sci.* 33:57–71.

Leffingwell, J. T., R. C. Spear, and D. Jenkins. 1975. The persistence of ethion and zolone residues on grape foliage in the central valley of California. *Arch. Environ. Contam. Toxicol.* 3:40–54.

Martin, J. T., and B. E. Juniper. 1970. *The cuticles of plants*. Edinburgh: Edward Arnold.

McEwen, F. L. 1977. Pesticide residues and agricultural workers—An overview. In *Pesticide management and insecticide resistance*, ed. D. L. Watson, and A. W. A. Brown, 37–49. New York: Academic Press.

McEwen, F. L., G. Ritcey, H. Braun, R. Frank, and B. D. Ripley. 1980. Foliar pesticide residues in relation to worker re-entry. *Pestic. Sci.* 11:643–50.

Nigg, H. N., L. G. Albrigo, H. E. Nordby, and J. H. Stamper. 1981. A method of estimating leaf compartmentalization of pesticides in citrus. *J. Agric. Food Chem.* 29:750–56.

Nigg, H. N., J. C. Allen, R. F. Brooks, G. J. Edwards, N. P. Thompson, R. W. King, and A. H. Blagg. 1977. Dislodgeable residues of ethion in Florida citrus and relationships to weather variables. *Arch. Environ. Contam. Toxicol.* 6:257–67.

Nigg, H. N., J. C. Allen, and R. W. King. 1979. Behavior of parathion residues in the Florida "Valencia" orange agroecosystem. *J. Agric. Food Chem.* 27:578–82.

Nigg, H. N., S. S. Brady, and I. D. Kelly. 1992. Dissipation of foliar dislodgeable residues of bendiocarb following application to azaleas. *Bull. Environ. Contam. Toxicol.* 48:416–20.

Nigg, H. N., R. D. Cannizzaro, and J. H. Stamper. 1986. Diflubenzuron surface residues in Florida citrus. *Bull. Environ. Contam. Toxicol.* 36:833–38.

Nigg, H. N., and J. H. Stamper. 1981. Comparative disappearance of dioxathion, malathion, oxydemetonmethyl and dialifor from Florida citrus leaf and fruit surfaces. *Arch. Environ. Contam. Toxicol.* 10:497–504.

Nigg, H. N., J. H. Stamper, and J. B. Knaak. 1984. Leaf, fruit, and soil surface residues of carbosulfan and its metabolites in Florida citrus groves. *J. Agric. Food Chem.* 32:80–85.

Nigg, H. N., J. H. Stamper, and R. M. Queen. 1984. The development and use of a universal model to predict tree crop harvester pesticide exposure. *Am. Ind. Hyg. Assoc. J.* 45:182–86.

Nilsson, U., T. Nybrant, M. Papantoni, and L. Mathiasson. 1996. Long-term studies of fungicide concentrations in greenhouses, part 2: Fungicide concentrations in air and on leaves after different exposure times and under different climate conditions. *J. Agric. Food Chem.* 44:2878–84.

Ontario Ministry of Agriculture and Food and Rural Affairs (OMAFRA). 1995. *Survey of pesticide use in 1993*. Toronto.

Popendorf, W. 1992. Reentry field data and conclusions. *Rev. Environ. Contam. Toxicol.* 128:71–117.

Popendorf, W. J., and J. T. Leffingwell. 1978. Natural variations in the decay and oxidation of parathion foliar residues. *J. Agric. Food Chem.* 26:437–41.

Popendorf, W. J., and J. T. Leffingwell. 1982. Regulating OP pesticide residues for farmworker protection. *Residue Rev.* 82:125–201.

Popendorf, W. J., and R. C. Spear. 1974. Preliminary survey of factors affecting the exposure of harvesters to pesticide residues. *Am. Ind. Hyg. Assoc. J.* 35:374–80.

Que Hee, S. S., and R. G. Sutherland. 1981. *The phenoxyalkanoic herbicides.* Vol. 1. Boca Raton, Fla: CRC Press.

Racke, K. D., and A. R. Leslie. 1993. Pesticides in urban environments, fate and significance. Washington, D.C.: *American Chemical Society.*

Riederer, M., and J. Schönherr. 1987. Covalent binding of chemicals to plant cuticles: Quantitative determination of epoxide contents of cutins. *Arch. Environ. Contam. Toxicol.* 15:97–105.

Roggenbuck, F. C., R. F. Burow, and D. Penner. 1994. Relationship of leaf position to herbicide absorption and organosilicone adjuvant efficacy. *Weed Technol.* 8:582–85.

Roggenbuck, F. C., D. Penner, R. F. Burow, and B. Thomas. 1993. Study of the enhancement of herbicide activity and rainfastness by an organosilicone adjuvant utilizing radiolabelled herbicide and adjuvant. *Pestic. Sci.* 37:121–25.

Sears, M. K., C. Bowhey, H. Braun, and G. R. Stephenson. 1987. Dislodgeable residues and persistence of diazinon, chlorpyrifos and isofenphos following their application to turfgrass. *Pestic. Sci.* 20:223–31.

Solomon, K. R., S. A. Harris, and G. R. Stephenson. 1993. Applicator and bystander exposure to home garden and landscape pesticides. In *Pesticides in urban environments*, ed. K. D. Racke and A. R. Leslie, 262–74. Washington, D.C.: American Chemical Society.

Spear, R. C., Y. S. Lee, J. T. Leffingwell, and D. Jenkins. 1978. Conversion of parathion to paraoxon in foliar residues: Effects of dust level and ozone concentration. *J. Agric. Food Chem.* 26:434–36.

Spencer, J. R., S. R. Bissell, J. R. Sanborn, F. A. Schneider, S. S. Margetich, and R. I. Krieger. 1991. Chlorothalonil exposure of workers on mechanical tomato harvesters. *Toxicol. Lett.* 55:99–107.

Staiff, D. C., S. W. Comer, and R. J. Foster. 1975. Residues of parathion and conversion products on apple and peach foliage resulting from repeated spray applications. *Bull. Environ. Contam. Toxicol.* 14: 135–39.

Stamper, J. H., H. N. Nigg, and W. Winterlin. 1981. Growth and dissipation of pesticide oxons. *Bull. Environ. Contam. Toxicol.* 27:512–517.

Sundaram, K. M., and A. Sundaram. 1994. Rain-washing of foliar deposits of Dimilin® WP-25 formulated in four different liquids. *J. Environ. Sci. Health* B29 (4):757–83.

Thompson, D. G., G. R. Stephenson, and M. K. Sears. 1984. Persistence, distribution and dislodgeable residues of 2,4-D following its application to turfgrass. *Pestic. Sci.* 15:353–360.

Tomlin, C. 1994. *The pesticide manual.* 10th ed. Surrey, UK: British Crop Protection Council.

Ware, G. W., N. A. Buck, and B. J. Estesen. 1983. Dislodgeable insecticide residues on cotton foliage: Comparison of ULV/cottonseed oil vs. aqueous dilutions of 12 insecticides. *Bull. Environ. Contam. Toxicol.* 31:551–58.

Ware, G. W., B. J. Estesen, and N. A. Buck. 1980. Dislodgable insecticide residues on cotton foliage: Acephate, AC222, 705, EPN, fenvalerate, methomyl, methyl parathion, permethrin, and thiocarb. *Bull. Environ. Contam. Toxicol.* 25:608–15.

Ware, G. W., and D. P. Morgan. 1976. Worker reentry safety, part 9: Techniques of determining safe reentry intervals for organophosphate-treated cotton fields. *Residue Rev.* 62:79–100.

Wicker, G. W., W. A. Williams, J. R. Bradley, Jr., and F. E. Guthrie. 1979. Exposure of field workers to organophosphorus insecticides: Cotton. *Arch. Environ. Contam. Toxicol.* 8:433–40.

Willis, G. H., L. L. McDowell, S. Smith, and L. M. Southwick. 1992. Foliar washoff of oil-applied malathion and permethrin as a function of time after application. *J. Agric. Food Chem.* 40:1086–1089.

Willis, G. H., L. L. McDowell, S. Smith, and L. M. Southwick. 1994a. Permethrin and sulprofos washoff from cotton plants as a function of time between application and initial rainfall. *J. Environ. Qual.* 23: 96–100.

Willis, G. H., L. L. McDowell, L. M. Southwick, and S. Smith. 1994b. Azinphosmethyl and fenvalerate washoff from cotton plants as a function of time between application and initial rainfall. *Arch. Environ. Contam. Toxicol.* 27:115–20.

Willis, G. H., S. Smith, L. L. McDowell, and L. M. Southwick. 1996. Carbaryl washoff from soybean plants. *Arch. Environ. Contam. Toxicol.* 31(2):239–43.

Winterlin, W. L., W. W. Kilgore, C. R. Mourer, and S. R. Schoen. 1984. Worker reentry studies for captan applied to strawberries in California. *J. Agric. Food Chem.* 32:664–72.

Winterlin, W., G. Walker, G. Hall, J. McFarland, and C. Mourer. 1980. Environmental fate of dialifor and for-
    mation of its oxygen analogue following application on grape vines in the San Joaquin Valley, California.
    J. Agric. Food Chem. 28:1078–83.

Wolfe, H. R., J. F. Armstrong, D. C. Staiff, S. W. Comer, and W. F. Durham. 1975. Exposure of apple thinners
    to parathion residues. *Arch. Environ. Contam. Toxicol.* 3:257–67.

Zweig, G., R. Gao, J. Witt, W. Popendorf, and K. Bogen 1984. Dermal exposure to carbaryl by strawberry
    harvesters. *J. Agric. Food Chem.* 32:1232–1236.

Zweig, G., R. Gao, and W. J. Popendorf. 1983. Simultaneous dermal exposure to captan and benomyl by
    strawberry harvesters. *J. Agric. Food Chem.* 31:1109–13.

Zweig, G., J. T. Leffingwell, and W. Popendorf. 1985. The relationship between dermal pesticide exposure by
    fruit harvesters and dislodgeable foliar residue. *J. Environ. Sci. Health.* B20 (1):27–59.

FOUR

DERMAL PENETRATION

Richard P. Moody

Product Safety Laboratory, Health Protection Branch, Health Canada, Ottawa, Ontario, Canada K1A 0L2

4-1 INTRODUCTION

The need for accurate dermal absorption efficacy data for determining the health risk of nondietary exposure to pesticides cannot be overemphasized (Franklin, Somers, and Chu 1989). The term "efficacy" is used to refer to the degree or magnitude of dermal absorption of pesticides although the term is more commonly used in reference to the beneficial aspects of transdermal drug delivery. The term is used here to refer to the transdermal efficacy of pesticide formulations containing an active ingredient.

Evaluation of the efficacy of pesticide transdermal absorption is directly dependent upon the accurate acquisition, transcription, reporting, and toxicological evaluation of percutaneous absorption data. This chapter reviews standard in vivo and in vitro test methods used in our laboratory and attempts to demonstrate where these methods may fail to provide accurate data. (For a more general review of in vitro and in vivo methods, see Bronaugh and Maibach [1991].) The efficacy of dermal absorption of a pesticide can be expressed quantitatively as the percentage of the total topically applied dose of the pesticide that becomes bioavailable via the dermal absorption route, including intercellular, intracellular, and transappendageal components. The theory concerning the mechanisms of absorption (e.g., chemical diffusion) through the skin tissue (Fig. 4-1) has been reviewed previously (Moody and Chu 1995).

This chapter focuses on the advantages and disadvantages of both the early methods of determining dermal absorption efficacy (e.g., in vivo animal testing) and the more recently developed in vitro and computer QSAR models. In the process, it is hoped that the reader will acquire at least a basic knowledge of some of the techniques that both laboratory and regulatory scientists use when evaluating dermal absorption data.

Dermal absorption studies undergo critical evaluation by both government and industry to ensure that the data acquired from these studies are applicable for predicting

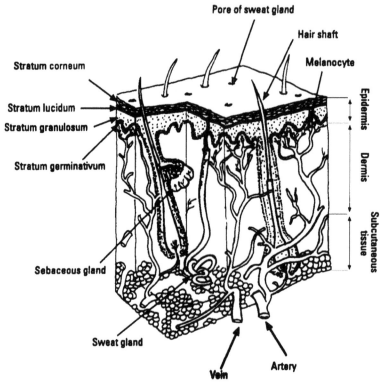

Figure 4-1 Microanatomy of human skin.
Source: Reprinted from Moody and Chu (1995).

human occupational and bystander systemic exposure to pesticides. At present, in vitro test data are not used directly by the Pest Management Regulatory Agency (PMRA) of Health Canada to derive exposure estimates for pesticide risk assessments. It is recognized by the PMRA, however, that in vitro data may be of some value (e.g., for bridging data gaps, characterizing formulation effects) and the PMRA is reviewing the acceptability of employing in vitro data for risk assessment (PMRA. 21 April 1997. letter to author). The acceptability of in vitro data recently has been given increasing scrutiny because these methods generally are considered to provide potentially humane and cost-effective alternatives to in vivo animal testing. Furthermore, the animal rights movement and growing public support have placed a stronger emphasis on the search for alternative methods to in vivo testing. In Europe, after January 1, 1998, a ban will be instituted on all animal testing that is conducted for the purpose of developing new cosmetics and their ingredients, unless validated alternative methods are not available (Griffin 1996).

This chapter examines the advantages and disadvantages of the methods employed, with particular emphasis on data originating from our own laboratory and this includes mention of our new Automated In Vitro Dermal Absorption (AIVDA) method that employs small autosampler vial inserts to hold skin tissue samples for dermal testing. A section concerning the comparison of in vivo and in vitro data is included to address issues pertaining to the validation of in vitro data. As well, a brief discussion of the use of QSAR methods together with a QSAR analysis of data recently originating from our laboratory is given. Finally, after considering model validation, this chapter concludes by recommending future research areas that are considered to be of particular importance to the rapidly evolving dermal absorption field.

4-2 EARLY IN VIVO ANIMAL MODELS

When our laboratory at Health Canada first commenced dermal absorption testing, in vitro methods employing isolated skin tissue were not commonly employed for assessing pesticide dermal bioavailability (i.e., systemic exposure). Hence our original investigations utilized established in vivo test methods that employed living animals (Moody and Franklin 1987; Moody, Kiedel et al. 1987; Moody, Benoit, et al. 1989; Moody, Grayhurst, and Ritter, 1989; Moody and Ritter, 1989; Moody et al. 1990). Several animal species were tested to determine which might be the best surrogate model for humans. Although the smaller animal species such as the rat and the rabbit were considered to be advantageous for technical and economic reasons (e.g., smaller urine volume for collection/analysis, more easily manipulated for dose applications, low cost of animal, food, and caging), the rhesus monkey was considered to be the best human surrogate because of the pioneering studies of Maibach at the University of California at San Francisco (UCSF) that were initiated in the late 1960s. In fact, the Health Canada test protocol for topical dose application, correction for nonurinary excretion via an intramuscular injection, and the use of ^{14}C-radiolabeled pesticides and liquid scintillation counting (LSC) analysis originated from the Maibach procedures. Table 4-1 gives a summary of our early data obtained by using in vivo tests, as well as some human data obtained from the Maibach laboratory (Moody et al. 1990).

Examining Table 4-1, it is difficult to obtain consistent species-related differences in skin permeability. A common paradigm is that rat skin is much more permeable to pesticides than human skin and that one can better approximate human skin permeability using monkey data. The assumption that rat skin is more permeable than human skin is a hazardous one because it can lead to false predictions of pesticide safety. For example, the data for 2,4-D amine in Table 4-1 shows only 20% skin absorption in rats but 31% and 58% absorption in monkeys and humans, respectively. Of the remaining compounds for both rat and monkey data in Table 4-1, aminocarb, DEET, and lindane gave similar percentage absorption in rats and monkeys whereas greater transdermal efficacy of fenitrothion and the two permethrin isomers (cis and trans ^{14}C-radiolabeled alcohol moiety) was obtained for rats. Hence, for the seven compounds in Table 4-1 for which both rat and monkey data are presented, only three showed rats as being more permeable than monkeys and two of these compounds were isomers. If we assume that the monkey is a good model of human skin permeability, then we could conclude here that the rat is less permeable than humans. If this were assumed to be the case, extrapolations of systemic pesticide exposure based on rat data could severely underestimate human exposure for compounds not fitting our low rat permeability assumption.

The premise that monkeys are a good predictor of human skin absorption, however, is not born out by the limited data in Table 4-1. In fact, for two (the acid form and iso-octyl

Table 4-1 Total mean percentage dermal absorption (±SD) of pesticides for different animal species including human

Pesticide	Dermal absorption (% ± SD)			
	Rat	Rabbit	Monkey	Human[a]
Aminocarb	88 ± 6		74 ± 4	
(Matacil)	n = 5		n = 8	
2,4-D (acid)		36 ± 11	29 ± 8	6 ± 2
		n = 3	n = 8	n = 6
				FA
2,4-D (amine)	20 ± 3	12 ± 4	31 ± 13	58 ± 23
[water]	n = 8	n = 8	n = 8	n = 6
2,4-D (iso-octyl)		50 ± 13	56 ± 15	6 ± 3
		n = 6	n = 6	n = 4
DEET	36 ± 8		33 ± 11	
	n = 8		n = 8	
Fenitrothion	84 ± 12		49 ± 4	
(Sumithion)	n = 5		n = 8	
Lindane	31 ± 10		34 ± 5	
	n = 5		n = 7	
Permethrin	46 ± 4		28 ± 6	
(cis-alcohol)[b]	n = 5		n = 8	
Permethrin	43 ± 5		21 ± 3	
(trans-alcohol)[b]	n = 5		n = 8	

Sources: Data on aminocarb and fenitrothion, Moody and Franklin (1987); on 2,4-D and derivatives, Moody et al. (1990); on DEET, Moody, Benoit, et al. (1989); on lindane, Moody and Ritter (1989); on permethrin, Sidon, Moody, and Franklin (1988).

Note: The percentage skin absorption is corrected for nonurinary recovery. The number (n) of animals used per test is indicated. Rats and rabbits were dosed on the back region. Monkey and human subjects were dosed on the forehead except for the human 2,4-D acid study in which the forearm (FA) was dosed. For all pesticides the dose application vehicle was acetone except for 2,4-D amine which was applied in water.

[a] Human data are from the Maibach UCSF laboratory.

[b] For brevity, the cis and trans ^{14}C-labeled cyclopropyl permethrin data are not included here but are given by Sidon, Moody, and Franklin (1988).

derivative of 2,4-D) of the three cases in which data for both monkeys and humans are given, the monkeys showed from about five- to ninefold greater skin absorption than humans. Given the high intersubject variation (SD = 23%) for the human data for the third compound (2,4-D amine), the monkey and human data were not significantly different (two-tailed Student's t-test; $P < .05$). Hence, the monkey data in Table 4-1 overpredict human absorption of these compounds. Of course, a new paradigm cannot be created on the basis of three data points, and the Maibach laboratory has reported data consistent with the monkey being a good human model (Wester and Maibach 1991), but there are often exceptions to the rules of general paradigms and these should be considered before fully accurate health risk and safety assessments are conducted.

Further to the monkey/human cross comparison, the Maibach laboratory has reported 17% and 9% absorption of DEET and lindane, respectively, for these compounds applied in acetone to the ventral forearm of human subjects (Feldmann and Maibach 1970, 1974). Table 4-1 shows 33% and 34% absorption for monkeys dosed on the forehead with DEET and lindane, respectively. Is this more evidence that the monkey could overpredict human skin permeability? To answer this question, it is necessary to consider the effect of anatomical site-related differences in skin permeability. Consistent

Table 4-2 Anatomic site-related differences in rhesus monkey skin permeability (mean percentage dermal absorption [±SD]) to pesticides

	Dermal absorption (%±SD)					
	Monkey			Human		
Pesticide	Forearm	Forehead	Palmar	Forearm	Forehead	Palmar
Aminocarb	37 ± 14	74 ± 4				
(Matacil)	n = 8	n = 8				
2,4-D (acid)	15 ± 10	29 ± 8		6 ± 2		
	n = 8	n = 8		n = 6		
2,4-D (amine)	6 ± 3	31 ± 13		7 ± 6*	58 ± 23	10 ± 12*
[water]	n = 8	n = 8		n = 4	n = 6	n = 4
2,4-D (iso-octyl)	40 ± 24	56 ± 15			6 ± 3	
	n = 8	n = 8			n = 4	
DEET	14 ± 5	33 ± 11	68 ± 9	17		
	n = 8	n = 8	n = 8			
Fenitrothion	21 ± 10	49 ± 4				
(Sumithion)	n = 8	n = 8				
Lindane	18 ± 4	34 ± 5	54 ± 26	9		
	n = 8	n = 7	n = 4			
Permethrin	9 ± 3	28 ± 6				
(cis-alcohol)	n = 8	n = 8				
Permethrin	12 ± 3	21 ± 3				
(trans-alcohol)	n = 8	n = 8				

Sources: Monkey data on aminocarb and fenitrothion, Moody and Franklin (1987); on 2,4-D and derivatives, Moody et al. (1990); on DEET, Moody, Benoit, et al. (1989); on lindane, Moody and Ritter (1989); on permethrin, Sidon, Moody, and Franklin (1988). Human data from the Maibach laboratory are included for comparison including the data (*) from Moody, Wester, and Maibach (1992).

Note: Acetone dose vehicle was employed in all cases except for 2,4-D amine where water was employed.

with the early report of Maibach et al. (1971) that skin permeability to topically applied compounds could vary severalfold in humans, our studies have reported anatomic site-related differences in rhesus monkeys. As shown in Table 4-2, the monkey forehead was consistently more permeable to pesticides than the forearm. In contrast with early reports from the Maibach laboratory that the human palmar region was not considered to be a relatively permeable site (Maibach et al. 1971), our monkey data showed for both DEET and lindane that the monkey palm (volar forepaw) was more permeable than either the forehead or the forearm (Table 4-2). Perhaps the intraspecies difference in the degree of cornification of the monkey palmar region versus other anatomic sites differs from that reported for humans. The human palm is known to have a thick stratum corneum in relation to other anatomic sites (Grasso and Lansdown 1972). Whatever the reason for this cross-species discrepancy, the paradigm that the human palmar region is less permeable to pesticides because of its thick protective stratum corneum was not validated by a more recent study in which the percentage absorption of 2,4-D amine in the human palmar and forearm sites was similar (Moody, Wester, and Maibach 1992). This point is stressed because pesticide exposure to field-workers commonly involves the hands, especially when protective gloves are not worn (Franklin, Somers, and Chu 1989) and the importance of transdermal exposure via the palmar route should not be underestimated.

To return to the question concerning whether the Maibach DEET and lindane human forearm data provide further evidence that monkey data may overpredict human exposure, it is apparent that we cannot directly cross-compare data for different anatomic sites. However, if we examine Table 4-2, it is evident that the monkey forearm data for DEET fit the general paradigm that the monkey is an accurate human model, but the lindane data do not (monkey forearm absorption of lindane is twice that of the human forearm). Thus for the four compounds considered here, there are two (2,4-D amine, DEET) in agreement and two (2,4-D acid, lindane) in disagreement with the monkey model being accurate for human percutaneous absorption. Unless a cross-species safety factor is introduced, it is inadvisable that human systemic exposure estimates be made on the basis of monkey data, at least for the type of pesticide compounds considered here. PMRA routinely applies an appropriate interspecies safety factor when assessing risk based on nonhuman toxicology or toxicokinetic data (PMRA. 1997. Letter to author, 21 April). "Margin of safety determinations are usually done (by the PMRA) on a case-by-case basis depending on the severity of the endpoint, strength of evidence and quality of data" (PMRA. 1997. Letter to author, 31 July). For a general description of the procedure used for pesticide registration in Canada, including the use of safety factors, see Ritter and Curry (1988).

As well as difficulty in cross-comparing the dermal absorption data obtained from different species and from different anatomic sites, a number of other variables hinder the practicality of conducting expensive, labor-intensive in vivo tests for predicting field-worker exposure to pesticides. It is not always obvious that the animal caging environment can differ substantially from that experienced in situ by the pesticide applicator, mixer/loader, home gardener, etc. Environmental factors such as temperature and wind speed (e.g., volatilization and persistence), rain (e.g., surface washing), humidity (e.g., skin hydration), and solar irradiation (e.g., photolysis), may affect the transdermal efficacy of pesticides. Because these factors are difficult to model in vivo, alternative methods are required.

4-3 ALTERNATIVE IN VITRO SKIN ABSORPTION MODELS

To circumvent the problems associated with in vivo testing, in vitro test methods, where the skin tissue is maintained separately from the animal during the dermal absorption test, were developed at Health Canada. As well as providing better control of environmental test conditions and a less laborious, more cost-effective procedure with much faster data acquisition, the in vitro method provides an alternative to the suffering that animals may experience during in vivo testing (Chu 1995). This latter provision is highly recommended from both an ethical and a scientific point of view because animals can suffer discomfort which can elicit physiological stress (e.g., rapid pulse) during in vivo testing. The rhesus monkeys in our tests were restrained in metal holding chairs for 24 h to prevent them from ingesting the topical dose. This form of stress was thought to be responsible for the high degree of intersubject variation obtained in the lindane dermal absorption (palmar dose site) study in which the restrained monkeys appeared to be unusually agitated (Moody and Ritter 1989). Stress from chair restraint, especially the first 24-h duration, has been reported to elevate serum lactate dehydrogenase levels in rhesus monkeys (Scott, Kosch, and Hilmas 1976). Even smaller test animals such as rats can undergo stress, for example, when the protective patch necessary to protect the dose site from ruboff of the pesticide was constricting. The fur-shaving procedure was also a stress factor because the fur needed to be shaved close to the skin surface to ensure good adhesion of the protective dose patch and the animals needed to be held tightly during shaving. In our studies, fur shaving was conducted immediately prior to the pesticide dose application. An attempt was made to find a less stressful dosing procedure where rats were dosed with lindane on the unshaved tail region. Although the results were promising and showed good agreement between the permeability of the rat tail and back regions to lindane, the ability of rat tail skin to provide a good model for a range of chemicals was not investigated (Moody, Grayhurst, and Ritter 1989b).

Any stress factor, even just the presence of animal care technicians during the dosing procedure, could elicit sufficient physiological change to alter the observed transdermal efficacy of pesticides. As discussed by Moody and Ritter (1989), a stress-induced, heart-rate elevation could affect the transdermal efficacy of pesticides by increasing the rate of blood flow in the dermal microvasculature and enhancing dermal clearance of the pesticide skin depot. Although it was suggested by Moody and Ritter (1989) that it would be necessary to monitor the effect of stress on the animals by employing a physiology monitoring system, the cost of this system was prohibitive. A brief attempt was made to train young monkeys so that chair restraint would not be required but this was not successful. Given the problems associated with in vivo tests, in vitro alternatives were evaluated to determine whether the data obtained could be used to predict in vivo data and hence to reduce the need for animal tests.

4-3-1 AIVDA Cell

This section details the progress from our early AIVDA aluminum flow-through cell method to the more recently developed Skin Permeation Tube Analysis (SPTA) AIVDA procedure. Detailed specifications of these methods have been reported in our published papers cited herein. This section summarizes our data and discusses some of the advantages and disadvantages of these in vitro methods in comparison with the more standard Bronaugh flow-through cell method. For review of in vitro dermal absorption cell methods (e.g., Franz finite, Bronaugh flow-through), skin preparation (e.g., dermatoming, skin viability), receiver solution composition (presence of serum albumin), and soforth, see Bronaugh and Maibach (1991) and Bronaugh (1996a).

The original AIVDA cell consisted of two aluminum plates that were clamped together to hold the dermatomed skin specimen between Teflon washers recessed into four donor/receiver chambers in the cell block (Fig. 4-2). The technical specifications are described by Moody and Martineau (1990). More recent AIVDA cell blocks had six donor/receiver chambers, the Teflon washers being replaced with aluminum clamping ridges. The AIVDA cell was designed to better control environmental conditions during the dermal absorption assay.

For example, the cell included quartz windows to permit investigation of the effect of solar ultraviolet irradiation (e.g., pesticide photolysis, sunburn of skin) on the absorption

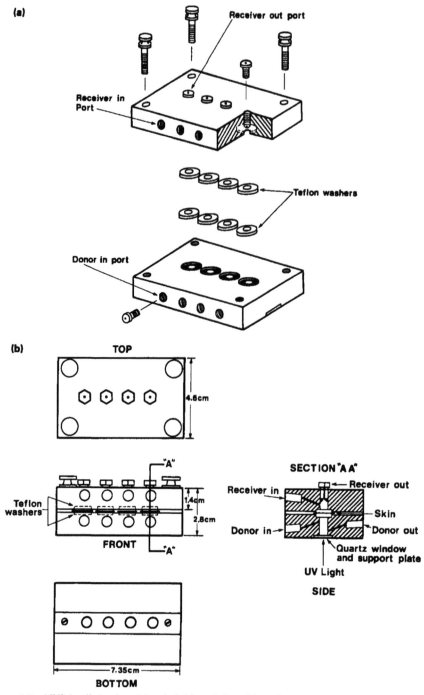

Figure 4-2 AIVDA cell chamber: (a) exploded frontal view; (b) top, front, bottom, and cross-sectional side view.

Source: Reprinted by permission of Moody and Martineau (1990).

process (Moody and Martineau 1990). A major advantage of the AIVDA cell was the small size of the exposed skin surface area ($0.2 \, cm^2$) and the simple clamping mechanism employed. The small exposure/dose area permitted the use of small, difficult-to-obtain human skin specimens, and the clamping mechanism permitted a rectangular skin strip to be cut with scissors and mounted over several receiver/donor wells at a time. This was less tedious than the need to punch out the large circular skin specimens ($0.64 \, cm^2$) required for the Bronaugh cells. A small skin exposure area in a flow-through type cell is, however, also a disadvantage because it can reduce the analytical sensitivity of the detection method. This latter point is addressed further in the section on the SPTA method.

As well as permitting standard finite dose applications that mimic exposure of field-workers to pesticide spray droplets, for example, the AIVDA cell was designed for investigating true, infinite, dose-type applications such as would be encountered by swimmers or bathers (Fig. 4-3). We recently reviewed the dermal exposure route for swimmers or bathers exposed to water contaminants in the Great Lakes (Moody and Chu 1995) and bathing exposure related to groundwater contamination (Moody [in press]). Other than one study that demonstrated dermal absorption of the commonly used swimming-pool chlorine stabilizer, cyanuric acid (Moody et al. 1993), most of our AIVDA studies employed the finite dose application. These data are summarized in Table 4-3 together with data on DDT from the Bronaugh cell. The technical specifications for the Bronaugh cell are given by Bronaugh and Stewart (1985).

The data in Table 4-3 are consistent with rat skin being more permeable than human skin. In general, the rat in vitro data agree with in vivo data, but the DEET in vitro data overestimate the in vivo data, whereas the reverse holds for the DDT Bronaugh cell data. The rat in vitro studies used skin from the same rats tested in vivo for each pesticide. Our early studies used a Ringer's-based receiver solution whereas a Hank's solution containing serum albumin was used later because this receiver had been reported to maintain skin viability (Bronaugh 1996a,b) and to enhance the dissolution of lipophilic skin permeants by the receiver solution. The issue of skin viability is considered later in this chapter.

SWIMMING POOL MODEL

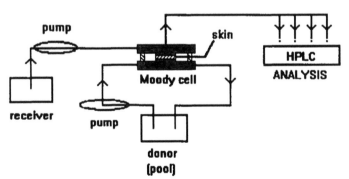

Figure 4-3 Use of the AIVDA cell for testing swimming or bathing exposure to pesticide water contaminants. The receiver solution flows under the skin to collect skin permeants. The donor swimming pool recirculates the pesticide-treated water over the exposed outer epidermal skin surface (i.e., the stratum corneum).
Source: Reprinted by permission of Moody et al. (1993) and Moody and Chu (1995).

The last two rows of Table 4-3 show that significantly less dermal absorption of FPE was obtained for rats tested in vivo with the FPE applied in a field formulation (Excel) than when this pesticide was applied in vivo in an acetone vehicle. Because vehicle effects are well documented in the literature, especially concerning the transdermal efficacy of pharmaceutical products, it is recommended that data used for assessing pesticide safety employ skin absorption data from studies in which the dose vehicle is the same as that of the end-use pesticide formulation. If the pesticide formulation is modified (e.g., diluted in water) prior to use, the dose vehicle should be modified accordingly so that field chemical exposure conditions are comparable.

Bronaugh flow-through cell. As reported by Bronaugh and Stewart (1985) (for recent method reviews, see Bronaugh [1996a,b]), this cell consists of a Teflon cell chamber equipped with a screw-in Teflon skin clamping part that also serves as a donor chamber. The skin specimen is held horizontally and the receiver solution is pumped under the untreated cut dermal surface of the skin and flows out into glass LSC vials held in a fraction collector. Sampling intervals and receiver flow rates are preselected. Our studies generally employed 2-h fraction collection intervals over the 24-h period following application of the pesticide, with a receiver flow rate of 1.5 ml/h. As discussed by Bronaugh, the flow must be maintained at a rate sufficient to wash out the cell receiver chamber. More rapid flow rates will dilute the pesticide concentration in the receiver. This dilution effect is not a problem for whole LSC vial radioactivity counting but does limit the analytical sensitivity if sample aliquot injection (e.g., liquid chromatography is used). A disadvantage of the Bronaugh-type cell concerns the potential for air gap formation at the skin–receiver solution interface following receiver outgassing. The

Table 4-3 Data from AIVDA cell tests except for DDT (Bronaugh [B] cell data) on skin permeability (mean percentage dermal absorption [±SD]) to pesticides

Pesticide	Receiver	Dermal absorption (% ±SD)		
		Rat in vitro	Rat in vivo	Human in vitro
2,4-D acid	Hank's	40 ± 5	49 ± 10	19 ± 2
		n = 4	n = 4	n = 4
DEET	Ringer's	65 ± 3	41 ± 11	28 ± 4
		n = 4	n = 4	n = 4
DDT	Hank's (B)	34 ± 11	73 ± 6	28 ± 3
		n = 4	n = 4	n = 4
Diazinon	Ringer's	47 ± 3	56 ± 1	20 ± 3
		n = 4	n = 4	n = 4
Fenoxaprop-ethyl (FPE)	Ringer's	44 ± 5	32 ± 7	
		n = 5	n = 6	
FPE (Excel)			18 ± 5	
			n = 6	

Sources: Data for 2,4-D, Moody, Nadeau, and Chu (1994a); for DEET, Moody and Nadeau (1993a); for diazinon, Moody and Nadeau (1993b); for DDT, Moody, Nadeau, and Chu (1994b); for FPE, Moody and Ritter (1992).

Note: The percentage absorption includes the percentage recovered from the skin depot (see text Section 4-4). Acetone dose vehicle employed in all cases except for FPE-Excel. Receiver solution used in indicated Ringer's or Hank's. For brevity, only data for rat and human skin are included; the studies also tested skin from guinea pigs, pigs, and a tissue-cultured skin. In vivo rat data are included for comparison.

design of the AIVDA cell with the receiver chamber located on top of the donor chamber minimizes this potential problem because air bubbles in the receiver solution should flow up and out of the receiver eluant port (Moody and Martineau 1990). Air gap formation is important because it could prevent the dissolution of the pesticide permeant into the receiver solution and lead to underestimates of systemic pesticide exposure. The potential for this problem is significant where the receiver solution needs to be aerated to ensure skin viability and a positive temperature gradient exists between the receiver reservoir and the heated cell.

Table 4-4 summarizes data obtained from our Bronaugh cell studies. Following a study with the lipophilic polycyclic aromatic hydrocarbon compound benzo[a]pyrene (B[a]P) that employed an acetone vehicle, tests with commercially formulated DEET and 2,4-D products were conducted.

It is apparent from the data in Table 4-4 that, in all cases, the rat in vitro data overestimate the rat in vivo data. In the one case in which we have comparative data for guinea pigs, the in vitro and in vivo B[a]P data are similar. As discussed in our published reports, the reason for the apparent discrepancy between the in vitro and in vivo data for rats may be due to a problem with interpreting the in vivo data. We have reported that a significant amount of the pesticide dose application was detected in the foam rubber patches used in vivo to protect the dose site from rub-off. These patches are placed over the dose site immediately following application of pesticide to the skin and may become contaminated with the dose application if the dose migrates laterally beyond the

Table 4-4 Data from Bronaugh cell tests of skin permeability (mean [n = 4] percentage dermal absorption [±SD]) to pesticides

| Pesticide | Dermal absorption (%±SD) | | | | |
	Rat in vitro	Rat in vivo	G-Pig in vitro	G-Pig in vivo	Human in vitro
B[a]P	95 ± 10	70 ± 8	51 ± 3	68 ± 9	23 ± 5 (32 year old) 43 ± 9 (50 year old)
2,4-D (CC)	30 ± 6	17 ± 5	21 ± 3		25 ± 6
2,4-D (WE)	26 ± 6	16 ± 6	16 ± 5		28 ± 3
DEET (OF)	50 ± 9	15 ± 3	46 ± 5		48 ± 3
DEET (DW)	49 ± 4	27 ± 4	37 ± 2		36 ± 4
DEET (MU)	44 ± 9	20 ± 7	19 ± 1		17 ± 1

Sources: Data on B[a]P, Moody, Nadeau, and Chu (1995a); on 2,4-D, Moody and Nadeau (1997); on DEET, Moody, Nadeau, and Chu (1995b).

Note: The percentage absorption includes the percentage recovered from the skin depot (see text). Acetone dose vehicle used for the benzo-[a]pyrene (B[a]P) tests. The B[a]P tests included human skin from both a 32-year-old and a 50-year-old subject. Formulated products of DEET (OFF! [OF], Deep Woods [DW], and Muskol [MU] containing 14%, 24%, and 95%, respectively) and 2,4-D amine (Clean Crop [CC] and Wilbur Ellis [WE], both containing 50% 2,4-D amine) were used. Receiver solution was Hank's HEPES buffered saline with serum albumin in all tests. For brevity only data for rat and human skin are included. In vivo rat and hairless guinea pig (g-pig) data are included for comparison.

application site (e.g., via capillary action). This artifact of the dosing procedure ranged as high as 76% of the applied dose in our study with formulated DEET products and could markedly contribute to a low value for dermal absorption being reported in vivo (Moody et al. 1995). Simply, if the dose runs off the skin and is irreversibly absorbed by the patch, it is not available for dermal absorption. This problem is expected to be more predominant for dose applications involving formulated pesticides because the acetone vehicle used in our in vitro and in vivo tests was evaporated from the skin surface by use of an electric fan to prevent runoff of the applied dose. Because the dose is contained in the donor chamber in vitro, it cannot run off the skin, but air drying of dose formulations employing acetone vehicles also was performed in vitro to maintain the same in vitro and in vivo dosing conditions.

Another way that pesticide exposure can be inadvertently misrepresented involves a failure to interpret dermal absorption data correctly. A good example of how this could readily occur also involves the data presented in our study with three commercial DEET formulations (Moody, Nadeau, and Chu 1995b). This DEET study reported a positive dose dependent *decrease* in the percentage of dermal absorption in vitro because lower percentage skin absorption was obtained for the more concentrated DEET products. An evaluation of these data could lead to the hazardously erroneous conclusion that it would be safer for the consumer to use the more concentrated DEET formulations. As discussed in our 1995 report, however, this was not the case. The total cumulative amount (mg/cm^2) of DEET absorbed into the skin *increased* with increasing concentration of DEET for the formulations tested. This apparent paradox was explained simply on the basis that $X\%$ absorption of a high-dose concentration would give greater cumulative absorption than $X\%$ absorption of a low-dose concentration (Moody, Nadeau, and Chu 1995b). In other words, the percentage absorption of a formulation containing a low DEET concentration can be greater than the percentage absorbed from a highly concentrated formulation, even though the cumulative total amount (mg/cm^2) of DEET absorbed is higher in the latter case. This point is stressed here because it is important to consider when evaluating product safety. The safety of DEET insect repellent formulations for consumer use has been addressed previously and concern has been expressed about the use of the highly concentrated DEET products, especially by children (Moody 1989). The generally higher body-surface-area-to-weight ratio for individuals such as children suggests that these individuals would obtain relatively higher total body dose of DEET per unit body weight under normal use conditions, where the entire exposed skin surface is commonly treated with mosquito repellent. Hence these individuals would be more at risk because the systemic dose (mg/kg body weight) would be higher than for individuals with low surface-area-to-body-weight ratios (see Moody [1989] and the references therein for further discussion). This theory concerning the vulnerability of individuals with low-surface-area to body-weight ratio is also applicable simply by logical extension to the swimming/bathing exposure model and should be considered when evaluating any form of whole-body exposure to pesticides and other environmental contaminants.

On returning to the paradigm that the rat will overestimate human skin permeability, note that, according to Table 4-4, this was true for only two of the six chemicals tested (for B[a]P and DEET [MU]) and that, for the remaining cases, the rat and human in vitro data were in agreement. For the guinea pig in vivo data, the B[a]P data overestimated the human in vitro data for the 23-year-old subject, whereas, for 2,4-D (WE), it underestimated the human data. For the remaining studies, the guinea pig and human in vitro data were comparable. The rat and guinea pig in vivo data for B[a]P were not significantly different.

New SPTA procedure. Having addressed some problems involved with current in vitro methods, a brief mention of our new SPTA procedure is warranted. The skin permeation tube or Moody/SPTA cell was designed to be used in a fully automated analytical method employing liquid chromatography (LC). Both the AIVDA and the Bronaugh cell procedures employ fraction collectors that leave the acquired receiver samples sitting in open LSC vials for the duration of the permeation study. Vials can be capped the next day but this still leaves the pesticide residues ample time to volatilize from the receiver fractions. As well as this potential for low percentage mass balance recovery, the capping/labeling and LSC analysis of the fractions are tedious. Further, as previously mentioned, the flowing receiver phase dilutes the pesticide permeant and this reduces analytical sensitivity for aliquot injection used, for example, in LC. The standard LSC whole fraction analysis does not chromatographically resolve the parent pesticide compound from its metabolites and degradation (e.g., hydrolytic, photolytic) products. Given that the current literature suggests that skin viability and dermal metabolites of pesticide permeants is important to consider (e.g., Moody, Nadeau, and Chu [1995a], Moir, Marwood, and Moody [1994]), a rapid automated LC procedure was developed in our laboratory.

The Moody/SPTA cell, as shown in Fig. 4-4, is basically an autosampler (AS) vial insert that was machined on a metal lathe to very precise dimensions to fit standard 2-ml glass AS vials. After trying various materials (e.g., brass, Teflon, stainless steel), we selected aluminum because it was the most cost-effective. The cell consists of an inner mounting tube that clamps the dermatomed skin specimen against the inner rim of the outer tube. The assembled Moody/SPTA cell is inserted into an AS vial. Either the epidermal surface can be dosed via micropipet before insertion of the cell into an empty AS vial for finite dose studies, or the cell holding the exposed skin surface can be inserted into an AS vial containing the donor dose solution for infinite swimming/bathing-type dose studies. As shown in Fig. 4-4, a small magnetic stir bar is used to swirl the donor pool during infinite-dose skin absorption bioassays. That this is, in effect, a true bioassay has very recently been confirmed in our laboratory with our ^{14}C-glucose/lactate skin viability test. This test, which uses ^{14}C-lactate as a specific biomarker for ^{14}C-glucose utilization by dermal tissue, has demonstrated that the skin specimen remains fully viable for at least 24 h in the Moody/SPTA cells (submitted).

There are several advantages to the Moody/SPTA cell. It requires only 5-mm-diameter dermatomed skin punches; thus small, difficult-to-obtain human skin specimens can be employed routinely. Although the exposed skin surface area of 0.07 cm^2 is much smaller than the Bronaugh cell (0.64 cm^2) used in our laboratory and even the AIVDA cell (0.2 cm^2), the Moody/SPTA cell uses a form of stop-flow technology in which the receiver sample sits still to permit the pesticide permeant to concentrate in the receiver for the duration of each sample interval. Consequently, the analytical sensitivity of this method is actually at least tenfold greater than the Bronaugh method on an aliquot injection basis (Moody 1996).

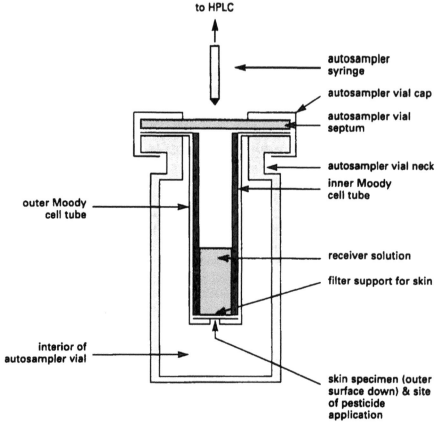

Figure 4-4   Moody/SPTA cell shown inserted in a glass AS vial. The inner tube of the cell is enclosed in the outer tube with the skin specimen clamped between the tubes and backed by a support filter. The AS syringe intermittently removes the receiver for LC analysis, and then refills the cell with fresh receiver solution. *Source:* Reprinted by permission of Moody (1997).

The miniaturization of the in vitro cell also greatly reduces the amount of hazardous and expensive radiotracer pesticides required for dose applications. The situation of the receiver solution on top of the skin specimen as in our AIVDA cell design prevents air bubbles from outgassing from the receiver solution and accumulating at the skin–receiver interface to form an air gap. Because the receiver solution is sealed within the AS vial (a serrated vial-cap septum is used to prevent gas pressure or vacuum buildup during sampling or refilling of receiver), the chance of volatilization of receiver pesticide and metabolites is minimized. The receiver solution is withdrawn entirely at each sample interval and then is refilled with fresh medium. This refill mechanism replenishes the receiver nutrient solution, ensuring the maintenance of optimal skin viability. The cell is swirled in an AS bar-code reader prior to the withdrawal of the receiver solution and mixing at each sampling interval. The AS carousel is kept heated with a water jacket to maintain the AS vials at skin temperature (32°C). Most important, the skin absorption bioassay is conducted using standard analytical equipment on a fully automated basis. In fact, the Hewlett Packard liquid chromatograph used in our laboratory was routinely monitored and controlled from a remote site via computer modem using the Reachout Remote Control computer software package (Ocean Isle Software, Vero Beach, Fla.). Remote control of the analytical environment provided a high degree of flexibility for experimental studies (e.g., the receiver solution can be changed to one with a different composition (e.g., pH, salt constituents, viability inhibitors) at preprogrammed intervals or even from a remote home site via modem after working hours. Because the AS carousel can hold up to 100 vials, this automated AIVDA method has the potential to run up to 50 Moody/SPTA cells concurrently with 50 receiver refill vials. The AIVDA and Bronaugh cells can only run six and seven concurrent cell replicates, respectively, per fraction collection arm. The Moody/SPTA cells are much less expensive to manufacture than the AIVDA cells and could be made available as prepackaged sterile disposable units. They are also ideal for testing the fragile synthetic tissue-cultured skins that have shown promise as human skin surrogates (Moody 1997).

There are a few disadvantages to the Moody/SPTA cell–AIVDA method. The mounting or insertion and microdosing of the dermatomed skin specimens take a certain degree of technical expertise. Serum albumin in high concentration (4% required by Hank's receiver solution) is not compatible with standard LC columns, thereby limiting the method to the analysis of less lipophilic skin permeants. A column switching system could be used (e.g., the CAPCELL PAK MF cartridge system currently available from Phenomenex, Torrance, Calif.) to negate the albumin problem, however, or another compound could be used to maintain the receiver permeant sink capacity. Last, and perhaps most limiting, is that, given the particular form of sequential sampling and immediate analysis used for SPTA, the sampling interval has to be kept longer than the cumulative (total vial) LC retention time (R_t) of the most strongly retained pesticide or metabolite during each analysis. As the number of vial replicates is increased, the sampling interval must be increased accordingly. In our laboratory, with the relatively short-retention-time compounds such as DEET and atrazine ($R_t < 5$ min), five replicate Moody/SPTA cells (four treated cells and one control) were run routinely and included an external standard injection at each 2-h sampling interval. In one study, 15 cells were run concurrently to test whether the use of serrated septa affected the absorption data obtained for DEET finite dose studies (Table 4-5, skin code 4). The data showed no significant effect of using septa versus studies conducted without septa. Also, in this 15-cell replicate run (Table 4-5), there was no major protective effect of a barrier cream tested for DEET. Most important, from Table 4-5 (skin code 5), was the observation that there was good agreement between the data obtained from the Moody/SPTA and Bronaugh cell tests that were conducted concurrently with skin specimens taken from the same rats. In conclusion, the new AIVDA method employing Moody/SPTA cells offers a rapid, cost-effective, and sensitive analytical procedure that is needed for conducting the large amount of tests required to validate the in vitro method as a viable alternative to in vivo testing.

4-4 IN VITRO/IN VIVO COMPARISONS

It is not the intent of this chapter to evaluate our own data from an in vitro/in vivo validation perspective. This already has been considered in our own reports cited herein. Of greater benefit is a brief consideration of the many experimental variables that can

Table 4-5 Results of in vitro Moody/SPTA and Bronaugh skin permeation cell tests, for average total % skin absorption ±SD of N,N-diethyl-m-toluamide (DEET) following topical application of Deep Woods insect repellent

Species (sex)	Site	Specimen storage condition	Test condition	Skin code	Total % absorption Moody cell	Total % absorption Bronaugh cell
Human (female)	breast	frozen		1	34 ± 4.5 (n = 4)	
Human (female)	abdomen	fresh		2		36 ± 4.2 (n = 4)
Human (female)	abdomen	fresh	skin sandwich	3	56 ± 9.1 (n = 4)	
Human (female)	abdomen	frozen	skin sandwich	4	48 ± 6.3 (n = 4)	
Human (female)	abdomen	frozen	Starburst septa	4	53 ± 8.0 (n = 4)	
Human (female)	abdomen	frozen	+barrier cream	4	39 ± 7.6 (n = 4)	
Rat (female)	back	fresh	screw cap, no septa	5	62 ± 12.8 (n = 4)	59 ± 8.1 (n = 3) (LC)
Rat (female)	back	fresh	screw cap, no septa	5		59 ± 9.1 (n = 3) (LSC)

Source: Reproduced from Moody (1997).

Note: The skin code designates different individual subjects, hence tests with code 4, for example, were conducted with skin from the same subject. The first two rows (skin codes 1,2) give data from 48-h tests and subsequent rows give data from 24-h tests. The Bronaugh cell data in row 2 (skin code 2) were from Moody, Nadeau, and Chu (1995b) and the test employed standard Hank's HEPES buffered receiver with 4% bovine serum albumin. The Moody cell data in rows 1 and 3 (skin code 1,3) were from tests with Hank's receiver with HEPES buffer diluted tenfold. The remaining tests were conducted with Hank's phosphate-buffered receiver solution. Moody cell tests were conducted with screw-cap AS vials without septa unless otherwise indicated. The study in row 3 (skin code 3) was conducted with the skin sandwiched between two support filters and data for efficacy testing of a barrier cream is given in row 6.

confound the method validation process. A few factors have been mentioned previously that are important to consider. For example, the failure of general paradigms of animal species-related dermal permeability and the importance of this when extrapolating to human systemic exposure for product safety evaluations. However, before species-related dermal permeabilities can be determined, experimental conditions used for conducting the skin absorption data must be evaluated. These include not only the laboratory conditions employed for both the in vitro and in vivo tests (e.g., ambient temperature, humidity, lighting, "wind" velocity, type of in vitro skin-test cell [AIVDA, Bronaugh, Moody/SPTA etc.]), type of dose patch used in vivo, but also the dermal physiological condition and the preparation method used for obtaining thin specimens for absorption bioassays. According to an early report by Maibach, in vitro skin testing at the time was considered to be no better than clamping a piece of shoe leather into a test cell (Maibach and Feldmann 1974). Given that thawed cadaver skin and distilled-water receiver solutions often were used in early in vitro studies, this perception was justifiable. However, since that time, much progress has been made in the area of preserving skin viability and physiological condition during the absorption tests. Several skin separation techniques have been advanced to provide human epidermal membranes for testing. An effective procedure, developed in the Bronaugh laboratory in Washington, D.C., is the use of a dermatome to prepare freshly cut specimens of standard thickness. This dermatome method was used extensively in our laboratory. Other methods, such as heat-stripping and enzyme cleavage, can be used, although more research needs to be done to ensure that these methods preserve skin viability and maintain the barrier integrity of the stratum corneum.

To focus on a specific case in point, Table 4-4 reports two values for in vitro absorption of B[a]P in human skin. The absorption through the specimen tested from a 50-year-old individual was about twice as high as that for the 32-year-old subject. Is this an age-related effect? Does human skin become more permeable with age? Are the elderly more at risk from swimming/bathing exposure to water contaminants? There are many experimental variables to consider before reaching any conclusion. Moody and Chu (1995) have discussed several factors that are important to consider when assessing swimming/bathing exposure. These include environmental (e.g., solar irradiance, water temperature, pH, and flow rate), physicochemical (e.g., dermaphilicity, adsorption or partitioning processes in water sediments, and surface slicks), and physiological (e.g., skin age, anatomic site, gender, tissue viability) factors. At least, in our own studies, a few of these variables can be eliminated. Our reported studies with human skin have employed only female Caucasian skin obtained fresh from a local hospital after surgical excision. The skin was transported to the laboratory in McCoy's nutrient medium and was dermatomed and mounted in the in vitro test cells within a few hours of surgery. The sole use of female Caucasian skin here eliminates race- and gender-related skin permeability differences from the evaluation. The use of fresh tissue coupled with recent tests in our laboratory that suggest that the skin remains viable in the Hank's receiver solution, and also in the glucose-supplemented Ringer's solution used for our AIVDA cell tests (submitted), support the hypothesis that the physiological condition of the skin was not altered by the excision and the in vitro test conditions. Because the human skins tested for B[a]P absorption were both from the abdominal site, we can rule out anatomical site-related skin permeability differences (Table 4-4), this leaving the possibility that age-related factors are important to consider. Because of ethical considerations, the in vivo human testing of B[a]P cannot be conducted to further examine the importance of factors such as age-related skin permeability. However, we can attempt to compare the rat and guinea pig in vivo B[a]P data with those obtained for humans in vitro to determine the usefulness of such in vivo animal models for providing surrogate human data.

The rat and guinea pig in vivo B[a]P data both overestimate the human in vitro data. This is not the case for the other pesticides tested in the Bronaugh cells (Table 4-4); the rat and guinea pig in vivo data either agreed with or underestimated the human in vitro data. This apparently unusual result for B[a]P may be due to the necessity for this highly lipophilic compound to be metabolized within the living skin to more dermaphilic permeant chemical species (Moody, Nadeau, and Chu 1995a). However, bringing in this assumption creates an apparent paradox. Because one might consider youthful skin to show greater metabolic activity, one might expect a faster rate of B[a]P cutaneous metabolism for the younger individual. Why then do we observe greater skin absorption of B[a]P for the older skin tissue? Obviously, a hypothesis based on the analysis of only two human specimens is likely to be invalid, and normal intersubject differences in human skin permeability are more than sufficient to account for a twofold difference. However, we have used this case to point out some of the complexities involved in cross comparing in vitro and in vivo data and trust that this will aid the pesticide safety evaluation process.

Before concluding this section, it is necessary to consider a still controversial assumption that is made to obtain the most conservative risk estimate of pesticide dermal exposure. This concerns the skin depot, or reservoir, effect. To ensure that pesticide systemic exposure estimates include a worst-case scenario, it is often necessary to assume that the skin depot will become fully bioavailable at some point in time. In calculating the percentage of dermal absorption, we have included the percentage recovery in the methanol washes and skin extractions of the dose region (i.e., the skin depot) as absorbed pesticide residues. This is a less conservative step from basically considering that all of a topical dose will be dermally absorbed because it includes the skin depot residues but ignores the percentage of the dose that may evaporate from the skin surface and the percentage washed off the skin with soap at 24 h posttreatment. However, an assumption of total bioavailability of the dermal depot may lead to overly restrictive regulations concerning pesticide use and safety. In this connection, Chu et al. (1996) have reported data consistent with the skin depot being bioavailable for three lipophilic compound—phenanthrene, B[a]P, and di-(2-ethylhexyl)pthalate—in hairless guinea pigs. To ensure that estimates of dermal absorption used for regulatory purposes are meaningful, the effects of skin depot residues are examined on a case-by-case basis by the PMRA. For instance, if sufficient data are available, excretion kinetics can be examined to characterize extent of dermal absorption over time.

In our own studies, a 24-h soap wash of the skin was conducted both in vitro and in vivo to remove nonabsorbed pesticide residues from the skin. A 24-h posttreatment soap-wash procedure was selected to provide a worst case for immediate attention to skin decontamination. The pesticide washed off the skin at 24 h is not considered to be a part of the skin depot in our studies. We mention this point because our recent studies have been exhibiting a very pronounced washing-in effect that appears to be attributed to the soap washing procedure (Moody, Nadeau, and Chu 1995b; Moody and Nadeau 1997). This is an important observation because it implies that soap washing of the skin, at least after a considerable delay from initial exposure, may actually wash the pesticide into the skin. Further, because the washed-in residues appear in the receiver fraction, they should be considered as being systemically bioavailable. The mechanism(s) for explaining this wash-in effect is not known at present. It simply may be due to a resolubilization of the pesticide dose on the skin surface or to an alteration in skin barrier integrity by the soap employed (e.g., skin delipidation). As is shown by the magnitude of this effect for the commonly used herbicide 2,4-D (Fig. 4-5), this phenomenon should be considered when evaluating the risk of pesticide exposure. In fact, it may be advisable not to recommend skin washing of pesticide workers for decontamination purposes, especially for individuals dermally exposed for sufficiently extended durations to permit a buildup of dermal depot levels.

Although in vivo data would include the total pesticide residues washed into the skin, because of long sampling intervals for urine and feces collections (e.g., 24 h), the secondary profoundly elevated receiver levels observed postwashing in vitro (Fig. 4-5)

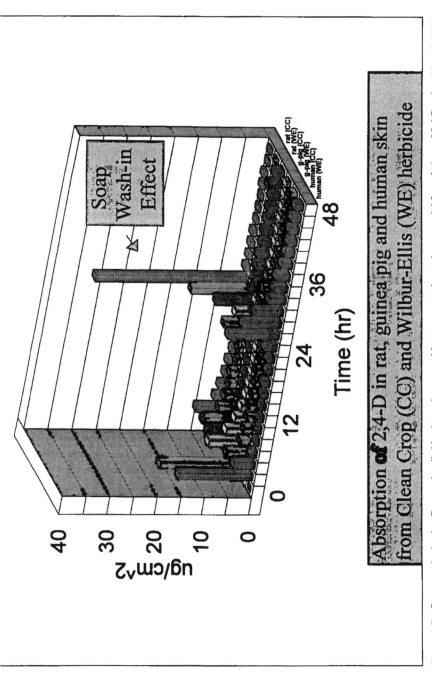

Figure 4-5 Cross-species in vitro Bronaugh cell skin absorption tests with two commonly used commercial formulations of 2,4-D amine. Note the high degree of skin absorption ($\mu g/cm^2$ of skin surface area) obtained following a 24-h soap wash of the skin for all species tested including human. The largest wash-in peak was for rat skin treated with the 2,4-DClean Crop (CC) and Wilbur-Ellis (WE) formulations.
Source: Data replotted by permission of Moody and Nadeau (1997).

Absorption of 2,4-D in rat, guinea pig and human skin from Clean Crop (CC) and Wilbur-Ellis (WE) herbicide

would not be observed in vivo (Moody and Nadeau 1997). Extended blood sampling in vivo would be necessary to provide this information. The in vitro method provides a humane, cost-effective alternative to the use of extended blood sampling.

4-5 QSAR COMPUTER MODELS

Given the numerous experimental and physiological factors that can operate to confound risk evaluations based on dermal absorption in vitro/in vivo data, computer-based Quantitative Structure-Activity Relationship (QSAR) models provide a useful tool for simplifying data extrapolations and correlations. We have reported good correlations of the logarithm of dermal permeability coefficient (K_p) values with the logarithm of octanol–water partition coefficients (K_{ow}) of 114 chemicals in our Health Canada database (Kirchner et al. 1997). These correlations were improved when compounds were assigned molar volume groupings and it was suggested that this type of correlation might be useful for predicting human skin permeability. Recently, we initiated three-dimensional QSAR analyses on data obtained from our laboratory using the Molecular Analysis Pro program (WindowChem Software™ Inc., Fairfield, Calif.). These analyses use three-dimensional molecular structure conformations and a wide range of algorithm-generated physicochemical descriptors and can be used to determine in vitro/in vivo correlations (IVIVCS). Three-dimensional QSAR multiple linear regression analyses of the rat in vitro data reported in Table 4-3 (AIVDA data) and Table 4-4 (Bronaugh data) gave an r^2 correlation coefficient of .73 between the percentage skin absorption in rats versus log K_{ow} and two other SARs (molecular volume and Kappa 2 Shape Index). The data for the eleven different pesticide formulations tested are plotted in Fig. 4-6. This correlation involved the in vitro absorption data that did not include the skin depot residues (Tables 4-3, 4-4). The correlation was not as good when the skin depot was included as percentage skin absorption ($r^2 = .68$; $n = 11$). Nevertheless, this level of correlation appears promising, especially because we are cross comparing data from two dermal penetration cells (AIVDA and Bronaugh) and for studies conducted with different dose vehicles (e.g., DEET in acetone versus three commercial DEET formulations). Excluding the two data points for 2,4-D amine that appear as outliers, (▲) in Fig. 4-6 gives a much improved correlation of $r^2 = .93$; however, valid reasons for excluding these data are not apparent. It is possible that a high degree of ionization of the dimethylamine salt of 2,4-D in its formulation vehicle (CC and WE) could have decreased the rate of skin absorption. Or perhaps, given that 2,4-D amine was the most hydrophilic compound (log $K_{ow} = 0.65$) tested in our laboratory, the lower degree of skin absorption obtained (Fig. 4-6) indicates the existence of a cutoff effect for hydrophiles. Given the small data set here ($n = 11$) and that the in vitro data in Fig. 4-6 were obtained using two different permeation cells (AIVDA and Bronaugh) and for pesticides applied in different vehicles, more data are required before we can conclude as to the validity of the correlation obtained. Future QSAR analysis of our data in combination with data obtained from other laboratories is recommended, especially the determination of IVIVCS. We note that, to ensure that IVIVCS are conducted properly, physiological factors must be accounted for. Physiologically based pharmacokinetic (PBPK) modeling is a potentially powerful

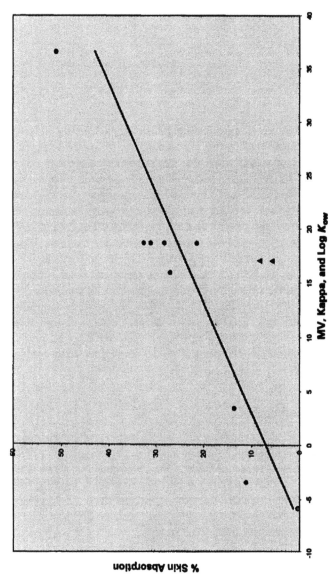

Figure 4-6 Three-dimensional QSAR analysis of rat in vitro AIDA and Bronaugh data. The percentage skin absorption data not including the percentage recovery in the skin depot are plotted versus values calculated from SAR values of molecular volume (MV), the κ2 shape index (Kappa) and log octanol–water partition coefficient (log K_{ow}) (see text). Data are given for the 11 pesticide formulations listed with rat in vitro data in Tables 4-3 and 4-4. The two points designated as ▲ were for the 2,4-D amine CC and WE formulations and may be outliers (see text).

tool for predicting systemic pesticide exposure via the dermal route and has been reviewed by McDougal (1996).

4-6 FUTURE DIRECTIONS

4-6-1 Model Validation

After considering the large number of experimental/physiological variables that can affect the interpretation/extrapolation of dermal absorption data for pesticide exposure risk assessment purposes, it is evident that future research should examine model validation, including studies to validate cross-species comparisons (e.g., rat/monkey/human) both in vivo and in vitro. To ensure the accuracy of these comparisons, the methodology used for in vitro and in vivo tests should be standardized. However, standardization should not confine scientists to the use of only one method. Method development is an important component of the progression of the scientific method. For example, if only one type of in vitro skin absorption test cell could be used, progress toward the development of better methods as shown herein would stop. Finally, and most important, both the in vivo and in vitro models need to be validated for their intended use, that being the prediction of exposure to pesticides in both the field and home environments. An initial attempt at cross comparing our own in vivo and in vitro data to actual field exposure studies with 2,4-D has been reported recently (Moody and Nadeau 1997). It is recommended that future reports of this nature should incorporate a field exposure component.

4-6-2 Research Recommendations

As well as model validation and further progress with in vivo and in vitro method development, the potential of computer-based QSAR and PBPK methods should not be neglected. Another area requiring intense research is the determination of the actual bioavailability of the skin depot under conditions best approximating field or home pesticide exposure. Research also is required to examine the importance of skin viability concerning the cutaneous metabolism of pesticides and how viability may alter the dermatopharmacokinetics of transdermal pesticide efficacy.

4-7 CONCLUSIONS

As methods have progressed from early animal testing methods to less animal-dependent, in vitro methods and, ultimately, to nonanimal-use computer-based methods such as QSAR, the three "Rs" (replacement, reduction, refinement) aspiration of the alternatives-to-animal-testing lobby will be realized. In the process, it is recommended that newer methods be well validated before they are used in place of standard methods for evaluating the safety of pesticide products. It should be stressed that the gold standard upon which the validation of in vitro data depends is the in vivo animal model. Future research is needed to investigate the use of IVIVCS in the pesticide hazard/risk assessment process.

4-8 ACKNOWLEDGMENTS

I am grateful to Dr. Claire Franklin for initiating the dermal absorption project at Health Canada, fellow coworkers, and many excellent students who provided technical support. I would like to thank the many researchers worldwide who provided research direction and technical advice. I also thank Dr. Conrad Watters and the staff of the Ottawa General Hospital, Ottawa, Ontario for kindly supplying fresh viable human skin tissue for our in vitro studies. I thank the Animal Resources Division of Health Canada for their excellent veterinary care. I am also very appreciative of financial assistance from the Multicenter Evaluation of in vitro Cytotoxicity via the Animal Defence League of Canada. I thank the Science and Technology Centre at Carleton University, Ottawa, for machining both the AIVDA cell and the SPTA cells. I thank Dr. Ih Chu for supporting this project. Finally, I thank Dr. James Quinn (Norgwyn Montgomery Software Inc., North Wales, Pa.) for technical support and advice concerning the use of the Molecular Analysis Pro program.

REFERENCES

Bronaugh, R. L. 1996a. *In vitro* percutaneous absorption. In *Dermatotoxicology*. 5th ed. Edited by F. N. Marzulli and H. I. Maibach, pp. 317–24. Washington, D.C.: Taylor and Francis.

Bronaugh, R. L. 1996b. *In vitro* viable skin model. In *Models for assessing drug absorption and metabolism*. Vol. 8 of *Pharmaceutical Biotechnology*, ed. R. T. Borchardt, P. L. Smith, and G. Wilson, pp. 375–86. New York: Plenum Press.

Bronaugh, R. L., and H. I. Maibach, eds. 1991. *In vitro percutaneous absorption: Principles, fundamentals, and applications*, p. 280. Boca Raton, Fla.: CRC Press.

Bronaugh, R. L., and R. F. Stewart. 1985. Methods for *in vitro* percutaneous absorption studies IV: The flow-through diffusion cell. *J. Pharm. Sci.* 74:64–67.

Chu, I. 1995. Alternative methods to animal testing: A Canadian Health Protection Branch perspective. *Alternatives Lab. Anim.* 23:257–61.

Chu, I., D. Dick, R. Bronaugh, and L. Tryphonas. 1996. Skin reservoir formation and bioavailability of dermally administered chemicals in hairless guinea pigs. *Fundam. Chem. Toxicol.* 34 (3):267–76.

Feldmann, R. J., and H. I. Maibach. 1970. Absorption of some organic compounds through skin in man. *J. Invest. Dermatol.* 54:339–404.

Feldmann, R. J., and H. I. Maibach. 1974. Percutaneous penetration of some pesticides and herbicides in man. *Toxicol. Appl. Pharmacol.* 28:126–32.

Franklin, C. A., D. A. Somers, and I. Chu. 1989. Use of percutaneous absorption data in risk assessment. *J. Am. Coll. Toxicol.* 8(5):815–27.

Grasso, P., and A. B. G. Lansdown. 1972. Methods of measuring, and factors affecting percutaneous absorption. *J. Soc. Cosmet. Chem.* 23:481–521.

Griffin, G. K. 1996. *Status of alternative methods in toxicity testing*, p. 124. Ottawa: Joseph Morgan Research Foundation.

Kirchner, L. A., R. P. Moody, E. Doyle, R. Bose, J. Jeffery, and I. Chu. 1997. The prediction of skin permeability using physicochemical data. *Alternatives Lab. Anim.* 25:359–70.

Maibach, H. I., R. J. Feldmann, T. H. Milby, and W. F. Serat. 1971. Regional variation in percutaneous penetration in man. *Arch. Environ. Health* 23:208–11.

Maibach, H. I., and R. J. Feldmann. 1974. Systemic absorption of pesticides through the skin of man. In *Occupational exposure to pesticides: Report to the Federal Working Group on Pest Management from the Task Group on Occupational Exposure to Pesticides*. Appendix B, pp. 120–27. Washington, D.C.: U.S. Government Printing Office.

McDougal, J. N. 1996. Physiologically based pharmacokinetic modelling. In *Dermatotoxicology*. 5th ed. Edited by F. N. Marzulli and H. I. Maibach, pp. 353–68. Washington, D.C.: Taylor and Francis.

Moir, D., T. M. Marwood, and R. P. Moody. 1994. *In vitro* cutaneous metabolism of DDT in human and animal skins. *Bull. Environ. Contam. Toxicol.* 52:474–78.

Moody, R. P. 1989. The safety of diethyltoluamide insect repellents (Letter to the Editor). *J. Am. Med. Assoc.* 262 (1):28–29.

Moody, R. P. 1993. *In vitro* dermal absorption of pesticides: A cross-species comparison including Testskin. *J. Cutaneous Ocular Toxicol.* 12 (2):197–202.

Moody, R. P. 1996. Automated *in vitro* dermal absorption (A*I*VDA): A new analytical method for determining transdermal absorption of environmental contaminants. Poster presented at 10th annual meeting of American Association of Pharmaceutical Scientists. 29–31 Oct. at Seattle, Wash.

Moody, R. P. 1997. Automated *in vitro* dermal absorption (A*I*VDA): A new *in vitro* method for investigating transdermal flux. *Alternatives Lab. Anim.* 25:347–57.

Moody, R. P. in press. Bathing water; percutaneous absorption of water contaminants. In *Dermal Absorption and Toxicity Assessment*; eds. M. Roberts and K. A. Walters; Drugs and the Pharmaceutical Sciences, Vol. 91. New York: Marcel Dekker.

Moody, R. P. submitted. Automated *in vitro* dermal absorption (AIVDA) II. Predicting skin permeation of atrazine with finite and infinite dose (swimming/bathing) exposure models. ATLA.

Moody, R. P., F. M. Benoit, D. Riedel, L. Ritter, and C. A. Franklin. 1989. Dermal absorption of the insect repellent (*N,N*-diethyl-*m*-toluamide) in rats and monkeys: Effect of anatomic site and multiple exposure. *J. Toxicol. Environ. Health* 26:137–47.

Moody, R. P., J. M. Carroll, and A. M. E. Kresta. 1987. Automated high performance liquid chromatography and liquid scintillation counting determination of pesticide mixture octanol/water partition rates. *J. Toxicol. Ind. Health* 3 (4):479–90.

Moody, R. P., and I. Chu. 1995. Dermal exposure to environmental contaminants in the great lakes. *Environ. Health Perspect.* 103 (9):103–14.

Moody, R. P., and C. A. Franklin. 1987. Percutaneous absorption of the insecticides fenitrothion and aminocarb in rats and monkeys. *J. Toxicol. Environ. Health* 20:209–18.

Moody, R. P., C. A. Franklin, L. Ritter, and H. I. Maibach. 1990. Dermal absorption of the phenoxy herbicides 2,4-D, 2,4-D amine, 2,4-D iso-octyl and 2,4,5-T in rabbits, rats, rhesus monkeys and humans: A cross-species comparison. *J. Toxicol. Environ. Health* 29:237–245.

Moody, R. P., M. Grayhurst, and L. Ritter. 1989. Evaluation of the rat tail model for estimating dermal absorption of lindane. *J. Toxicol. Environ. Health* 28:317–26.

Moody, R. P., and P. Martineau. 1990. An automated *in-vitro* dermal absorption procedure. Part 1. Permeation of ^{14}C-labelled *N,N*-diethyl-*m*-toluamide through human skin and effects of short-wave ultraviolet radiation on permeation. *Toxicol. in vitro* 4 (3):193–99.

Moody, R. P., and B. Nadeau. 1993a. An automated *in vitro* dermal absorption procedure. Part 3. *In vivo* and *in vitro* comparison with the insect repellent DEET (*N,N*-diethyl-*m*-toluamide) in mouse, rat, guinea pig, pig, human and tissue-cultured skin. *Toxicol. in vitro* 7 (2):167–76.

Moody, R. P., and B. Nadeau. 1993b. *In vitro* dermal absorption of pesticides. Part 4. Cross-species *in vivo* and *in vivo* comparison with the organophosphorus insecticide, diazinon, in rat, guinea pig, pig, human and tissue-cultured skin. *Toxicol. in vitro* 7 (2):167–76.

Moody, R. P., and B. Nadeau. 1997. *In vitro* dermal absorption of two commercial formulations of 2,4-Dichlorophenoxyacetic acid dimethylamine (2,4-D amine) in rat, guinea pig, and human skin. *Toxicol. in vitro* 11:251–62.

Moody, R. P., B. Nadeau, and I. Chu. 1994a. *In vitro* dermal absorption of pesticides. Part 5. Cross-species *in vivo* and *in vitro* comparison with the herbicide, 2,4-Dichlorophenoxyacetic acid, in rat, guinea pig, pig, human and tissue-cultured skin. *Toxicol. in vitro* 8 (6):1219–1224.

Moody, R. P., B. Nadeau, and I. Chu. 1994b. *In vitro* dermal absorption of pesticides. Part 6. Cross-species *in vivo* and *in vitro* comparison with the organochlorine insecticide, DDT, in rat, guinea pig, pig, human and tissue-cultured skin. *Toxicol. in vitro* 8 (6):1225–32.

Moody, R. P., B. Nadeau, and I. Chu. 1995a. *In vivo* and *in vitro* dermal absorption of benzo[a]pyrene in rat, guinea pig, human and tissue-cultured skin. *J. Dermatol. Sci.* 9:48–58.

Moody, R. P., B. Nadeau, and I. Chu. 1995b. *In Vitro* dermal absorption of *N,N*-diethyl-*m*-toluamide (DEET) in rat, guinea pig and human skin. *In Vitro Toxicol.* 8 (3):263–275.

Moody, R. P., B. Nadeau, S. Macdonald, and I. Chu. 1993. *In vitro* skin absorption of ^{14}C-cyanuric acid in a simulated swimming pool. *Bull. Environ. Contam. Toxicol.* 50:12–18.

Moody, R. P., D. Riedel, L. Ritter, and C. A. Franklin. 1987. The effect of DEET (*N,N*-diethyl-*m*-toluamide) on dermal persistence and absorption of the insecticide, fenitrothion, in rats and rhesus monkeys. *J. Toxicol. Environ. Health* 22 (4):51–59.

Moody, R. P., and L. Ritter. 1989. Dermal absorption of the pesticide lindane (1α, 2α, 3β, 4α, 5α, 6β-hexachlorocyclohexane) in rats and rhesus monkeys: Effect of anatomical site. *J. Toxicol. Environ. Health* 28:161–69.

Moody, R. P., and L. Ritter. 1992. An automated *in vitro* dermal absorption procedure. Part 2. Comparative *in vivo* and *in vitro* dermal absorption of the herbicide, Fenoxaprop-ethyl (HOE 33171) in rats. *Toxicol. in vitro* 6 (1):53–59.

Moody, R. P., R. Wester, and H. I. Maibach. 1992. Dermal absorption of the phenoxy herbicide 2,4-D dimethylamine in humans: Effect of DEET and anatomic site. *J. Toxicol. Environ. Health* 36:241–50.

Ritter, L. and P. Curry. 1988. Regulation of pesticides in Canada. *Toxicol. Ind. Health* 4 (3):331–340.

Scott, S. K., P. C. Kosch, and D. E. Hilmas. 1976. Serum lactate dehydrogenase of normal, stressed and yellow fever virus-infected rhesus monkeys. *Lab. Anim. Sci.* 26 (3):436–42.

Sidon, E. W., R. P. Moody, and C. A. Franklin. 1988. Percutaneous absorption of *cis*- and *trans*-permethrin in rhesus monkeys and rats: Anatomic site and interspecies variation. *J. Toxicol. Environ. Health* 23:207–16.

Wester, R. C., and H. I. Maibach. 1991. *In vivo* animal models for percutaneous absorption. In *Percutaneous absorption mechanisms-methodology-drug delivery*. 2d ed. Edited by R. L. Bronaugh and H. I. Maibach, pp. 221–238. New York: Marcel Dekker.

OCCUPATIONAL MONITORING

Herbert N. Nigg

Citrus Research and Education Center, University of Florida, Lake Alfred, FL 33850

5-1 INTRODUCTION

Occupational monitoring of agricultural workers for pesticide exposure has been linked inextricably with the use of parathion for pest control. Parathion was synthesized sometime between 1940 and 1944 by Gerhard Schrader (Holmstedt 1963). Schrader's research thrust from 1936 to 1945 at I. G. Farben was phosphorus compounds for use as insecticides. Besides parathion, Schrader's laboratory also synthesized TEPP, DFP, OMPA, paraoxon, and the chemical warfare agents tabun, sarin, and soman (Holmstedt 1963).

Parathion use in agriculture began after World War II. In 1949, a loader-mixer was killed by parathion in Lake Placid, Florida (Griffiths, Stearns, and Thompson 1951). Blood cholinesterase monitoring and poison incident investigations began in 1950 among Florida citrus workers and continued in 1951 (Griffiths, Stearns, and Thompson 1951; Griffiths, Williams, et al. 1951). Based on cholinesterase measurements and symptoms, Griffiths, Williams, et al. (1951) determined that the most hazardous task was a worker applying pesticides with a handgun; the next most hazardous task was mixing for an airblast sprayer, followed by the applicator. Illnesses were reduced by reducing the length of exposure, bathing, and changing clothes daily (Griffiths, Williams, et al. 1951). Griffiths, Williams, et al. (1951) stated there was no danger from vapors of parathion, based on the work of Stearns et al. (1951), but they made a clear recommendation that citrus groves not be reentered for 7 days for discing, chopping, or irrigation and for 2 weeks for picking and pruning. Grob, Garlick, and Harvey (1950) were perhaps the first to report pesticide illnesses of fruit harvesters, although their emphasis was on the deaths and symptoms of plant workers and applicators.

What these early workers did not know was that the oxon metabolites and iso compounds of the organophosphate pesticides complicated their studies. Metcalf and

March (1953) showed that technical parathion contained from 5.0 to 20% of the S-ethyl oxon isomer (termed isoparathion), a much more toxic chemical than parathion (tenfold in the rat). US patent 2,601,219 was issued for the S-ethyl isomer of parathion as an insecticide. Other organophosphate insecticide formulations contained up to 50% of the S-ethyl oxon isomer (Metcalf and March 1953).

Over forty-five years ago, Carman et al. (1952) recognized the adverse potential of field-worker exposure to pesticide residues. Subsequent acute organophosphate intoxications of field-workers and the resulting political pressure led to regulations by the Occupational Safety and Health Administration and the US Environmental Protection Agency. These historical regulations and the worker reentry situation have been the subject of reviews (Milby 1974; Gunther et al. 1977).

The Milby report listed eighteen worker reentry cases from 1949 to 1971 in California citrus (Milby 1974). Gunther et al. (1977) cited forty-seven reentry incidents from 1949 to 1976 in California citrus. Peoples, Maddy, and Smith (1978) reported thirty-eight cases of occupational illness in California from 1974 to 1976. Approximately fourteen of these cases were due to exposure to aldicarb residues. Peoples, Maddy, and Topper (1978) reported that 351 cases of occupational illness caused by parathion exposure were seen by California physicians in 1975. Seventeen of these cases can be regarded as worker reentry incidents. Knaak et al. (1978) reported a case of 118 worker illnesses from dialifor exposure in a California grape vineyard in 1976. Davies, Cassady, and Raffonelli (1973) reported two cases (24 field-workers) involving parathion in Florida sweet corn. In 1975, one death was reported among North Carolina workers reentering a parathion-treated tobacco field (Wicker et al. 1979).

A major summary of reentry incidents from 1966 to 1979 classified 86 of 25,500 pesticide incident reports as reentry incidents (US EPA 1979). Of these eighty-six incidents, forty-seven were field-worker reentry incidents (Table 5-1). Nine states reported incidents. California had the highest number with thirty-four incidents, followed by North Carolina with three. Sulfur led with nine cases and parathion was involved in six. All of the sulfur cases were from California, usually involving eye irritation. Twenty-four different pesticides, covering most pesticide classes, were implicated in

Table 5-1 Agricultural reentry incidents by state, 1966 to 1979

State	No. of incidents	No. of persons
California	34	62
Hawaii	1	1
Illinois	1	79
Indiana	2	77
Florida	1	17
Michigan	2	6
North Carolina	3	5
North Dakota	1	2
Washington	2	4
Total	47	253

Source: US EPA (1979).

Table 5-2 Pesticides implicated in agricultural reentry incidents

Pesticide	No. of incidents
Aldicarb	2
Azinphos-methyl	2
Benomyl	1
Carbaryl	1
Carbofuran	1
Chlorothalonil	1
Copper sulfate	1
Demeton	1
Dichlorvos	1
Difolatan	2
Dimethoate	3
Disulfoton	1
Endosulfan	2
Lime	1
Malathion	2
Methidathion	1
Methomyl	2
1-Naphthalene acetic acid	1
Naled	2
Oryzalin	1
Oxydemeton-methyl	1
Parathion	6
Phosalone	1
Sulfur	9
Unknown	1
Total incidents	47
Total pesticides	25

Source: US EPA (1979).

these reported incidents (Table 5-2). This summary was incomplete because the 1976 dialifor incident in California (118 persons) (Knaak & Peoples et al. 1978) and the cases reported by Peoples, Maddy, and Smith (1978) and Peoples, Maddy, and Topper (1978) were not included.

Based on the record in Table 5-3, worker reentry incidents can be placed in either of two classes. One class occurs shortly after application, with workers reentering a treated area by mistake or by design and being exposed to toxic and fast-acting parent compounds. Additionally, these residues may be readily available to the worker because the field is wet from the application itself or from dew, or the residue is in a confined space such as a greenhouse or mushroom facility. The second type of incident involves reentry into a treated area after a period of time usually considered safe, i.e., 5 to 120 days after application. This type of reentry incident has been limited primarily to California and to organophosphate insecticides, usually parathion and dialifor.

A compilation of reentry incidents from 1949 to 1955 described cases of organophosphate intoxication among workers in pears, apples, grapes, citrus, and hops (Quinby,

Table 5-3 Reentry-incident pesticides and human symptoms

Compound type	Health complaints
Sulfur	Symptoms of sulfur exposure were dermatitis and eye irritation. Incidents generally occurred within 1 to 4 days after application, but have occurred up to 3 weeks after application during mowing operations.
Carbamates (carbaryl, benomyl, carbofuran, aldicarb, methomyl)	Symptoms of exposure were nausea, dizziness, blisters, dermatitis, malaise, sweating, tearing eyes, breathing difficulty, and chest tightness. Usually, incidents have occurred within 1 to 2 days after application except for aldicarb (up to 4 days after application).
Chlorinated hydrocarbons (chlordane, endosulfan, and chlorothalonil)	Symptoms of exposure were dermatitis, eye irritation, nausea, and dizziness. Entry into treated areas was from 0 to 1 day after application.
Organophosphates (parathion, naled, phosalone, dialifor, oxydemeton-methyl, dimethoate, azinphos-methyl, methidathion)	Symptoms of exposure were headache, nausea, vomiting, dizziness, malaise, abdominal pain, and weakness. The organophosphates were implicated in reentry incidents from 0 to 120 days after application.

Walker, and Durham 1958; Quinby and Lemmon 1958). Workers were picking, thinning, cultivating, and irrigating. Most cases occurred 8 days after application; one case occurred 33 days after application. Although these were descriptive cases, there is no doubt that they were real organophosphate intoxications. The first inkling of why pesticide residues made workers ill 8 or more days after application came in 1964 (Milby, Ottoboni, and Mitchell 1964). Ninety-four orchard workers became ill even though the parathion residue present should not have made them ill. Paraoxon was postulated to be the culprit. One residue sample contained more paraoxon than parathion. Sweet-corn workers in North Carolina showed significant depression of red-blood-cell (RBC) and serum cholinesterase after parathion exposure. The 24-h residues were 0.26 $\mu g/cm^2$ parathion, 0.35 pg/cm^2 paraoxon and the 48-h residues were 0.16 $\mu g/cm^2$ parathion, 0.23 $\mu g/cm^2$ paraoxon (Wicker, Williams, and Guthrie 1979). No clinical symptoms were observed. The simple expedient of wearing gloves appeared to reduce cholinesterase depression (Wicker, Williams, and Guthrie 1979).

Field studies have had difficulty in reproducing the reentry pesticide illnesses discussed above. A North Carolina study of peach thinners exposed to foliar dislodgeable residues of 2.36 pg/cm^2 of parathion with 0.13 pg/cm^2 of paraoxon on fruit and 0.03 pg/cm^2 paraoxon on leaves showed no significant depression of either RBCs or serum cholinesterase (Wicker, Williams, and Guthrie 1979). A study of California peach harvesters over 4 weeks, where both phosalone and azinphos-methyl had been applied, showed no depression of RBC cholinesterase (Popendorf et al. 1979). The dermal doses that these subjects received ranged from 122 to 232 mg phosalone per week, 26 mg azinphos-methyl during the first week only, and 2 to 3.4 mg phosalone oxon per week. Dislodgeable residues of azinphos-methyl on plant surfaces were 0.40 $\mu g/cm^2$ for the first week and nondetectable thereafter. Phosalone ranged from 2 to 7 pg/cm^2 and phosalone oxon from 0.02 to 0.2 pg/cm^2. Apple thinners received parathion exposures ranging from 5 to 12 mg/h, depending on reentry time, with dislodgeable residues of 1 to

5 $\mu g/cm^2$ on the leaf surface (Wolfe et al. 1975). No ill effects on apple thinners were noted. Peach thinners in California showed no change in RBC cholinesterase following initial exposure to 2.6 pg/cm² azinphos-methyl on reentry. No residue was found in workers' shirts, but urinary metabolites of azinphos-methyl were detected at levels up to 20 ppm by day 5 of exposure (Richards et al. 1978). Agricultural hand labor was studied in Washington state using cholinesterase determinations and urinary metabolite analyses (Robbins, Nash, and Comer 1977). No organophosphate residues were detected in urine, and cholinesterase values were normal. In a study of peach thinners exposed to azinphos-methyl, the workers reentered the orchard only 12 to 18 h after the application because of a misunderstanding (Kraus et al. 1977). Residues were about 2.6 $\mu g/cm^2$ of azinphos-methyl with very low azinphos-methyl oxon residues. Low urinary excretion of metabolites and 15% or less cholinesterase depression were noted over 5 exposure days. In a study of California citrus trimmers and pruners exposed to phosalone during reentry, results were unremarkable (Knaak, Maddy et al. 1978). Reentry was at 14 and 21 days with exposure to 2.6 and 3.6 pg/cm², respectively, on each reentry day. In a major study of peach thinners, lettuce harvesters, and artichoke harvesters, low or no urinary metabolites were observed and cholinesterase values were within the normal ranges (Richards et al. 1978; Kraus et al. 1977; Kilgore 1977). Phosdrin and methomyl had been applied to lettuce, and parathion and methyl parathion to artichokes. Dislodgeable residues were less than 1 pg/cm². A human-subject experiment to determine tobacco reentry after a monocrotophos application showed no urinary metabolites and insignificant cholinesterase depression when reentry was 48 h or greater (Guthrie et al. 1976).

Dialifor was more persistent in the San Joaquin central valley than anticipated and vineyards varied considerably in their dialifor content (Knaak, Peoples et al. 1978). Vineyards had been sprayed 15 to 40 days earlier. Cholinesterase values were reduced 49 to 58% in workers who became ill. Analyses of trousers, patches, and skin swabs indicated a higher level of residue on the worker than on the foliage. Skin swabs contained about ten times the level of foliage. Unfortunately, analyses for dialifor oxon residues were not performed until later (Winterlin et al. 1978). Analyses for dialifor oxon on a row-by-row basis showed possible spot treatment of rows, and dialifor oxon residues were as much as one-third of the dialifor residues. It was speculated that dialifor oxon contributed to this reentry incident (Winterlin et al. 1978). Cotton scouts in North Carolina showed occasional RBC and serum cholinesterase depressions of 20 to 50% over eight growing seasons after exposure to toxaphene–methyl parathion, EPN–methyl parathion, monocrotophos, or azinphos-methyl 1 to 7 days after application (Wicker, Williams, Bradley et al. 1979). These depressions were linked to early reentry of fields (prior to 48 h), entering dew-wet fields and poor personal hygiene practices such as wearing contaminated clothing more than one day (Wicker, Williams, and Guthrie et al. 1979; Wicker and Guthrie 1980).

A reentry study 12 h after application in Arizona cotton exposed volunteers to methyl parathion, ethyl parathion, or monocrotophos for 5 h and produced no clinical signs of poisoning. However, cholinesterase depression averaged 14%, and both ethyl and methyl parathion were found in the blood as well as *p*-nitrophenol (PNP) in the urine. These workers were not judged to be in jeopardy because cholinesterase depression was less

Table 5-4 Historical harvest and reentry times in days for California and Florida citrus

Pesticide	Harvest		Reentry	
	Calif.	Fla.	Calif.	Fla.
Azinphos-methyl	7(28)[a]	7(28)	30	1
Carbophenothion	30	14 or 30	14	2
Demeton	21	21	5	2
Diazinon	21	21	5	0
Dimethoate	15	15	4	0
Dioxathion	0	0	30	0
Ethion	0(21)[b]	0(21)	30	1
Malathion	7	7	1	0
Methidathion	14	14	30	0
Sulfur	0	0	1	0
Parathion	30	RNR[c]	30, 45, or 60[d]	RNR

Sources: Florida Cooperative Extension Service (1981) and California Cooperative Extension (1980–1982).
[a] 28 days for two applications.
[b] 21 days for lemons.
[c] RNR = registered but not recommended—2-day national reentry time; 14- or 30-day harvest depending on application rate.
[d] Dependent on application rate.

than 30% and the PNP excretion was less than 4 mg following exposure (Ware, Morgan et al. 1975).

The conclusion of two worker-reentry reviews was that the worker reentry problem differs in various regions (Milby 1974; Gunther et al. 1977). Carbamate and organophosphate pesticide use doubled from 1964 to 1976 (Ridgeway et al. 1978). By weight, carbamates represented 19% and organophosphates 50% of insecticides in use in 1976, up from 10% and 25%, respectively, in 1964. Of all pesticides in use worldwide, 21% were in cotton and 32% in fruits and vegetables, crops that employ scouts or hand labor. These use patterns should have resulted in more hazard and thus more cases of field-worker intoxication, but there have been no reports of increased pesticide-related illnesses in field-workers during this period. However, in a review of worldwide pesticide poisonings, cases were estimated at 0.5 million in 1972 and 25 million in 1990 (Levine and Doull 1992). These are poisonings from all circumstances (suicide, workplace, accidents, etc.) but, nonetheless, these are disturbing estimates.

The lack of an increase in reentry incidents can be attributed partly to harvest times: the time that a grower must wait after a pesticide application before harvesting the crop (Table 5-4). The control of food quality through residue monitoring was established in 1947 and reenforced in 1954 (Gunther 1969). Harvest times are usually longer than reentry times and do not vary regionally (Table 5-4). This is not true for reentry times. For instance, parathion in citrus had a 30-day national harvest time and a 2-day national reentry time, but up to a 60-day reentry time in California. Azinphos-methyl-treated fruit could be harvested in 7 days, groves could be reentered in 1 day nationally, but not for 30 days in California (Table 5-4). These differences reflect the experience of agricultural labor with pesticide exposure in various regions.

Four primary factors apparently have led to an incident 5 or more days after application: a dusty work environment, use of a sufficiently toxic organophosphate, conversion of the parent compound to its oxon metabolite, and dry conditions. A dusty working environment has been recognized since 1952 as a key element for transferring pesticides from leaf, fruit, and soil surfaces to a field laborer (Carman et al. 1952). The type of dust, that is, soil type, also influences the rate of conversion of an organophosphate to its more toxic oxon (Gunther et al. 1977). Dry conditions are necessary for the persistence of these oxon residues over long periods of time (Gunther et al. 1977).

Florida citrus is grown in primarily sandy soils with clay content of 2% or less (Nigg 1979). Clay is defined as particles of <2 μm. The Florida silt fraction contains less than 8% of particulates 2 to 50 μm. California soils have over 10% of particulates <2 μm and 28 to 57% of particulates 2 to 50 μm (Adams, Iwata, and Gunther 1976; Iwata, Westlake, and Gunther 1973; Iwata et al. 1975). The limit for particulate matter swept up by grove operations and wind, that is, airborne particulates that settle on foliage, is about 50 μm. These data indicate a five- to eightfold difference between Florida and California in the potential for dusty leaves and the transfer of soil surface residues to workers directly or via plant surfaces.

California agriculture has a history of using parathion and this compound has been implicated in almost every case of delayed reentry illness (0–120 days), with the exception of dialifor (Milby 1974; Gunther et al. 1977; Knaak, Peoples et al. 1978). Florida had difficulty with parathion applicator intoxication over 25 years ago which substantially reduced the use of parathion (Griffiths, Stearns, and Thompson 1951; Griffiths, Williams, et al. 1951). Also, about 30 years ago, Florida citrus growers discovered that a decreased use of organophosphates assisted pest management by increasing the natural enemies of scale insects.

In a study of ethion on grape foliage in the California central valley, ethion dioxon was above the level of the parent compound after 7 days and the monoxon and dioxon combined were approximately three times the level of ethion after 28 days (Leffingwell, Spear, and Jenkins 1975). A similar study in Florida citrus under various weather conditions found little ethion dioxon and less than 10% of the dislodgeable residue was the monoxon (Nigg, Allen, Brooks, Edwards, et al. 1977). Of thirty-two citrus groves in Tulare County, California, substantial paraoxon residues were found in thirty-one groves (Spear et al. 1975). On day 2 after application, one grove contained twice as much paraoxon as parathion, and paraoxon was about twice as long-lived as parathion. Experiments in Florida citrus over a two-year period under various weather conditions showed that dislodgeable paraoxon levels were always well below the level of parathion and disappeared at about the same rate (Nigg et al. 1978; Nigg, Allen, and King 1979). Similar data have been reviewed (Gunther et al. 1977). The point here is that oxon levels have differed regionally in the United States and, in fact, regionally within California (Gunther et al. 1977).

It is well established that environmental conditions modify the behavior of pesticides (Nigg, Allen, Brooks, Edwards, et al. 1977). Long-term (5 to 120 days) worker-reentry-related illnesses have been limited primarily to the hot and dusty central California valley. Because these incidents have been linked to the oxon metabolites and these metabolites reach higher levels in dry areas, rainfall and dew appear to be additional major components for the regionality of worker reentry incidents in the United States. Davies, Cassady, and Raffonelli (1973) proposed a regional definition in 1973.

Arizona citrus belts	Two wet seasons (July to September, December to February); mean annual precipitation 3 in. in Yuma area, 7 in. in Phoenix area, 11 in. in Tucson area. Normal harvest seasons dry.
California citrus belts	One wet season (November to March); mean annual precipitation 20 in. in central district, 11 in. in southern district. Except for navel oranges, normal harvest seasons dry.
Florida citrus belts	No prolonged dry season; mean annual precipitation 52 in, higher than 52 in south of Orlando; wet summers (8 in. in Orlando in July, 7 in. in Miami in June). Normal harvest seasons wet.

Beyond these differences in rainfall, most areas of the United States, excepting desert areas, have frequent dew condensation at night. In Florida, which may be the extreme, dew presence ranges from 2 h during the dry winter to 12 h in the wet summer. In extensive and repeated studies of organophosphate insecticides and their oxon metabolites in Florida citrus, the oxons have always been in low quantity (below the parent) and have dissipated rapidly (Nigg, Allen, and Brooks 1977; Nigg, Allen, Brooks, Edwards, e al. 1977; Nigg et al. 1978; Nigg, Allen, and King 1979; Nigg and Stamper 1980). Rainfall also has been linked to oxon dissipation in California (Gunther et al. 1977) and moisture to dissipation of oxons on soil surface particulates (Spencer et al. 1975).

5-2 MONITORING TECHNIQUES

5-2-1 Enzymes in Blood

One of the earliest biological monitoring studies was undertaken by Quinby, Walker, and Durham (1958). Cholinesterase activity was measured in aerial applicators together with residues on worker clothing and in respirator filters. In spite of physical complaints by the pilots, either normal or only slightly depressed cholinesterase values were reported. Cholinesterase values were compared with the normal range for the U.S. population rather than the pilots' own individual normal values. Kay et al. (1952) measured cholinesterase levels in orchard parathion applicators. These were compared with cholinesterase levels taken from the same workers during nonspray periods. Plasma cholinesterase for workers reporting physical symptoms was 16% lower during the spray period. The corresponding reduction for symptomless workers was about 13%. RBC cholinesterase was depressed 27% for the symptomatic group versus 17% for the nonsymptomatic group, but these means were not statistically different. Roan et al. (1969) measured plasma and erythrocyte cholinesterase and serum levels of ethyl and methyl parathion in aerial applicators. Serum levels of the parathions could not be correlated with cholinesterase levels. However, serum residue levels did correlate with the urine concentration of PNP. Drevenkar et al. (1983) measured plasma and erythrocyte cholinesterase levels and urine concentrations of organophosphate and carbamate pesticides in plant workers who formulated these materials. No correlation existed between urinary metabolites and cholinesterase depression.

Even in tightly controlled animal experiments, correlations between dose, cholinesterase depression, and urinary metabolites have been difficult to obtain. Bradway, Shafik, and Loros (1977) fed seven different organophosphate compounds at two different doses to rats. Animals were removed from exposure after day 3 and blood and urine were collected and cholinesterase was measured for the next 10 days. The average percentages of the total doses excreted in urine over the 10 days were 12% dimethoate, 10% dichlorvos, 11% ronnel, 57% dichlofenthion, 66% carbophenothion, 40% parathion, and 50% leptophos. Very little of this excretion occurred beyond day 3 postexposure. Intact residues of ronnel, dichlofenthion, carbophenothion, and leptophos were found in fat on day 3 and day 8 postexposure. Correlations of cholinesterase activity with blood residues and urine metabolite levels were poor. The overriding general conclusion from the human and animal studies was that cholinesterase inhibition contains too many variables, known and unknown, to be useful for estimating dose. Whole blood rarely has been used to monitor dose in humans despite pleas from pesticide chemists, toxicologists, and pharmacologists (Moseman and Enos 1982).

Wilson and Henderson (1992) reviewed the techniques used to measure blood esterases and their problems. A solvable and important problem with these measurements is that there is no biochemically based standardized assay available.

There are other problems with blood monitoring. It is difficult and probably not possible with the present committee oversight of human subjects to conduct human experiments as above. Workers are averse to blood testing; noninvasive methods are much preferred. Blood has been used in epidemiological studies (Davis 1980; Keil et al. 1973; Klemmer, Rashad, and Mi 1973) but an original dose probably never can be inferred. The rapid clearance of newer, biodegradable pesticides from blood may, in fact, eliminate blood as an exposure monitoring tool for pesticide dose and undoubtedly will eliminate blood as an epidemiological monitoring tool.

Lewalter and Korallus (1986) reviewed erythrocyte protein conjugates as a monitoring technique for pesticide exposure. This technique appears promising for estimating dose from pesticides that form erythrocytic adducts. There are a variety of biomarkers for pesticide exposure monitoring (Brewster, Hulka, and Lavy 1992). Unfortunately, none of these biomarker techniques has been validated for the estimation of an internal pesticide dose (Brewster, Hulka, and Lavy 1992).

5-2-2 Noninvasive Techniques

Monitoring techniques for the occupational exposure of agricultural workers were based on the techniques of Batchelor and Walker (1954) and Batchelor, Walker, and Elliott (1956). In a series of studies, Durham, Wolfe, and coauthors modified and refined these techniques to produce methods that are in use today (Wolfe et al. 1959; Wolfe, Durham, and Batchelor 1961; Durham and Wolfe 1962, 1963; Wolfe, Durham, and Armstrong 1963, 1965, 1967, 1970; Wolfe, Armstrong, and Durham 1966, 1974; Durham, Wolfe, and Quinby 1965; Wolfe and Armstrong 1971; Wolfe and Williams 1972; Durham, Wolfe, and Elliott 1972; Wolfe 1972; Wolfe et al. 1972a, b; Wolfe, Armstrong, Staiff,

Comer, and Durham et al. 1975; Wolfe, Staiff, and Armstrong 1978). By far the most referenced of these studies are Durham and Wolfe (1962, 1963). The 1962 reference outlines the basic procedures for monitoring dermal, hand, and respiratory exposure. The 1963 reference contains a correlation between dermal exposure to parathion and urinary excretion of PNP. These techniques and their utility were reviewed by Davis (1980), but the original articles contain the detailed information necessary for replication.

The drawbacks to the Durham and Wolfe methods are that the respiratory dose is estimated from a respirator collection pad or a personal air monitor, the material and design of the dermal collection devices needs assessment before an experiment, and dermal exposure is estimated from collection devices that may inaccurately represent exposure. For example, Serat, Van Loon, and Serat (1982) recognized that pesticides might be lost (volatilize) from a collection device during the sampling period. This deficiency can be assessed with a short loss experiment before human experiments are begun. Kirchner et al. (1994, 1996) have compared the capture efficiencies of materials from which dosimeters are made. We should remember, however, that the goal is to estimate a dose, either to the skin and lungs or an internal dose.

There are various ways of estimating pesticide dose. *Environmental monitoring* includes measurement of ambient levels of pesticide in the worker's environment and passive dosimetry to estimate the quantity that comes into contact with the worker. Dermal patches and personal air samplers have been and are used to estimate dermal and inhalation exposure in pesticide workers. *Biological monitoring* includes the measurement of parent compounds, their metabolites or an indicator of response (e.g., acetylcholinesterase inhibition) in a biological sample such as urine, blood, sweat, saliva, exhaled breath, hair, or nails. If human biological samples are negative for a given pesticide, what can be concluded in reality about dose? If they are positive, can dose be reliably quantified? These same questions apply to any method for estimating dose.

Here, exposure to pesticide means the deposit of compound onto the clothing or skin of the worker, the latter being dermal exposure. Dose is the amount that enters the body, including skin sequestration. The majority of the dose to an agricultural worker is usually through transdermal absorption, but other routes such as respiratory or oral ingestion also play a role. A biological indicator is any biological sample taken from the worker that contains reliable evidence that a pesticide dose has been received. Some of the biological methods we review here have been combined with nonbiological methods for measuring exposure. Gunther et al. (1977), Davis (1980), Nigg and Stamper (1982), Popendorf and Leffingwell (1982), and Nigg and Stamper (1985) provide reviews of nonbiological methods.

For applicator/mixer-loader groups, the type of equipment used, the number of tanks applied per unit time, the concentration of the tank mix, the loading method, and the protective clothing worn affect the dermal exposure process. Especially important are personal work habits. This has been known for years and is described in many published reports (Davis 1980; Nigg, Stamper, and Queen 1984).

Regardless of crop type, the production rate of harvesters appears to be related to their exposure. The worker's production rate is related to the amount of contact with the plant, a subject that has been studied using movies and time analysis (Wicker and Guthrie 1980) and estimated with surveys (Wicker et al. 1980). The production rate can be quantified as the number of boxes picked, crates loaded, tassels removed, and soforth, and is confounded with residue levels in affecting exposure.

One of the reasons for measuring worker pesticide exposure is to aid in determining the safety of the product. To do this, it must be known whether the amount of pesticide to which the worker is exposed may lead to a toxic effect. The toxic effect of a pesticide is evaluated through testing in animals by establishing the relationship of a physiological effect with dose. To extrapolate the animal data to humans, there must be an equivalent estimate of dose. Two approaches have been used to do this in pesticide workers. One involves the measurement of dermal exposure (most prevalent route) with a correction for the estimated percentage of percutaneous penetration. The other approach relates urinary metabolite level(s) to dose.

Urinary Monitoring. Excellent studies of 2,4,5-T and 2,4-D in human blood and urine were conducted by Gehring et al. (1973), Kohli et al. (1974), and Sauerhoff, Braun, et al. (1977). An oral dose of 2,4,5-T was excreted in the urine of five human volunteers with a half-life of 23 h and $88.5 \pm 5.1\%$ of the administered dose was recovered. Its half-life in plasma was also 23 h (Gehring et al. 1973). 2,4-D was taken orally and also excreted unchanged in urine (75% recovered in 96 h) (Kohli et al. 1974). There was marked individual variation in when the peak of 2,4-D occurred in plasma. Sauerhoff, Braun, et al. (1977) administered a single 5-mg/kg oral dose of 2,4-D to five human male volunteers. It disappeared from plasma with an average half-life of 11.6 h and from urine with an average half-life of 17.7 h. From 88 to 106% of the dose was recovered in urine over 144 h; over 50% was recovered in 24 h. Silvex 2-(2,4,5-trichlorophenoxy) propionic acid was administered to seven men and one woman at 1 mg/kg. Biphasic disappearance from plasma/urine with half-lives of $4.0 \pm 1.9/16.5 \pm 7.3$ h and $5.0 \pm 1.8/25.9 \pm 6.3$ h, respectively, were determined and 65% of the dose was excreted in urine in 24 h (Sauerhoff, Chenoweth, et al. 1977). Human exposure to phenoxy herbicides has been reviewed by Lavy (1987). These studies are the best we have available. Yet even this type of study cannot be used to estimate dose to a pesticide-exposed worker, as will be readily apparent.

The first difficulty in using urine to estimate a pesticide dose is urinary output (Table 5-5). Creatinine measurements are assumed to normalize differences in the urinary excretion of a compound. Differences in creatinine excretion can arise from changes in fluid intake, pathology, or physiological changes such as sweating. To be useful, day-to-day excretion rate of creatinine in urine should be constant regardless of the volume of urine output, food intake, or physical activity, and should reflect only small interindividual variation. In a large study of creatinine excretion, Alessio et al. (1985) found that men excreted more creatinine than women, younger subjects more than old, and individuals varied in their 24-h creatinine output from 9.2 to 79.4%, perhaps because of fluid intake. It appears that great care must be taken if creatinine is to be used to normalize urine samples. Perhaps the best study of adjusting urine concentrations of excreted substances was that by Araki et al. (1986). These authors studied the excretion of lead, inorganic mercury, zinc, and copper in urine. They concluded that urinary flow-adjusted concentration could be used to monitor all substances in urine, creatinine included. Creatinine, urinary specific gravity, and timed excretion plus urinary flow had

Table 5-5 Human variability in daily urine volume

Subject	Volume (mL/day)	Reference
Newborn infants	17–34	Lentner (1981)
Males, 6–10 years	459 ± 175	Lentner (1981)
Males, 10–14 years	605 ± 294	Lentner (1981)
Females, 6–10 years	274 ± 125	Lentner (1981)
Females, 10–14 years	441 ± 305	Lentner (1981)
Males, mature	1,360 ± 443	Lentner (1981)
Females, mature	1,000 ± 430	Lentner (1981)
Males, working in heat	896 ± 18	Nigg et al. (1991)
Males, long-distance runners	600	Costill (1981)
Male control	1,300	

limited uses. It appears necessary to use 24-h, total urine collection when urine is used to estimate the pesticide dose.

Urine monitoring to determine dose also depends on certain assumptions and disparate pieces of available data. The excretion kinetics of the pesticide must be known for both dermal and oral administration. If the total dose is excreted by small animals in 24–48 h, the same may or may not be true of humans. If the dose is excreted over a period of a week, a simple test correlating dermal exposure with an *immediate* effect on urine pesticide levels will not work. The use of prolonged sampling periods over which mean values of urinary pesticide excretion rates are calculated presents problems also. It is common to see a 100% coefficient of variation in residues among urinary excretion samples. As a consequence, a major change in urinary pesticide levels would have to occur before a significant difference could be confirmed statistically. Part of this unfortunate variation no doubt results from reckless work practices (e.g., spills, sloppy work practices, poor hygiene). If the dermal dose excretion kinetics are known, if urine can be shown to be reflective of dose, if 24-h urines are collected, and if sample numbers are based on the expected variation, then urine could be used to monitor dose.

Significant storage in body tissue may occur. After *multiple* oral doses of dicofol, this organochlorine rapidly reached a plateau in urine, whereas excretion levels steadily increased in the feces (Brown, Hughes, and Viriyanondha 1969). The excretion kinetics available for most compounds have resulted from studies using only one, or a very few *oral* doses. Workers, however, receive daily dermal and respiratory doses. There are no animal data reflecting these continuous and multiple routes of exposure. There are kinetic models that could represent pesticides recycled into the bile (Colburn 1984). These models do not account for the differences in the exposure route for workers, the frequency of dose, or the subsequent excretion differences of workers compared to small animals.

Urinary metabolites of pesticides have been used to indicate absorption and excretion. Swan (1969) measured paraquat in the urine of spraymen; Gollop and Glass (1979) and Wagner and Weswig (1974) measured arsenic in timber applicators. Lieban et al. (1953) measured PNP in urine after parathion exposure, as did Durham, Wolfe, and Elliott (1972). The chlorobenzilate metabolite, dichlorobenzilic acid, was detected in citrus workers (Levy, Brady, and Pfaffenberger 1981); phenoxy acid herbicides were detected in farmers (Kolmodin-Hedman, Hoglund, and Akerblom 1983); and organophosphate metabolites were detected in the urine of mosquito-treatment applicators (Kutz and Strassman 1977). Davies et al. (1979) used urinary metabolites of organophosphate compounds and carbamates to confirm poisoning cases. These studies documented exposure, but no quantitative estimate of dose was made.

Other studies have used air sampling and have monitored hand exposure in combination with urinary levels (Cohen et al. 1979), and air sampling plus cholinesterase inhibition plus urine levels (Hayes, Wise, and Weir 1980). Nonbiological methods, combined with measurement of urinary metabolites, also have been used to compare worker exposure for different pesticide application methods (Wojeck et al. 1983; Carman et al. 1982) as well as to monitor exposures to formulating plant workers (Comer et al. 1975) and to homeowners (Franklin et al. 1981).

Several studies have used nonbiological methods in an attempt to correlate urine levels with an exposure estimate (Wojeck et al. 1981, 1983; Franklin et al. 1981; Staiff et al. 1975; Lavy, Shepard, and Mattice 1980; Lavy et al. 1982). Lavy, Shepard, and Mattice (1980), and Lavy et al. (1982) found no correlation of dermal exposure with urinary metabolites of 2,4-D or 2,4,5-T. Wojeck et al. (1983) found no paraquat in urine and consequently no relationship between dose and urine levels. However, the *group daily mean* concentration of urinary metabolites of ethion and the group *mean* total estimated dermal dose to ethion on the same day correlated positively, with significance at the 97% confidence level (Wojeck et al. 1981). For arsenic, the cumulative total estimated dose and daily urinary arsenic concentration (lead arsenate exposure) correlated positively, with significance at the 99% confidence level (Wojeck et al. 1982). Franklin, et al. (1981) found a positive correlation between 48-h excretion of azinphos-methyl metabolites and the *amount* of active ingredient sprayed. A significant correlation could not be made, however, between 48-h excretion and an exposure estimate using dosimeters. In the Franklin et al. (1981) experiment, a fluorescent tracer was added to the spray mixture. Qualitatively, unmonitored areas (face, hands, neck) also received significant exposure, perhaps leading to a weak correlation between the exposure estimate and urinary metabolites. Franklin, Muir, and Moody (1986) also studied excretion of dimethylthiophosphate (DMTP, a metabolite or azinphos-methyl) in orchard workers in two other areas in Canada and again found that urinary data more closely reflected the amount of pesticide applied. Grover et al. (1986) also obtained a correlation ($R = .82$) between the amount of 2,4-D applied and urinary excertion. Total deposition of 2,4-D estimated with dosimeters and urinary excretion correlated at only $R = .47$. Winterlin et al. (1984) monitored the dermal exposure of applicators, mixer-loader, and strawberry harvesters to captan. Although the applicator/mixer-loader group showed higher estimated dermal exposure, no metabolite was detected in their urine; harvester urine had detectable levels. Leng et al. (1982) reviewed the excretion of 2,4,5-T from humans exposed in the field and found that it was slower with dermal field exposure than with oral doses. Kinetic excretion models of 2,4,5-T indicated that the absorbed dose did not exceed 0.1 mg/kg daily of body weight. This dose estimation was well below the exposure estimates made from dermal collection patches.

Part of the problem appears to be the excretion kinetics and excretion routes of pesticides after dermal exposure, the complexity of which may render useless any simple correlation between exposure and urinary metabolites. Funckes, Hayes, and Hartwell

(1963) exposed the hands and forearms of human volunteers to 2% parathion dust. During exposure, the volunteers breathed pure air and placed their forearm and hand into a plastic bag that contained the parathion. This exposure lasted 2 h and was conducted at various temperatures. There was an increased excretion of PNP with increasing exposure temperature. Importantly, PNP still could be detected in the urine 40 h after this dermal exposure. Feldmann and Maibach (1974) dosed human volunteers with ^{14}C-radiolabeled pesticides intravenously (i.v.) and followed urinary excretion for 5 days using total urine collection. The same volunteers subsequently were exposed dermally on the forearm with the same ^{14}C-labeled pesticides, and urinary metabolites were again followed for 5 days. The percentage of pesticide excreted after forearm (dermal) dosing then was corrected if i.v.-related excretion was incomplete. For instance, the estimated dermal absorption of a pesticide was doubled if only 50% of the i.v. dose was excreted in urine. The compounds administered were ^{14}C-azodrin, ethion, guthion, malathion, parathion, baygon, carbaryl, aldrin, dieldrin, lindane, 2,4-D and diquat. Wester et al. (1984) applied ^{14}C-labeled paraquat dichloride to the legs, hands, and forearms in a similar experiment. Urinary excretion of ^{14}C was followed over 5 days. The highest corrected recovery from urine in either study was carbaryl at approximately 74% of the dermally applied compound (forearm); the lowest was for paraquat at 0.23% (hand). These dermally applied ^{14}C-labeled pesticides were recovered in urine at rates of 0.3% (diquat) to 19.6% (baygon) of the dermally applied dose. Individuals varied in their absorption of chemicals. Different anatomic sites absorbed chemicals at different rates in both humans and animals (Maibach 1982; Skinner and Kilgore 1982). Occlusion of the anatomic site also increased absorption of chemicals (Feldmann and Maibach 1974; Stoughton 1965; Riley, Kemppainen, and Norred 1985).

Dedek (1980) presented an interesting case for the effect of solubility of pesticides in solvents and skin penetration. Polar compounds penetrated better when applied in nonpolar solvents and nonpolar compounds penetrated better in polar solvents. Wester and Noonan (1980) reviewed the relevance of animal models for percutaneous absorption. The pig and the monkey were most nearly like humans in this respect. These authors warned that a clear understanding of the species and methods used is a prerequisite to extrapolation to humans.

In another human experiment, Kolmodin-Hedman, Hoglund, Swenson, Akerblom (1983) applied methyl-chlorophenoxy acetic acid (MCPA) to the thigh. Plasma MCPA reached a maximum in 12 h and MCPA appeared in the urine for 5 days with a maximum in about 48 h. Given orally, urinary MCPA peaked in urine in 1 h with about 40% of the dose excreted in 26 h. Using three male and two female volunteers, Akerblom, Kolmodin-Hedman, and Hoglund (1983) performed three experiments with MCPA. In one experiment, the hands and forearms were dipped into a 2% solution. The dried solution was washed off after 30 min. In a second experiment, oral MCPA (15 μg/kg) was given. In a third experiment, 10 mL of a 10% aqueous solution of commercial MCPA formulation was placed on a 10-cm^2 pad which was applied to the thigh for 2 h. Urine and blood were monitored. MCPA was not detected in blood or urine with the hand and forearm dip. With oral MCPA, the peak concentration occurred at about 6 h in urine and 1 h in plasma. In the MCPA thigh absorption study, the urine peak occurred at 24–48 h.

Nolan, Freshour, et al. (1984) dosed six male volunteers orally with amino-3,5,6-trichloropicolinic acid (picloram). They were exposed dermally in a separate experiment. In 72 h, over 90% of the oral dose was excreted in the urine. The dermal application was poorly absorbed (0.2%) and still was being excreted in urine at 72 h. Nolan, Rick, et al. (1984) studied the excretion of chlorpyrifos in human urine after an oral or a dermal application. Total urine was collected from 24 to 48 h prior to administration through 120 h after administration; $70 \pm 11\%$ ($n = 6$) of the oral dose was recovered in urine, and $1.28 \pm 0.83\%$ ($n = 6$) of the dermal dose was recovered in urine. Blood levels peaked in 6 h after the oral dose and in 24 h after the 5mg/kg dermal dose. The urine elimination half-life was about 27 h for both oral and dermal doses.

Morgan et al. (1977) dosed humans orally with methyl and ethyl parathion. PNP, which appears in urine after exposure to either pesticide, was excreted at 37% of the theoretical whereas the alkyl phosphate metabolites were excreted at 50 +% of the theoretical in 24 h. Thus, about 50% of the oral dose administered to humans was not recovered in urine in 24 h.

Rats were administered azinphos-methyl dermally and intramuscularly to determine whether a linear response existed between the amount of azinphos-methyl applied dermally and the excretion of DMTP (Franklin, Greenhalgh, and Maibach 1983). A linear relationship appears to exist in humans. This suggests that DMTP levels can be multiplied by the ratio of azinphos-methyl exposure to DMTP level in urine. This turned out to be a 10-to-1 ratio. Using data from several field applications, it appears that the urinary DMTP-times-10 result is more closely related to the amount sprayed than the dermal exposure estimated from patches (Franklin, Muir, and Moody 1986). Franklin, Muir, and Moody (1986) concluded that without adequate pharmacokinetic information on the parent compound, urine is useful only as a screening tool for indicating exposure or as a predictor of overexposure.

These experiments illustrate the excretion differences between dermal, intramuscular, and oral administration; the excretion differences between compounds; and problems concerning the monitoring of urinary metabolites (Franklin, Greenhalgh, and Maibach 1983). A very discriminating experimental design would be necessary to model dermal exposure and absorption versus urinary metabolite levels.

Saliva Monitoring. The use of saliva for monitoring chemicals in humans has been reviewed (Nigg and Wade 1992). Saliva has the advantages of being easy to collect and analyze for the parent molecule. The level of the parent compound in saliva and the free parent molecule in plasma has been correlated for a variety of drugs and metals. However, the binding of a chemical to proteins and other biological molecules makes dose determination difficult (Nigg and Wade 1992).

Skin Monitoring. Feldmann and Maibach (1974) suggested that the effect of sweating on dermal penetration should be documented, and this was perhaps the first suggestion that sweat might be important to the human pesticide absorption/excretion process. Let us assume that all problems associated with monitoring urine for pesticides and metabolites have been solved. Further, assume that the dermal exposure of several workers is exactly the same in each of 2 weeks. Can we expect the mean level of urinary metabolites during the second week to equal that during the first week? Rosenberg, Queen, and Stamper (1985) used sweat patches to collect dichlorobenzilic acid (a metabolite of chlorobenzilate) from the backs of chlorobenzilate applicators. Urinary analyses also were performed. The proportion found in sweat and urine was approximately 1:1. Sell, Maitlen, and Aller (1982) conducted a human exposure experiment with 2,4-D and monitored urine and sweat. Of the total 2,4-D excreted over 1 week after a single dermal application to the hand, 15% was excreted in urine and 85% was found in sweat (average for three workers). If a pesticide and/or metabolite is excreted in sweat, the sweat/urine excretion ratio may change with the human physiological response to heat. Consider how little we know about this process. Could a dermal exposure, subsequently manifested in sweat, simply be pesticide recycled at the skin level? Is this then an "internal" dose? Does a sweat collection device also collect interstitial fluid? If it collected only interstitial fluid, we could quickly estimate an internal dose. Peck, Conner, Bolden, et al. (1986) and Peck et al. (1987) have studied the transdermal collection of drugs and parathion in small animals. A model for chemical disposition in skin is available (Peck, Conner, Scheuplein, et al. 1986), and outward migration of chemicals through skin has been studied (Peck, Lee, and Becker 1981; Peck et al. 1988). The cumulative internal dose of several compounds, including parathion, was correlated with amount of *parent* compound collected by a transdermal collection device (activated charcoal in a gel, applied to the skin). Collecting a chemical across a worker's skin is noninvasive and appears ripe for development.

The perceived problem of quantifying sweat makes it a less attractive medium to monitor. However, cystic fibrosis is diagnosed with a sweat test which depends on quantification, and Webster (1983) has reviewed the problems of sweat quantification. Quantification may not be a serious problem locally, but the rate of sweating and total daily production would have to be estimated. There are estimates available of fluid loss due to sweating. A nationally ranked, 47-kg runner sweated at 1.2–1.3 l/h (reduction in body wt; 5+%) over 10 miles and 82 min (Costill 1977a,b). Twelve unacclimatized males were tested at 30.0 ± 0.7°C wet bulb globe temperature (WBGT), 1.0 ± 1 m/s wind velocity on a treadmill (Armstrong et al. 1985). Subjects walked and climbed for 30 min and rested sitting for 30 min, over a 6-h period. They sweated at a rate of about 425 ml/h while walking. This rate decreased to 125 ml/h while sitting. Their cumulative sweat loss over 6 h averaged about 3,300 ml. This loss was offset by water intake over the 6 h. Seven men exercising at 50% of maximum oxygen uptake averaged 2.19-l water loss through sweat in 2 h (Costill et al. 1981). Marathon runners in a warm environment may lose 2.8 l/h through sweat and incur 8% loss of body weight through sweat, resulting in a 13–14% loss of body water (Costill 1977a,b). In spite of quantification difficulties, the study of this route of pesticide excretion and its inclusion in exposure estimate is warranted.

When the excretion of pesticide metabolites and parent compounds in human sweat is understood, we may be able to determine our possible underestimation of exposure or dose through the use of urine only. If urinary excretion kinetics following dermal exposure are ignored, the interpretation of the results of a urine experiment might be 100% wrong. If, in addition, sweat excretion is ignored, additional 100% mistakes appear possible.

Table 5-6 Excretion rates of biological fluids and estimated internal dose after a single dermal exposure

Fluid	Fluid excretion rate	Estimated pesticide conc.	Pesticide excretion rate	% of Total
Urine + swallowed saliva (excreted in urine)	900 ml/day[a]	50 ng/ml	45 μg/day	47
Sweat[b]	6,400 ml/day	8 ng/ml	51 μg/day	53
Sebum	3.8 mg/day	8 pg/mg	$3 \times 10^{-5}\mu$g/day	—
Total			96 μg/day	
	Total Dose = 96 μg, or 0.0014 mg/kg			

Note: Excretion rates are based on Light steady work by a 70 kg adult and temperatures above 32°C.
[a] Nigg et al. (1991), Lentner (1981).
[b] Over 8-h period, 0.6 l/h; 16-h period, 0.1 l/h (Armstrong et al. 1985; Costill 1977a,b; Costill et al. 1981).

5-3 EXPOSURE (DOSE) ESTIMATION

Table 5-6 presents a hypothetical case of pesticide excretion in a worker in a hot environment. Typical adult daily fluid elimination rates were assigned to urine, sweat, and sebum. From typical pesticide concentrations within those fluids following a single dermal exposure, daily pesticide excretion rates were estimated for each of the three elimination pathways. A total internal dose then was inferred under the assumption that 100% of the dose was excreted during the day. If the internal dose were estimated from urinary data alone, the result would have been roughly one-half of that calculated.

The rough calculations of Table 5-6 suggest a real problem. Table 5-7 presents the i.v. data of Feldmann and Maibach (1974) in a different form. Because the original published half-lives were excretion-*rate* half-lives, the original data were transformed to percentage of dose unaccounted-for by multiplying excretion rate by hours, producing a *cumulative* total for each of the eight time periods monitored and subtracting this value from 100. These decreasing percentages then were subjected to a first-order linear

Table 5-7 Projected urinary excretion of pesticides

Pesticide	Projected period for excretion of total i.v. dose				First-order correlation coefficients ($n = 8$)
	50% excretion period (h)		90% excretion period (h)		
Azodrin	76	[160, 104][a]	253	[200, 345]	0.966
Ethion	187	[129, 337]	620	[429, 1,120]	0.913
Guthion	68	[58, 83]	266	[192, 276]	0.984
Malathion	73	[37, 1,284]	241	[124, 4,264]	0.727
Parathion	134	[90, 258]	445	[300, 857]	0.901
Baygon	99	[51, 1,263]	329	[171, 4,194]	0.735
Carbaryl	1,670	[1,063, 3,894]	5548	[3,532, 12,936]	0.868
2,4-D	17	[16, 19]	58	[53, 64]	0.997
Diquat	204	[129, 493]	679	[429, 1,637]	0.863

Source: Adapted from Feldmann and Maibach (1974).
[a] 95% confidence interval (h).

regression analysis. Correlation coefficients from those analyses are presented. The linear regression lines allowed us to predict statistically the times at which 50% (and 90%) of the i.v. dose would be eliminated in urine. For each of the two times, 95% confidence internals were calculated. Except for 2,4-D, the projections of Table 5-7 had to be extended beyond the range of the data. This procedure adds to the uncertainty in the projections, as evidenced by some of the larger confidence intervals.

We encourage investigators in this area to conduct urinary excretion experiments to nondetection. Elimination- or excretion-*rate* half-lives are virtually useless for predicting dose if only a small percentage of the dose is accounted for with data. The same *rate* half-life could result, regardless of whether 10% or 100% of the dose was excreted. Mathematical representation might present problems with urinary excretion of oral doses and will certainly be a problem with dermally applied pesticides that penetrate human skin poorly and are excreted in a non-first-order manner. Problems that have not been resolved with any human excretion of pesticides include whether the excretion rate or route is dose dependent. We encourage, in any event, mass-balance or near mass-balance experiments. We should, as scientists, account for the rest of the dose. Do Feldman and Maibach's data suggest other routes of excretion? Are some of these compounds or their metabolites excreted in sweat? Are some residues excreted in saliva either as parent or metabolite(s) and only partially reabsorbed and excreted in urine or stored in adipose tissue or metabolized and excreted as CO_2? Only for 2,4-D does it appear that urine monitoring would be adequate.

The first problem of modeling the excretion process, as in Table 5-6, lies with the oral (stomach) dosing of the rats in most kinetic studies. As previously discussed, the kinetics of excretion are different for oral versus dermal doses. Because rats do not sweat, rat data, even following a dermal exposure, may not be applicable. We do not presently have the data to make an estimate as in Table 5-6. The basic data on human excretion of organic molecules are missing and it seems that the rat and mouse models really apply only to rats and mice.

5-3-1 Statistical Considerations

The question must be addressed as to how many replicates (subjects × days) will be necessary to confirm statistically a difference between means in an exposure experiment. Two parameters must be estimated first. What difference in means is expected to arise from the data and what variation among replicates is expected? A large estimate for the first expectation or a small estimate for the replicate variation would reduce the number of necessary replicates per mean. Our experience has shown that protective clothing can reduce mean exposure pad concentrations by about 90% (Nigg and Stamper 1983). The use of gloves may reduce hand-wash residues by about 85% (Nigg, Stamper, and Queen 1986). The coefficient of variation among replicates is typically about 100% for exposure-pad and hand-wash residues (Nigg, Stamper, and Queen 1984; Nigg, and Stamper 1983). This variation also could subsequently affect the variation in parent/ metabolite quantities in biological samples. Standard statistical procedures show that, on the basis of the above estimates, two means are significantly different at the 95% confidence level by taking at least ten replications per mean (Dowdy and Weardon

1983). The (approximate) calculation is $n > 8$ (100% − 90%), or $n > 10$ replications. For gloves, the corresponding calculation would be $n > 8$ (100% − 85%), or $n > 11$ replications. These numbers represent an absolute minimum based on the above two expected values. An increase in n of 50% to provide some margin for error, in the above cases to $n > 15$ and $n > 17$, is certainly warranted in view of the guesswork involved about sample means and variances.

If more than two means are to be compared concurrently, as with comparing residues at various body locations, the situation becomes more complicated. Although it is now harder to generalize, the optimum number of replications per mean can be estimated roughly by the same calculation, with the final result that the means are grouped into significantly different categories, at some confidence level.

Whether to utilize, for example, three subjects for six sampling periods each, or six subjects for three sampling periods each, to obtain, for example, eighteen replications is usually dictated by factors other than statistical ones. A good rule to follow is not to overload the design too heavily in favor of either variable. If many subjects are used for a very few sampling days, and the data indicate large differences from subject to subject, the available number of replicates now decreases to the number of sampling days alone. This type of experiment makes a very unreliable statement about each of many subjects and no valid overall conclusion may emerge.

Remember that statistical calculations show significant differences between means, but do not estimate the difference. Suppose one mean is 90% less than the other, as above, but that one wishes to validate the claim that 50% of that 90% is statistically significant. The approximate calculation for the requisite number of replications per mean is now $n > 8 [100\% \div (90\% − 50\%)]^2 = 50$, or $n > 75$ with the safety factor.

5-3-2 Biological Sampling Combined with Exposure Monitoring

From another perspective, suppose 24-h urines were collected from twenty workers during two different periods: one where workers wore protective clothing and one where they did not. The urinary excretion means for the two periods would have to differ by 48% to be statistically different at the 95% level, assuming a coefficient of variation of 75%. For *field* studies using fewer than twenty subjects (replicates), differences in urinary excretion means would have to exceed 48% for a statistical difference to be validated.

A paramount consideration is the amount of chemical removed by a collection device that subsequently is *not* excreted in urine because it never became part of the internal dose. If twenty 103 sq cm (4 in. × 4 in.) pads are used for monitoring a worker, about 2,000 cm^2 of body surface area would be covered. The average human has a surface area of ca. 20,000 cm^2. If 10% of the body surface area is covered, will this reduce the urinary metabolite by 10%? Hardly. First, the pads are worn no longer than 4 h, usually only 1 or 2 h, over the 8-h workday. The obvious percentage reduction is thus 5%. This 5% difference could be reduced by incomplete percutaneous absorption. We challenge any researcher to show a statistically valid 5% difference in urinary excretion in any field experiment.

The use of a hand rinse or cotton gloves to monitor hand exposure could invalidate urinary results. Suppose workers who normally work ungloved and do not customarily wash their hands participate in an exposure study. Requiring that they wear gloves in order to monitor hand exposure might reduce their total dermal exposure, and hence their urinary metabolite level, by 21%. This is based on the assumption that hand exposure originally constituted 50% of the total dermal exposure and that the gloves, worn for half of the work day, gave 85% protection. Requiring hand washes might lead to a similar reduction. However, for workers who normally wear gloves or who regularly wash their hands, hand rinses or cotton-glove monitors may not affect the experimental outcome. For example, suppose we measured 1,000 units of a pesticide urinary metabolite in an experiment in which hand exposure contributed 50% of the urinary metabolite. If we removed (rinse or gloves) hand residue after 2 h, we would remove one-quarter of the hand contribution for an 8-h day. The new urinary value would be 875 units, $1,000 - (0.25 \times 500)$, or a urinary metabolite reduction of 12.5%. If we monitored hands during an ungloved period, measured 1,000 urinary metabolite units, and otherwise totally excluded other sources of exposure, and then protected the hands with gloves, what then? Gloves have afforded 85% protection (Nigg, Stamper, and Queen 1986). Now we see only 150 urinary units. If we removed one-quarter with a hand rinse, we would see about 113 urinary units. We could easily describe this difference statistically, but it is obvious that the urinary data would be uninterpretable.

Otherwise unexplainable sources of variation within replicate specimens might become evident if the work practices of the subjects are considered, but this may require a period of careful observation in the field extending over a week or longer.

5-4 REGULATORY ASPECTS

Pesticide assessment guidelines, Subdivision U, Applicator Exposure Monitoring (US EPA 1987) are the current regulations for occupational monitoring for pesticide registration requirements. These regulations are available from the National Technical Information Service, U.S. Department of Commerce, Springfield, VA 22161. Although ten participants are listed in this document, only four (Davis, Hickey, Nigg, Ware) had field experience.

Some of the studies discussed here in detail were used in the development of Subdivision U to determine that monitoring of biological fluids was a very risky approach to determine dose; that use of dermal pads, plus hand rinses, plus personal air monitors or respirators were more reliable approaches. As previously discussed, both approaches have problems, but passive dosimetry tends to overestimate exposure (err on the side of safety) whereas biological monitoring may underestimate exposure by 100% or more.

Out of Subdivision U came the generic database for exposure, termed the Pesticide Handlers Exposure Database (PHED). This was a valiant effort to avoid costly field experiments for the registration of pesticides. The requirements for PHED include quality assurance (QA) and good laboratory practices (GLP) and rank the reliability of a study.

Unfortunately, every study referenced in this chapter is a poor candidate for PHED or simply is not used because it was conducted before GLP and QA requirements were in place. So, in an attempt to standardize studies and save money, the best studies were scrapped. The studies that are included in PHED are company-sponsored studies of short duration with volunteers (students, employees, nonfarmworkers) as subjects. These subjects can be as careful as possible and do not reflect the real business of applying pesticides. This practice underestimates the exposure of agricultural labor to pesticides.

Subdivision U allows the use of biological monitoring if the method can be shown to be accurate. Based on the preceding discussion and the cost of validating a biological method, no biological method has been validated for PHED as of this writing. With the possible overestimation of exposure with passive dosimetry and the underestimation of exposure in PHED studies, perhaps the average protects workers in reality, but no one knows for certain.

The exposure monitoring requirements of Subdivision U were driven by the fact that better methods were not available. Thus, other, but validated, methods are acceptable. The *estimated* exposures then can be used in an estimated-permissible-concentration (EPC) approach to risk assessment (Williams et al. 1994). The estimated exposure dose may be termed RFD (oral or dermal reference dose) and RFC (inhalation reference concentration). The EPC process of risk assessment depends on accurate exposure estimates and few of these have been made (Williams et al. 1994).

Exposure modeling is another possible approach to estimate exposure, but models must be tested in the field. Van Hemmen (1992) and Popendorf (1992) reviewed the modeling of pesticide exposure. The primary difficulties in pesticide exposure models appears to be the differences in the databases from which they were produced (Van Hemmen 1992). Even under tightly controlled conditions, exposure-model estimates may be only 40% accurate (Carlton and Flynn 1997a,b).

Another approach to exposure monitoring for risk assessment is mean testing (Hewett 1997a,b). This approach requires dosimeters that permit averages to be generated over a long period of time. This has not been done for PHED or for any agricultural pesticide exposure situation.

5-5 CONCLUSIONS

The excretion kinetics of pesticides and pesticide metabolites in sweat and urine, and perhaps saliva, after a dermal exposure must be understood before any biological fluid can be validated as a dose estimator. When we understand *human* exposure-route absorption and excretion-route kinetics and their interrelationships, biological fluid monitoring will allow an accurate calculation of pesticide exposure and internal dose. We should remember, however, that few biological monitoring experiments in the literature were designed to monitor dose. The uncertainty in a dose estimate based on biological monitoring generally results from the lack of basic information on relevant human biochemistry and physiology. Future biological monitoring experiments designed to estimate dose must consider the possible application of human physiological data never collected.

Noninvasive methods lead necessarily to an overestimation of external dose. This estimated dose then is used as the basis for an estimated internal dose which also may be overestimated. It seems prudent, however, that with the uncertainties inherent in biological monitoring, the noninvasive methods provide the safest regulations for the worker.

One of the most common (and perhaps easiest) statements to make in a review is that more research is needed on the subject of the review. We operate with estimation methods that are based on short-time experiments (1–4 h). Exposure models are either inaccurate or have not been field tested. Data used in most models do not reflect the real world. If we are serious about worker exposure to chemicals, consistent and persistent research programs should be created and funded, programs whose ultimate goal is an accurate estimation of dose.

REFERENCES

Adams, J. D., Y. Iwata, and F. A. Gunther. 1976. Worker environment research. 4. The effect of dust derived from several soil types on the dissipation of parathion and paraoxon dislodgable residues on citrus foliage. *Bull. Environ. Contam. Toxicol.* 15:547–54.

Akerblom, M., B. Kolmodin-Hedman, and S. Hoglund. 1983. In *Human welfare environment,* ed. J. Miyamoto, 227–32. New York: Pergamon Press.

Alessio, L., A. Berlin, A. Dell'Orto, F. Toffoletto, and I. Ghezzi. 1985. Reliability of urinary creative as a parameter used to adjust values of urinary biological indicators. *Int. Arch. Occup. Environ. Health* 55: 99–106.

Araki, S., K. Murata, H. Aono, S. Yanagihara, Y. Niinuma, R. Yamamoto, and N. Ishihara. 1986. Comparison of the effects of urinary flow on adjusted and non-adjusted excretion of heavy metals and organic substances in "healthy" men. *J. Appl. Toxicol.* 6:245–51.

Armstrong, L. E., R. W. Hubbard, P. C. Szlyk, B. S. Matthew, and I. V. Sils. 1985. Voluntary dehydration and electrolyte losses during prolonged exercise. *Aviat. Space Environ. Med.* 765–70.

Batchelor, G. S., and K. C. Walker. 1954. Health hazards involved in use of parathion in fruit orchards of North Central Washington. *Arch. Ind. Health* 10:522–29.

Batchelor, G. S., K. C. Walker, and J. W. Elliott. 1956. Dinitroorthocresol exposure from apple-thinning sprays. *Arch. Ind. Health* 13:593–96.

Bradway, D. E., T. M. Shafik, E. M. Loros. 1977. Comparison of cholinesterase activity, residue levels, and urinary metabolite excretion of rats exposed to organic phosphorus pesticides. *J. Agric. Food Chem.* 25:1353–58.

Brewster, M. A., B. S. Hulka, and T. L. Lavy. 1992. Biomarkers of pesticide exposure. *Rev. Environ. Contam. Toxicol.* 128:17–42.

Brown, J. R., H. Hughes, and S. Viriyanondha. 1969. Storage, distribution and metabolism of 1,1-bis(4-chlorophenyl)-2,2,2-trichloroethanol. *Toxicol. Appl. Pharmacol.* 15:30–37.

Carlton, G. N., and M. R. Flynn. 1997a. A model to estimate worker exposure to spray paint mists. *Appl. Occup. Environ. Hyg.* 12:375–82.

Carlton, G. N., and M. R. Flynn. 1997b. Field evaluation of an empirical-conceptual exposure model. *Appl. Occup. Environ. Hyg.* 12:555–61.

Carman, G. E., F. A. Gunther, R. C. Blinn, and R. D. Garmus. 1952. The physical fate of parathion applied to citrus. *J. Econ. Entom.* 45:767–77.

Carman, G. E., Y. Iwata, J. L. Pappas, J. R. O'Neal, and F. A. Gunther. 1982. Pesticide applicator exposure to insecticides during treatment of citrus trees with oscillating boom and airblast units. *Arch. Environ. Contam. Toxicol.* 11:651–59.

Cohen, B., E. Richler, E. Weisenberg, J. Schoenberg, and M. Luria. 1979. Sources of parathion exposures for Israeli aerial spray workers, 1977. *Pestic. Monit. J.* 13:81–86.

Colburn, W. A. 1984. Pharmacokinetic analysis of concentration-time data obtained following administration of drugs that are recycled in the bile. *J. Pharm. Sci.* 73:313–17.

Comer, S. W., D. C. Staiff, J. F. Armstrong, H. R. Wolfe. 1975. Exposure of workers to carbaryl. *Bull. Environ. Contam. Toxicol.* 13:385–91.

Costill, D. L. 1977a. Sweating: Its composition and effects on body fluids. *Ann. NY Acad. Sci.* 301:160–74.

Costill, D. L. (moderator). 1977b. Sweating: Discussion. *Ann. NY Acad. Sci.* 301:183–88.

Costill, D. L., R. Cote, W. J. Fink, and P. Van Handel. 1981. Muscle water and electrolyte distribution during prolonged exercise. *Int. J. Sports Med.* 2:130–34.

Davis, J. E. 1980. Minimizing occupational exposure to pesticides: Personnel monitoring. *Residue Rev.* 75: 33–50.

Davies, J. E., J. C. Cassady, and A. Raffonelli. 1973. The pesticide problems of the agricultural worker. In *Pesticides and the environment: A continuing controversy.* Vol. 2. ed. W. B. Deichman, 223–31. New York: Stratton.

Davies, J. E., H. F. Enos, A. Barquet, C. Morgade, and J. X. Danauskas. 1979. Pesticide monitoring studies. The epidemiologic and toxicologic potential of urinary metabolites. In *Toxicology and occupational medicine.* Vol. 4. ed. W. B. Deichmann, 369–80. New York: Elsevier.

Dedek, W. 1980. Solubility factors affecting pesticide penetration through skin and protective clothing. In *Field worker exposure during pesticide application,* ed. W. F. Tordoir and E. A. H. van Heenstra, 47–50. New York: Elsevier.

Dowdy, S., and S. Weardon. 1983. Inference about two variances. In *Statistics for research,* 192–98. New York: Wiley.

Drevenkar, V., B. Stengl, B. Tralcevic, and Z. Vasilic. 1983. Occupational exposure control by simultaneous determination of *N*-methylcarbamates and organophosphorous pesticide residues in urine. *Int. J. Environ. Anal. Chem.* 14:215–30.

Durham, W. F. 1965. Pesticide exposure levels in man and animals. *Arch. Environ. Health* 10:842–46.

Durham, W. F., and C. H. Williams. 1972. Mutagenic, teratogenic, and carcinogenic properties of pesticides. *Ann. Rev. Entomol.* 17:123–48.

Durham, W. F., and H. R. Wolfe. 1962. Measurement of the exposure of workers to pesticides. *Bull. WHO.* 26:75–91.

Durham, W. F., and H. R. Wolfe. 1963. An additional note regarding measurement of the exposure of workers to pesticides. *Bull. WHO.* 29:279–81.

Durham, W. F., H. R. Wolfe, and J. W. Elliot. 1972. Absorption and excretion of parathion by spraymen. *Arch. Environ. Health* 24:381–87.

Durham, W. F., H. R. Wolfe, and G. E. Quinby. 1965. Organophosphorus insecticides and mental alertness. *Arch. Environ. Health* 10:55–66.

Feldmann, R. J., and H. I. Maibach. 1974. Percutaneous penetration of some pesticides and herbicides in man. *Toxicol. Appl. Pharmacol.* 28:126–72.

Florida Cooperative Extension Service. 1981. Florida citrus spray guide. University of Florida, IFAS, Gainesville, FL.

Franklin, C. A., R. A. Fenske, R. Greenhalgh, L. Mathieu, H. V. Denley, J. T. Leffingwell, and R. C. Spear. 1981. Correlation of urinary pesticide metabolite excretion with estimated dermal contact in the course of occupational exposure to guthion. *J. Toxicol. Environ. Health* 7:715–31.

Franklin, C. A., R. Greenhalgh, H. I. Maibach. 1983. Correlation of urinary dialkyl phosphate metabolite levels with dermal exposure to azinphos-methyl. *Pesticide chemistry human welfare and the environment,* ed. J. Miyamoto and P. C. Kearny, 221–26. Vol. 4 of *Pesticide residues and formulation chemistry.* New York: Pergamon Press.

Franklin, C. A., N. I. Muir, and R. P. Moody. 1986. The use of biological monitoring in the estimation of exposure during the application of pesticides. *Toxicol. Lett.* 33:127–36.

Funckes, A. J., G. R. Hayes, and W. V. Hartwell. 1963. Urinary excretion of paranitrophenol by volunteers following dermal exposure to parathion at different ambient temperatures. *J. Agric. Food Chem.* 11: 455–57.

Gehring, P. J., C. G. Kramer, B. A. Schwetz, J. Q. Rose, and V. K. Rowe. 1973. The fate of 2,4,5-trichlorophenoxy-acetic and (2,4,5-T) following oral administration to man. *Toxicol. Appl. Pharmacol.* 26:352–61.

Gollop, B. R., and W. I. Glass. 1979. Urinary arsenic levels in timber treatment operators. *NZ Med. J.* 89: 10–11.

Griffiths, J. T., C. R. Stearns, Jr., and W. L. Thompson. 1951. Parathion hazards encountered spraying citrus in Florida. *J. Econ. Entomol.* 44:160–63.

Griffiths, J. T., J. W. Williams, C. R. Stearns, and W. L. Thompson. 1951. Health status of parathion when used on citrus in 1951. *Proc. Fl. State Hort. Soc.* 64:79–82.

Grob, D., W. L. Garlick, and A. M. Harvey. 1950. The toxic effects in man of the anticholinesterase insecticide parathion (*p*-nitrophenyl diethyl thionophosphate.) *Bull. Johns Hopkins Hosp.* 87:106–29.

Grover, R., C. A. Franklin, N. I. Muir, A. J. Cessna, and D. Riedel. 1986. *Toxicol. Lett.* 33:73–83.

Gunther, F. A. 1969. Insecticide residues in California fruits and products. *Residue Rev.* 28:1–127.

Gunther, F. A., Y. Iwata, G. E. Carman, and C. A. Smith. 1977. The citrus reentry problem: Research on its causes and effects, and approaches to its minimization. *Residue Rev.* 67:1–139.

Guthrie, F. E., J. J. Domanski, A. L. Chasson, D. E. Bradway, and R. J. Monroe. 1976. Human subject experiments to estimate reentry periods for monocrotophos-treated tobacco. *Arch. Environ. Contam. Toxicol.* 4:217–25.

Hayes, A. L., R. A. Wise, and F. W. Weir. 1980. Assessment of occupational exposure to organophosphates in pest control operators. *Ind. Hyg. Assoc. J.* 41:568–75.

Hewett, P. 1997a. Mean testing. 1. Advantages and disadvantages. *Appl. Occup. Environ. Hyg.* 12:339–46.

Hewett, P. 1997b. Mean testing. 2. Comparison of several alternative procedures. *Appl. Occup. Environ. Hyg.* 12:347–55.

Holmstedt, B. 1963. Structure-activity relationships of the organophosphorus anticholinesterase agents. Chap. 9 in Handbuch der Experimentellen Pharmakologie. ed. O. Eichler, A. Farah and Koelle (G.B.), 428–85. Berlin: Springer-Verlag.

Iwata, Y., M. E. Dusch, W. E. Westlake, and F. A. Gunther. 1975. Behavior of five organophosphorus pesticides in dust derived from several soil types. *Bull. Environ. Contam. Toxicol.* 14:49–56.

Iwata, Y., W. E. Westlake, and F. A. Gunther. 1973. Persistence of parathious in six California soils under laboratory conditions. *Arch. Environ. Contam. Toxicol.* 1:84–96.

Kay, K., L. Monkman, J. P. Windish, T. Doherty, J. Pare, and C. Raciot. 1952. Parathion exposure and cholinesterase response of Quebec apple growers. *Ind. Hyg. Occup. Med.* 6:252–62.

Keil, J. E., C. B. Loadholt, S. H. Sandifer, W. Weston, R. H. Gadsden, C. G. Hames. 1973. Sera DDT elevation in black components of two southeastern communities: Genetics or environment? In *Pesticides and the environment: A continuing controversy.* Vol. 2. ed. W. B. Diechmann, 203–213. New York: Intercontinental Medical Book Corp.

Kilgore, W. (coordinator). 1977. Human physiological effects of organophosphorous pesticides in a normal agriculture field population. A preliminary report. 2. Scientific aspects, p. 55. Food Protections. Toxicology Center, University of California, Davis.

Kirchner, L. M., R. A. J. Taylor, R. A. Downer, and F. R. Hall. 1994. Calibrating the pesticide capture efficiency of passive dosimeters. *Pestic. Sci.* 46:61–69.

Kirchner, L. M., R. A. J. Taylor, R. A. Downer, and F. R. Hall. 1996. Comparison of the pesticide capture efficiency of potential passive dosimeter materials. *Bull. Environ. Contam. Toxicol.* 57:938–45.

Klemmer, H. W., M. N. Rashad, and M. P. Mi. 1973. Age, sex, and race effects on the distribution of organochlorine pesticide residues in serum. In *Pesticides and the environment: A continuing controversy.* Vol. 2. ed. W. B. Diechmann, 53–61. New York: Intercontinental Medical Book Corp.

Knaak, J. B., K. T. Maddy, M. A. Gallo, D. T. Lillie, E. M. Craine, and W. F. Serat. 1978. Workers reentry study involving phosalone application to citrus groves. *Toxicol. Appl. Pharmacol.* 46:363–74.

Knaak, J. B., S. A. Peoples, T. J. Jackson, A. S. Fredrickson, R. Enos, K. T. Maddy, J. B. Bailey, M. E. Dusch, F. A. Gunther, and W. L. Winterlin. 1978. Reentry programs involving the use of dialifor on grapes in the San Joaquin Valley of California. *Arch. Environ. Contam. Toxicol.* 7:465–81.

Kohli, J. D., R. N. Khanna, B. N. Gupta, M. M. Dhar, J. S. Tandon, and K. P. Sircar. 1974. Absorption and excretion of 2,4-dichlorophenoxyacetic acid in man. *Xenobiotica* 4:97–100.

Kolmodin-Hedman, B., S. Hoglund, and M. Akerblom. 1983. Studies on phenoxy acid herbicides. (MCPA, dichloroprop, mecoprop, and 2,4-D) in agriculture. *Arch. Toxicol.* 54:257–65.

Kolmodin-Hedman, B., S. Hoglund, A. Swenson, and M. Akerblom. 1983. Studies on phenoxy acid herbicides. 2. Oral and dermal uptake and elimination in urine of MCPA in humans. *Arch. Toxicol.* 54:267–73.

Kraus, J. F., D. M. Richards, N. O. Borhani, R. Mull, W. W. Kilgore, and W. Winterlin. 1977. Monitoring of grape harvesters for evidence of cholinesterase inhibition. *Arch. Environ. Contam. Toxicol.* 5:471–85.

Kutz, F. W., and S. C. Strassman. 1977. Human urinary metabolites of organophosphate insecticides following mosquito adulticiding. *Mosq. News* 37:211–18.

Lavy, T. L. 1987. *Human exposure to phenoxy herbicides.* Washington, D.C.: Veterans Administration, Agent Orange Projects Office.

Lavy, T. L., J. S. Shepard, and J. D. Mattice. 1980. Exposure measurements of applicators spraying (2,4,5-trichlorophenoxy) acetic acid in the forest. *J. Agric. Food Chem.* 28:626–30.

Lavy, T. L., J. D. Walstad, R. R. Flynn, and J. D. Mattice. 1982. (2,4-dichorophenoxy) acetic acid exposure received by aerial application crews during forest spray operations. *J. Agric. Food Chem.* 30: 375–81.

Leffingwell, J. T., R. C. Spear, and D. Jenkins. 1975. The persistence of ethnion and zolone residues on grape foliage in the Central Valley of California. *Arch. Environ. Contam. Toxicol.* 3:40–54.

Leng, M. L., J. C. Ramsey, W. H. Brown, and T. L. Lavy. 1982. Review of studies with 2,4,5-trichlorophenoxyacetic acid in humans including applicators under field conditions. In *Pesticide residues and exposure.* ed. J. Plimmer, 133–56. ACS Symposium Series, no. 182. Washington, D.C.: American Chemical Society.

Lentner, C., ed. 1981. *Geigy scientific tables, units of measure, body fluids, composition of the body, nutrition, 8th ed.,* Vol. 1, 53–122. Medical Education Division, Ciba Geigy Corp. West Coldwell, New Jersey. 53–122.

Levine, R. S., and J. Doull. 1992. Global estimates of acute pesticide moridity and mortality. *Rev. Environ. Contam. Toxicol.* 129:29–50.

Levy, K. A., S. S. Brady, and C. Pfaffenberger. 1981. Chlorobenzilate residues in citrus-worker urine. *Bull. Environ. Contam. Toxicol.* 27:235–38.

Lewalter, J., and U. Korallus. 1986. Erythrocyte protein conjugates as a principal of biological monitoring for pesticides. *Toxicol. Lett.* 33:153–65.

Lieban, J., R. K. Waldman, and L. Krause. 1953. Urinary excretion of paranitrophenol following exposure to parathion. *Ind. Hyg. Occup. Med.* 7:93–98.

Maibach, H. I. 1982. *Animal models in dermatology.* Edinburgh: Churchill Livingstone.

Metcalf, R. L., and R. B. March. 1953. The isomerization of organic thionophosphate insecticides. *J. Econ. Entomol.* 46:288–94.

Milby, T. H. (chairman). 1974. Occupational exposure to pesticides: Report to the Federal Working Group on Post Management from the Task Group on Occupational Exposure to Pesticides. Federal Working Group on Pest Management Wash., D.C. 155 p.

Milby, T. H., F. Ottoboni, and H. W. Mitchell. 1964. Parathion residue poisoning among orchard workers. *J. Am. Med. Assoc.* 189:351–56.

Morgan, D. P., H. L. Hetzler, E. F. Slach, and L. I. Lin. 1977. Urinary excretion of paranitrophenol and alkyl phosphates following ingestion of methyl or ethyl parathion by human subjects. *Arch. Environ. Contam. Toxicol.* 6:159–73.

Moseman, R. F., and H. F. Enos. 1982. Utility of urine and blood analysis for the assessment of man's exposure to chemicals. *J. Environ. Sci. Health* A17:519–23.

Nigg, H. N. 1979. Comparison of pesticide and particulate recoveries with the vacuum and dislodgeable surface pesticide residue techniques. *Arch. Environ. Contam. Toxicol.* 8:369–81.

Nigg, H. N., J. C. Allen, and R. F. Brooks. 1977. Weather and pesticide residues. *Proc. Int. Soc. Citriculture* 2:437–41.

Nigg, H. N., J. C. Allen, R. F. Brooks, G. J. Edwards, N. P. Thompson, R. W. King, and A. H. Blagg. 1977. Dislodgeable residues of ethion in Florida citrus and relationships to weather variables. *Arch. Environ. Contam. Toxicol.* 6:257–67.

Nigg, H. N., J. C. Allen, and R. W. King. 1979. Behavior of parathion residues in the Florida "Valencia" orange agroecosystem. *J. Agric. Food Chem.* 27:578–82.

Nigg, H. N., J. C. Allen, R. W. King, N. P. Thompson, G. J. Edwards, and R. F. Brooks. 1978. Dislodgeable residues of parathion and carbophenothion in Florida citrus. A weather model. *Bull. Environ. Contam. Toxicol.* 19:578–88.

Nigg, H. N., and J. H. Stamper. 1980. Persistence of phenthoate (cidial) and phenthoate oxon on fruit, leaf, and soil surfaces and in air in Florida citrus. *Chemosphere* 9:343–50.

Nigg, H. N., and J. H. Stamper. 1982. Regional consideration in worker reentry. In *Pesticide residues and exposure.* ed. J. Plimmer, 59–73. ACS Symposium Series, no. 182, Washington, D.C.: American Chemical Society.

Nigg, H. N., and J. H. Stamper. 1983. Exposure of spray applicators and mixer-loaders to chlorobenzilate miticide in Florida citrus growers. *Arch. Environ. Contam. Toxicol.* 12:477–82.

Nigg, H. N., and J. H. Stamper. 1985. Field studies: Methods overview. In *Dermal exposure related to pesticide use*. ed. R. C. Honeycutt, G. Zweig, N. Ragsdale, 95–98. ACS Symposium Series, no. 273, Washington D.C.: American Chemical Society.

Nigg, H. N., J. H. Stamper, S. N. Deshmukh, and R. M. Queen. 1991. 4,4'-Diclorobenzilic acid urinary excretion by dicofol pesticide applicators. *Chemosphere* 22:365–73.

Nigg, H. N., J. H. Stamper, and R. M. Queen. 1984. The development and use of a universal model to predict tree crop harvester pesticide exposure. *Am. Ind. Hyg. Assoc. J.* 45:182–86.

Nigg, H. N., J. H. Stamper, and R. M. Queen. 1986. Dicofol exposure to Florida citrus applicators: Effects of protective clothing. *Arch. Environ. Contam. Toxicol.* 15:121–43.

Nigg, H. N., and S. E. Wade. 1992. Saliva as a monitoring medium for chemicals. *Rev. Environ. Contam. Toxicol.* 129:95–119.

Nolan, R. J., N. L. Freshour, P. E. Kastl, and J. H. Saunders. 1984a. Pharmacokinetics of picloram in male volunteers. *Toxicol. Appl. Pharmacol.* 76:264–69.

Nolan, R. J., D. L. Rick, N. L. Freshour, and J. H. Saunders. 1984b. Chlorpyrifos: Pharmacokinetics in human volunteers. *Toxicol. Appl. Pharmacol.* 73:8–15.

Peck, C. C., D. P. Conner, B. J. Bolden, R. G. Almirez, L. M. Rowland, T. E. Kwiatkowski, B. A. McKelvin, and C. R. Bradley. 1987. Outward transdermal migration of theophylline. In *Pharmacology and the skin*. Vol. 1, 201–08. Basel, Switzerland: Karger.

Peck, C. C., D. P. Conner, B. J. Bolden, R. G. Almirez, T. E. Kingsley, L. D. Mell, M. G. Murphy, V. E. Hill, L. M. Rowland, D. Ezra, T. E. Kwiatkowski, C. R. Bradley, and M. Abdel-Rahim. 1988. Outward transcutaneous chemical migration: Implications for diagnostics and dosimetry. *Skin Pharmacol.* 1:14–23.

Peck, C., D. Conner, B. Bolden, T. Kingsley, L. Mell, T. Kwiatkowski, V. Hill, D. Ezra, A. Dubois, M. A. Rahim, and G. Murphy. 1986. Transdermal chemical collection: A novel research tool. Paper presented at 3rd Annual Symposium of the Skin Pharmacology Society, 1–2 August, at Karolinska Institute, Stockholm.

Peck, C., D. Conner, R. Scheuplein, R. Dedrick, and J. McDougal. 1986. Physiologically based pharmacokinetic modeling of chemical disposition in skin. Paper presented at 3rd Annual Symposium of the Skin Pharmacology Society, 1–2 August, at Karolinska Institute, Stockholm.

Peck, C. C., K. Lee, and C. E. Becker. 1981. Continuous transepidermal drug collection: Basis for use in assessing drug intake and pharmacokinetics. *J. Pharmacokinet. Biopharm.* 9:41–58.

Peoples, S. A., K. T. Maddy, and C. R. Smith. 1978. Occupational exposure to temik (aidicarb) as reported by California physicians for 1974–76. *Vet. Human Toxicol.* 20:321–24.

Peoples, S. A., K. T. Maddy, and J. Topper. 1978. Human occupational illness problems due to ethyl parathion exposure in California in 1975. *Vet. Human Toxicol.* 20:327–29.

Popendorf, W. 1992. Reentry field data and conclusions. *Rev. Environ. Contam. Toxicol.* 128:71–117.

Popendorf, W. J., and J. T. Leffingwell. 1982. Regulating of pesticide residues for farmworker protection. *Residue Rev.* 82:125–201.

Popendorf, W. J., R. C. Spear, J. T. Leffingwell, J. Yager, and E. Kahn. 1979. Harvester exposure to zolone (phosalone) residues in peach orchards. *J. Occup. Med.* 21:189–94.

Quinby, G. E., and A. B. Lemmon. 1958. Parathion residues as a cause of poisoning in crop workers. *J. Am. Med. Assoc.* 166:740–46.

Quinby, G. F., K. C. Walker, and W. F. Durham. 1958. Public health hazards involved in the use of organic phosphorous insecticides in the delta area of Mississippi. *J. Econ. Entomol.* 51:831–38.

Richards, D. M., J. F. Kraus, P. Kurtz, N. O. Borhani, R. Mull, W. Winterlin, and W. W. Kilgore. 1978. A controlled field trial of physiological responses to organophosphate residues in farm workers. *J. Environ. Pathol. Toxicol.* 2:493–512.

Ridgeway, R. L., J. C. Tinney, J. T. MacGregor, N. J. Starler. 1978. Pesticide use in agriculture. *Environ. Health Perspect.* 27:103–12.

Riley, R. T., B. W. Kemppainen, and W. P. Norred. 1985. Penetration of aflatoxins through isolated human epidermis. *J. Toxicol. Environ. Health* 15:769–77.

Roan, C. C., D. P. Morgan, N. Cook, and E. H. Paschal. 1969. Blood cholinesterases, serum parathion concentrations and urine *p*-nitrophenol concentrations in exposed individuals. *Bull. Environ. Contam. Toxicol.* 4:362–69.

Robbins, A. L., D. F. Nash, and S. W. Comer. 1977. A monitoring study of workers in a Central Washington orchard. *Bull. Environ. Contam. Toxicol.* 17:233–40.

Rosenberg, N. M., R. M. Queen, and J. H. Stamper. 1985. Sweat-patch test for monitoring pesticide absorption by airblast applicators. *Bull. Environ. Contam. Toxicol.* 35:68–72.

Sauerhoff, M. W., W. H. Braun, G. E. Blau, and P. J. Gehring. 1977. The fate of 2,4-dichlorophenoxyacetic acid (2,4-D) following oral administration to man. *Toxicology* 8:3–11.

Sauerhoff, M. W., M. B. Chenoweth, R. J. Karbowski, W. H. Braun, J. C. Ramsey, P. J. Gehring, and G. E. Blau. 1977. Fate of silvex following oral administration to man. *J. Toxicol. Environ. Health* 3: 941–52.

Sell, C. R., J. C. Maitlen, and W. A. Aller. 1982. Perspiration as an important physiological pathway for the elimination of 2,4-diclorophenoxyacetic acid from the human body. Paper presented at American Chemical Society meeting, 1 April, at Las Vegas, NV.

Serat, W. F., A. J. Van Loon, and W. H. Serat. 1982. Loss of pesticides from patches used in the field as pesticide collectors. *Arch. Environ. Contam. Toxicol.* 11:227–34.

Skinner, C. S., and W. W. Kilgore. 1982. Percutaneous penetration of [14C] parathion in the mouse: Effect of anatomic region. *J. Toxicol. Environ. Health* 9:483–90.

Spear, R. C., W. J. Popendorf, J. T. Leffingwell, and D. Jenkins. 1975. Parathion residues on citrus foliage. Dust and composition as related to worker hazard. *J. Agric. Food Chem.* 23:808–10.

Spencer, W. F., M. M. Cliath, K. R. Davis, R. C. Spear, and W. J. Popendorf. 1975. Persistence of parathion and its oxidation to paraoxon on the soil surface as related to worker reentry into treated crops. *Bull. Environ. Contam. Toxicol.* 14:265–72.

Staiff, D. C., S. W. Comer, J. F. Armstrong, and H. R. Wolfe. 1975. Exposure to the herbicide, paraquat. *Bull. Environ. Contam. Toxicol.* 14:334–40.

Stoughton, R. B. 1965. *Toxicol. Appl. Pharmacol.* 7:1–6.

Swan, A. A. B. 1969. Exposure of spray operators to paraquat. *Br. J. Ind. Med.* 26:322–29.

U.S. Department of Agriculture (USDA) Cooperative Extension Service. 1980–1982. Treatment Guide for California Citrus Crops. Leaflet 2903. Division of Agricultural Science, University of California, Berkeley.

U.S. Environmental Protection Agency (US EPA). 1979. Pesticide Incident Monitoring System Report No. 260. Health Effects Branch, Office of Pesticide Program. US EPA, Washington D.C.

U.S. Environmental Protection Agency (US EPA). 1987. Pesticide assessment guidelines, Subdivision U, Applicator exposure monitoring (Reinart, J. C., A. P. Nielsen, J. E. Davis, K. D. Hickey, H. N. Nigg, G. W. Ware, C. Lunchick, O. Hernandez, F. L. Smith, and D. M. Mazzetta [participants]). PB87-133286, Washington, D.C.

Van Hemmen, J. J. 1992. Estimating worker exposure for pesticide registration. *Rev. Environ. Contam. Toxicol.* 128:43–54.

Wagner, S. L., and P. Weswig. 1974. Arsenic in blood and urine of forest workers. *Arch. Environ. Health* 28:77–79.

Ware, G. W., D. P. Morgan, B. J. Estesen, and W. P. Cahill. 1975. Establishment of reentry intervals for organophosphate-treated cotton fields based on human data. Part 3. 12 to 72 hours post-treatment exposure to monocrotophos, ethyl and methyl parathion. *Arch. Environ. Contam. Toxicol.* 3:289–306.

Webster, H. L. 1983. Laboratory diagnosis of cystic fibrosis. *Crit. Rev. Clin. Lab. Sci..* Vol. 18:313–338. Boca Raton, Fla.: CRC Press.

Wester, R. C., and P. K. Noonan. 1980. Relevance of animal models for percutaneous absorption. *Int. J. Pharm.* 7:99–110.

Wester, R. C., H. I. Maibach, A. W. Bucks, and M. B. Aufere. 1984. *In vivo* percutaneous absorption of paraquat from hand, leg, and forearm of humans. *J. Toxicol. Environ. Health* 14:759–62.

Wicker, G. W., and F. W. Guthrie. 1980. Worker-crop contact analysis as a means of evaluating reentry hazards. *Bull. Environ. Contam. Toxicol.* 24:161–67.

Wicker, G. W., P. E. Stinner, F. E. Reagan, and F. E. Guthrie. 1980. Mail survey to determine amounts of and potential work force exposure to foliarly-applied insecticides in North Carolina. *Bull. Entomol. Soc. Am.* 26:156–61.

Wicker, G. W., W. A. Williams, J. R. Bradley, and F. E. Guthrie. 1979. Exposure of field workers to organophosphorous insecticides: Cotton. *Arch. Environ. Contam. Toxicol.* 8:443–40.

Wicker, G. W., W. A. Williams, and F. E. Guthrie. 1979. Exposure of field workers to organophosphorous insecticides: Sweet corn and peaches. *Arch. Environ. Contam. Toxicol.* 8:175–82.

Williams, C. A., H. D. Jones, R. W. Freeman, M. J. Wernke, P. L. Williams, S. M. Roberts, and R. C. James. 1994. The EPC approach to estimating safety from exposure to environment chemicals. *Regulat. Toxicol. Pharmacol.* 20:259–80.

Wilson, B. W., and J. D. Henderson. 1992. Blood esterase determinations as markers of exposure. *Rev. Environ. Contam. Toxicol.* 128:55–69.

Winterlin, W., W. W. Kilgore, C. Mourer, R. Mull, G. Walker, J. Knaak, and K. Maddy. 1978. Dislodgeable residues of dialifor and phosalone and their oxygen analogs following a reported worker-injury incident. *Bull. Environ. Contam. Toxicol.* 20:255–60.

Winterlin, W. L., W. W. Kilgore, C. R. Mourer, and S. R. Schoen. 1984. Worker reentry studies for captan applied to strawberries in California. *J. Agric. Food Chem.* 32:664–72.

Wojeck, G. A., H. N. Nigg, R. S. Braman, J. H. Stamper, and R. L. Rouseff. 1982. Worker exposure to arsenic in Florida grapefruit spray operations. *Arch. Environ. Contam. Toxicol.* 11:661–67.

Wojeck, G. A., H. N. Nigg, J. H. Stamper, and D. E. Bradway. 1981. Worker exposure to ethion in Florida citrus. *Arch. Environ. Contam. Toxicol.* 10:725–35.

Wojeck, G. A., J. F. Price, H. N. Nigg, and J. H. Stamper. 1983. Worker exposure to paraquat and diquat. *Arch. Environ. Contam. Toxicol.* 12:65–70.

Wolfe, H. R. 1972. Safety problems related to exposure of workers in pesticide formulation plants. In *Industrial production and formulation of pesticides*. UN pub. no. E.72.11.B.5, 137–44. U.S. Environmental Protection Agency. United Nations Industrial Development Organization, Vienna.

Wolfe, H. R., and J. F. Armstrong. 1971. Exposure of formulating plant workers to DDT. *Arch. Environ. Health* 23:169–76.

Wolfe, H. R., J. F. Armstrong, and W. F. Durham. 1996. Pesticide exposure from concentrate spraying. *Arch. Environ. Health* 13:340–44.

Wolfe, H. R., J. F. Armstrong, and W. F. Durham. 1974. Exposure of mosquito control workers to fenthion. *Mosq. News* 34 (3):263–67.

Wolfe, H. R., J. F. Armstrong, D. C. Staiff, and S. W. Comer. 1972a. Exposure of spraymen to pesticides. *Arch. Environ. Health* 25:29–31.

Wolfe, H. R., J. F. Armstrong, D. C. Staiff, and S. W. Comer. 1972b. The use of protective clothing and equipment for prevention of exposure to pesticides. *Proceedings of the National Conference on Protective Clothing and Safety Equipment for Pesticide Workers*, Atlanta, Georgia, May 1–2 1972 Federal Working Group on Post Management, Wash., D.C. 155–64.

Wolfe, H. R., J. F. Armstrong, D. C. Staiff, and S. W. Comer. 1975. Potential exposure of workers to parathion through contamination of cigarettes. *Bull. Environ. Contam. Toxicol.* 13 (3):369–76.

Wolfe, H. R., J. F. Armstrong, D. C. Staiff, S. W. Comer, and W. F. Durham. 1975. Exposure of apple thinners to parathion residues. *Arch. Environ. Contam. Toxicol.* 3:257–67.

Wolfe, H. R., W. F. Durham, and J. F. Armstrong. 1963. Health hazards of the pesticides endrin and dieldrin. *Arch. Environ. Health* 6:458–64.

Wolfe, H. R., W. F. Durham, and J. F. Armstrong. 1967. Exposure of workers to pesticides. *Arch. Environ. Health* 14:622–33.

Wolfe, H. R., W. F. Durham, and J. F. Armstrong. 1970. Urinary excretion of insecticide metabolites. *Arch. Environ. Health* 21:711–16.

Wolfe, H. R., W. F. Durham, and G. S. Batchelor. 1961. Health hazards of some dinitro compounds. *Arch. Environ. Health* 3:468–75.

Wolfe, H. R., D. C. Staiff, J. F. Armstrong. 1978. Exposure of pesticide formulating plant workers to parathion. *Bull. Environ. Contam. Toxicol.* 20:340–43.

Wolfe, H. R., K. C. Walker, J. W. Elliott, and W. F. Durham. 1959. Evaluation of the health hazards involved in house-spraying with DDT. *Bull. WHO.* 20:1–14.

SIX

REVIEW OF VIDEO IMAGING TECHNIQUES FOR ESTIMATING DERMAL EXPOSURE TO PESTICIDES

Donna Houghton

Novartis Crop Protection Canada Inc., Research Park, University of Guelph, Guelph, Ontario, Canada N1G 4Z3

Bruce A. Archibald

Food Laboratory Services Division, University of Guelph, Guelph, Ontario, Canada, N1H 8J7

Keith R. Solomon

Center for Toxicology, University of Guelph, Guelph, Ontario, Canada N1G 1Y4

6-1 INTRODUCTION

Workers in many industries, including agriculture, are exposed to chemicals that may pose a health risk. Dermal contamination has been recognized as a significant source of exposure. Over the years, dermal exposure to pesticides has been evaluated using a variety of techniques. The dermal patch technique, developed by Durham and Wolfe (1962), has been used by numerous researchers to estimate dermal exposure to chemicals in agricultural and indoor structural scenarios (Comer et al. 1975; Gold, Leavitt, and Ballard 1981; Gold et al. 1984; Bonsall 1985; Roberts and Camann 1989). Disadvantages of this technique, including underestimation of exposure, prompted the development of additional equipment and techniques such as surface dislodgeability sleds (Vaccaro 1990), rollers (Hsu et al. 1990; Lewis et al. 1991; Lewis, Fortmann, and Camann 1994; Ross et al. 1991) and rubbing devices (Kells and Solomon 1995; Houghton, Solomon, and Harris 1996; Houghton 1997), treated-surface wipes (Naffziger, Sprenkel, and Mattler 1985; Fenske et al. 1991), hand presses followed by hand wipes or rinses (Davis 1980; Hsu et al. 1990; Lewis et al. 1991; Lewis, Fortmann, and Camann 1994; Geno et al. 1996; Houghton, Solomon, and Harris 1996; Houghton 1997), and dosimeter clothing analysis (Chester and Ward 1983; Ross et al. 1990). Hand wipe and rinse methods were reviewed recently by Fenske (1993) and McArthur (1992).

Although these newer methods have advantages over the patch technique, disadvantages, including the lack of similarity between skin and cotton cloth, differences in absorptivity of these matrices, and the need to use large volumes of solvent to extract dosimeter clothing, prompted research in the area of image analysis. Since the late 1970s, several researchers have developed, modified, and used video imaging techniques to assess dermal exposure to pesticides mixed with nontoxic fluorescent tracer surrogates (Ashbaugh 1977; Chester and Ward 1983; Fenske 1984, 1987, 1988a–d, 1989, 1990; Fenske, and Elkner 1990; Fenske, Leffingwell, and Spear 1985, 1986; Fenske, Wong, et al. 1986; Fenske, Horstman, and Bentley 1987; Fenske, Hamburger, and Guyton 1987; Black 1993; Archibald 1993; Archibald, Solomon, and Stephenson 1994a,b, 1995; Methner and Fenske 1994a,b; Roff 1994, 1995, 1996a–c; Bierman, Brouwer, and van Hemmen 1995; Houghton, Solomon, and Harris 1996; Houghton 1997). The tracers used with this technique are fluorescent whitening agents (FWAs) having applications in the textile, paper, and detergent industries. This group of dyes gives a blue-white fluorescence after absorption of near-ultraviolet light.

The basic principle of the video imaging technique is that by mixing a pesticide with a fluorescent tracer that is visible only under long-wave ultraviolet (UV) light, tracer deposition and hence pesticide can be measured using a video camera and a specialized computer, imaging board, and program designed to measure changes in fluorescence on skin between preexposure and postexposure images.

Once preliminary laboratory work has been conducted to establish dislodgeability ratios for the tracer–pesticide mixture, chemical compatibility, lack of phytotoxicity to plants (in agricultural situations), lack of antagonism of pesticide efficacy, and lack of target staining (indoor residential and commercial applications), this technique is rapid, noninvasive, avoids the drawbacks of previous methods, and provides a visual indication of the anatomical distribution of pesticides on skin. The technique has tremendous visual impact and would be an excellent qualitative tool for extension or teaching purposes. Fluorescent tracers have the potential to be used as surrogates to mimic pesticide exposure in human exposure studies of new chemicals, thereby eliminating subject exposure to pesticide active ingredients. The disadvantages of the technique are that the chemical properties of the tracer used (e.g., water solubility) must be matched closely with those of the pesticide to ensure that the two behave similarly, required preliminary laboratory testing can be extensive and time-consuming, skilled technicians are required to analyze the images, and quantitative assessments can be challenging because of inherent variability in the technique.

This chapter reviews basic concepts of image analysis, the toxicology of FWAs, the development of video imaging techniques for estimating dermal exposure to fluorescent tracers used as surrogates for other chemicals, and the results of studies employing this technique to evaluate dermal exposure to pesticides in agricultural and indoor structural scenarios.

6-2 FUNDAMENTALS OF IMAGE ANALYSIS

Prior to discussing video imaging techniques for assessing dermal exposure to various substances, including pesticides, it is important for the reader to understand the basic concepts of image analysis. Images are two-dimensional representations of three-dimensional objects. Cones in the human eye possess three color receptors (red, green, and blue) and only a narrow range of wavelengths of the electromagnetic spectrum (400–700 nm) is visible to the human eye (Russ 1992; Imaging Research Inc. 1996). Electronic sensors, however, have been developed that permit the detection of UV and infrared light, radio waves, and x-rays. Human vision is primarily qualitative rather than quantitative. We are poor judges of image color and brightness. If image features are to be quantified, for example, the brightness of a fluorescent tracer in an image of human skin, image analysis tools are required. Computer-assisted image analyzers are accurate and time-efficient, provide many complex analysis features, and are the best option for researchers assessing the extent and distribution of dermal exposure using image analysis.

6-2-1 Image Analysis Software

Computer-assisted image analyzers consist of a computer equipped with an image processor; one or two monitors, depending on the system; a solid-state video camera; and image analysis software. There are two main categories of software: turnkey and macro-based systems (Imaging Research Inc. 1996). Turnkey systems are designed for a researcher's specific applications. No programming is required. Macro-based systems require some user programming and a simplified programming feature is supplied to end users to generate macros for their particular needs. Turnkey software has advantages for quantitative analyses because its measurements are guaranteed repeatable. Reputable turnkey software has been validated by extensive testing and is supported by an extensive publication record. Scientists around the globe, including those working in basic research and employed with regulatory agencies, can be assured that measurements are standardized internationally (Imaging Research Inc. 1996). For example, an image analyzed with the same system in Britain, Canada, and Tokyo will yield identical results. The major disadvantage of turnkey systems is that they are much less flexible than macro-based systems and often are not easily adapted for certain applications.

Despite the advantages of turnkey software, most commercial image analysis software is macro-based, most likely because programmers are not privy to all customer needs. Macro-based software allows end users to design customized packages to meet their specific requirements. Although macro-based software is useful for unique applications, it can be disadvantageous when used for quantitative analyses because macros written may not be thoroughly validated and may provide inaccurate data. As a result, data from the same image may not be comparable from one laboratory to another. *It is critical to consult the manufacturer's technical staff and test macros for accuracy before analyzing any images.* Developing an accurate and efficient image analysis software necessitates cooperation among a team of end users, research scientists, engineers, and computer programmers who are experts in image analysis techniques. Scientists must choose a system based on the science used to develop it and the level of expertise available for technical support once the software has been purchased (Imaging Research Inc. 1996). Since few companies possess this level of expertise, many researchers have encountered problems with software performance or obtaining satisfactory customer support.

6-2-2 Computer Equipment Requirements

The image processor, supplied with an image analysis system, is a circuit board or board set specialized for handling large image files and designed to interface to cameras. The image processor, under the control of the host computer, is responsible for image acquisition, transformation, and display. Image processors use large memory buffers to store images and specialized hardware designed to increase the speed of certain operations. A frame buffer is a bank of random access memory located within the imaging system. Large amounts of fast image memory are incorporated in standard imaging systems. In a modular image processor, the imaging components can transfer data very rapidly by communicating with each other using a dedicated and fast data path called the imaging bus. Data can be transferred more quickly in this manner than via the host-computer bus. The bandwidth of the imaging bus determines the data manipulation limit of the imaging system. At the time of writing, most manufacturers used a 15- to 20-MHz imaging bus, although more expensive systems use bus bandwidths of 40 MHz or more.

6-2-3 Image Acquisition and Processing

Images are acquired using solid-state video cameras equipped with a charge-coupled device (CCD) chip. This chip contains an array of individual picture element (pixel) sensors that record incident illumination. Currently, arrays of greater than 300,000 sensors can be packed into an area less than 1 cm^2. Each sensor acts as a photon counter. A charge transfer (of electrons) is passed from one sensor to the next, along each row of the chip, and is collected in an isolated well (Russ 1992). The signal readout from each row of detectors on the chip passes through an amplifier that produces the analog voltage. Camera noise, which appears in images, is caused by variations in the sensitivity of the photon sensors in the CCD chip.

The first step in image processing is digitization, which is the transformation of a live image into discrete digital values that can be interpreted by the computer. During this process, the analog voltage signal from the camera is digitized using an 8-bit flash analog-to-digital converter (ADC). This chip rapidly samples and measures the analog voltage in less than 100 ns producing a number value that represents brightness. The number is immediately stored in the image processor memory and another measurement taken, so that a series of brightness values are obtained along each scan line. Once location and density, or color, of each pixel are digitally coded, processing the image involves manipulating numbers. (A pixel is a single picture element in image memory.) Each pixel has a memory address that corresponds to a specific (x,y) location in a Cartesian coordinate system (e.g., 3,1 or 0,4) and a density, or brightness value, reported as a gray level. On the newer systems used for dermal pesticide exposure analysis, monochrome densities have been measured producing a number value from 0 (black) to 255 (the whitest white) representing brightness. (An 8-bit ADC gives 2^8, or 256, brightness levels). Higher-density precision is available with advanced imaging systems used for medical imaging (e.g., 12-bit resolution systems give 2^{12}, or 4,096 gray levels).

In true color systems, each of red, green, and blue (RGB) is assigned 8 bits for a total of 24 bits. From these RGB values, which replace the gray level density values of the monochrome systems, any combination of hue and intensity can be created. To date, all of the systems used to measure dermal exposure to pesticides on skin have used monochrome gray value measurements (Fenske, 1988d; Fenske, Leffingwell, and Spear 1986; Fenske, Wong et al. 1986; Fenske, Hamburger, and Guyton 1987; Archibald

1993; Archibald, Solomon, and Stephenson 1994a,b, 1995; Black 1993; Groth 1994; Methner and Fenske 1994a,b; Roff 1994, 1996a–c; Houghton, Solomon, and Harris 1996; Houghton 1997). The 8-bit system corresponds to the most common organization of computer memory into bytes, such that one byte of storage retains the brightness value from one pixel.

The image analysis software used for measuring dermal exposure must be capable of performing general fluorescence measurements including area, number of fluorescent targets, and fluorescence intensity. Unlike the analysis of images of tissue sections, body surfaces are not planar; therefore, body images taken to assess exposure to fluorescent tracers require correction for four effects that fall into two categories (Roff 1994):

1. Effects relating to surface illumination:
   a. Lambert's cosine law—a surface tilted away from the perpendicular to a light source receives less illumination.
   b. Inverse square law—a surface placed farther away from a light source receives less illumination.
2. Effects relating to apparent surface size:
   a. Foreshortening—according to a cosine law, a surface tilted away from the perpendicular to the camera appears smaller.
   b. Diminution—a surface placed farther from the camera appears smaller, according to the inverse square law.

These laws apply to pinpoint sources of light, which have not been used with video imaging techniques until recently. Fluorescent tubes normally are used as the source illumination. Roff (1994) added a long-wave UV flashgun (pinpoint light source) to his imaging equipment, making it possible to correct for the cosine and inverse square laws. Prior to Roff's developments, software used for dermal exposure assessment made anthropometric adjustments to correct for the fact that the human body does not consist solely of planar surfaces. The software used by Fenske (1984, 1987, 1988a–c, 1990), Fenske, Leffingwell, and Spear (1985, 1986), Fenske, Wong et al. (1986), Fenske, Birnbaum, and Cho (1990), Fenske and Elkner (1990), Black (1993), Archibald, Solomon, and Stephenson (1994a,b, 1995), Bierman, Brouwer, and van Hemmen (1995), and Methner and Fenske (1994a,b) used conical and cylindrical shapes to represent various body regions. Estimated curvatures were used to correct fluorescence intensity and surface area of tracer deposits before using standard curves to quantify mass. There are other methods for measuring surface morphology and usually they require projection of raster lines or grids onto the image. If grid lines are not used, stereoscopic methods using two cameras can achieve a similar result. All of these methods, however, involve substantial image processing to calculate target curvature and are therefore complex.

The imaging system must be capable of performing density and spatial measurements. The latter are relatively simplistic for most image software; however, intensity/density measurements are difficult because fluorescence can change over time and accurate, visually uniform calibration standards are difficult to create. Systems must be capable of accurate densitometric measurement. Fluorescence imaging densitometry is a particularly complex methodology. When using video imaging techniques for evaluating dermal exposure, a variation of densitometry actually is being performed to measure fluorescent tracer intensity, or brightness, on skin. In densitometry, which is a common measurement in fluorescence microscopy, the following terms and equations are used (Imaging Research Inc. 1996):

Incident illumination: light falling on a specimen
Transmittance: light passing through a specimen, or (transmitted light)/(incident light)
Reflectance: light reflected by a specimen, or reflected light/incident light

Reflectance helps make targets visible; however, transmittance is more commonly measured in life-science densitometry. Transmittance decreases as a specimen absorbs more of the incident illumination; therefore, there is an inverse relationship between transmittance and darkness.
Transmittance and reflectance are transformed to opacity using the following equation:

$$\text{Opacity} = \frac{1}{\text{transmittance}} = \frac{\text{incident light}}{\text{reflected or transmitted light}}$$

Density is a value often preferred to transmittance, reflectance, or opacity and is obtained by taking the common logarithm of opacity:

$$\text{Density} = \log_{10} \text{opacity} = \log_{10} \frac{\text{incident light}}{\text{transmitted or reflected light}}$$

Density, like opacity, increases as a specimen darkens but has the advantage that it corresponds to our visual perceptions of how dark a specimen is. For example, a density of 1.0 is approximately twice as dark as a density of 0.5. For a given specimen, the estimate of density is only repeatable under tightly controlled conditions (e.g., identical conditions of illumination).
Density reference standards for microscopy and computer-assisted tomography (CAT) scans, for example, are composed of a transparent substance, usually glass or film, containing various quantities of light-absorbing materials. When working with live human subjects, however, known masses of fluorescent tracer standards must be applied directly to skin. This is necessary because tracers fluoresce differently on different skin tones (Fenske, Birnbaum, and Cho 1990; Archibald 1993; Archibald, Solomon, and Stephenson 1994a; Black 1993; Groth 1994).
Optical density is the logarithm of the ratio of transmitted-to-incident illumination and is a logarithmic function of brightness. *It is brightness that is measured, not optical density, when using video imaging techniques to estimate dermal exposure.* Most scanning densitometers are equipped with a mechanism that calibrates density to a diffuse density standard, thereby eliminating the need for external standards. The technology to internally calibrate cameras has yet to be elucidated; therefore, calibration to an external standard is required to ensure repeatable density and brightness measurements. The system also must be calibrated for the camera-to-target distance to ensure that specimen sizes are measured accurately. Image analysis programs contain a distance calibration feature.

The dynamic range of an image analysis system refers to extremes in the range of brightness that can be measured, from just below the sensor's sensitivity threshold, to that which is too bright and saturates the sensor. The dynamic range is digitized into discrete steps or levels of brightness. As previously discussed, each level corresponds to a shade of gray in the image and is called a gray level. The higher the gray level, the brighter or lighter the shade of gray (the closer it is to pure white or a gray level of 256). Roff (1994, 1995, 1996a–c) measured intensity in units called "least significant bits" (l.s.b's). This is discussed in Section 6-6-4 on the Fluorescent Interactive Video Exposure System (FIVES).

Since incident illumination measurements are not a practical feature of camera-based image analysis systems, these analyzers use gray levels or relative optical density (ROD) as density measurement parameters because they do not require measurement of incident illumination.

Gray level transmittance (GLT) and reflectance (GLR), are calculated as follows:

$$GLT \text{ or } GLR = \frac{\text{observed gray levels}}{\text{maximum possible number of gray levels}}$$

GLT and GLR are converted to ROD as follows:

$$ROD = \log_{10} \frac{1}{\text{gray-level transmittance or reflectance}}$$

$$= \log_{10} \frac{\text{maximum possible number of gray levels}}{\text{observed gray levels}}$$

Since there is no measurement of incident illumination when using a camera, the observed gray (brightness) and ROD values will change with different lighting conditions, hence the importance of stabilizing light intensity during the course of any study using video imaging techniques to evaluate dermal exposure to fluorescent tracers used as surrogates for other compounds. With this technique, researchers *simply measure brightness of the tracer on skin*.

6-2-4 Camera Requirements

The image analyzer must interface to a sophisticated digital camera (preferably a cooled CCD camera) and be capable of handling image acquisition, processing, and calibration at levels of precision (ideally 8 or 12 bits) that often exceed those of standard video-based systems. Most image analysis systems produce medium resolution images (640 × 480 pixels) which are adequate for most biological science applications. More expensive systems operate at high resolution (1024 × 1024 pixels).

Specialized low-light cameras, or regular security video cameras with the gain feature turned off, are required for fluorescence imaging. Black (1993), Archibald, Solomon, and Stephenson (1994a,b, 1995), Methner and Fenske (1994a,b), Groth (1994), and Bierman, Brouwer, and Van Hemmen (1995) used the latter type of camera, but newer integrating cameras such as those used by Roff (1994, 1996b,c), Houghton, Solomon, and Harris (1996), and Houghton (1997) have considerable advantages. Some skin types prove to be a significant challenge for the video imaging technique. Fair-skinned individuals are extremely bright under near-UV light, whereas individuals whose skin contains more melanin pigment emit very little fluorescence and are difficult to distinguish from the image backdrop. An integrating camera allows the operator to brighten or darken an image as required, similar to the effect obtained by changing the shutter speed on a 35-mm camera, except that a video camera achieves this effect by integrating the signal from the camera to the computer. Integrating cameras accumulate incident illumination over time as directed by the operator. By increasing the number of integrations, the signal can be sent less frequently, which results in a situation similar to adding images, causing them to be brightened. Alternatively, integrating less frequently produces a darker image. This is a very valuable feature for improving accuracy because an image can be adjusted such that none of the tracer deposits is bright enough to cause quenching or saturation of the system (a brightness level or gray value greater than 256). This allows the gray value of deposits to be accurately measured. Often integrating cameras are cooled to reduce camera noise. Electronic noise is caused by temperature and power fluctuations on the performance of more than 245,000 sensors that produce the image and because the sensors do not respond identically. System noise must be accounted for when quantifying tracer deposits using fluorescent imaging techniques. Integrating cameras provide better image quality when used for acquiring static images than intensified cameras that amplify incident illumination by an adjustable factor. Integrating cameras are now much more affordable than when earlier work was published; therefore, they should be a mandatory piece of equipment in any laboratory using a fluorescent imaging technique for evaluating dermal exposure to tracers as surrogates for other compounds.

6-2-5 Image Segmentation and Mapping

The imaging system must be capable of discriminating valid targets (deposits of tracer) from background skin to allow counting and mapping. This ability is known as image segmentation. Thresholding is a simple method of selecting pixels within a range of gray levels in the target image as belonging to the area of interest and rejecting all other pixels to the background (Imaging Research Inc. 1996). Setting density thresholds is not always sufficient for accurate segmentation and counting, due to the frequent presence of irrelevant portions of an image that contain densities similar to those of the target and because thresholding is not consistent from one operator to another, nor for one operator over a period of time. Unfortunately, less expensive image software, such as that used by Houghton, Solomon, and Harris (1996) and Houghton (1997), rely on threshold setting. Reasonably accurate results were obtained with this system, most likely due to placing a standard curve in every image and having only one operator threshold all study images. Variability, however, would have been reduced with a more sophisticated software package.

There are techniques that can be used to improve segmentation accuracy such as the use of spatial filters to help distinguish targets from the background and the use of color. The system must also be capable of performing morphometry, or measuring shape. This allows measurement of proportional areas (area fraction) and object diameters. Measurements can be presented as pixel counts or metric measurements (e.g., cm^2).

The system must permit tracing, or mapping of objects, to enable measurement of the areas of interest. Mapping refers to selecting pixels in an image by location rather than brightness and is conducted by using the computer mouse to trace around target areas. Mapping without prior thresholding introduces human error because people tend to draw just outside the actual boundary, making dimensions larger than they should be. This was a source of error with methods other than those used by Roff (1994, 1996a–c) and Houghton, Solomon, and Harris (1996).

6-2-6 Tracer Quantification and Standard Curve Development

Consistency of lighting intensity is absolutely critical to obtain accurate quantitative results, as discussed by Houghton (1997). Some means of stabilizing fluctuations in light intensity caused by changes in line voltage must be achieved or variability in illumination will dramatically affect results. To quantify tracer deposits on skin using video imaging techniques, calibration standards containing various masses of fluorescent tracer were pipetted over a standardized area on the skin of subjects, the brightness measured using the image analysis system, the postexposure image adjusted for background skin fluorescence and each standard was equated to a specific fluorescence intensity (gray level) (Fenske, Leffingwell, and Spear 1986; Fenske, Birnbaum, and Cho 1990; Archibald 1993; Archibald, Solomon, and Stephenson 1994a,b, 1995; Black 1993; Methner and Fenske 1994a,b; Roff 1994, 1995, 1996a–c; Groth 1994; Houghton, Solomon, and Harris 1996; Houghton 1997). The methods used to adjust for background fluorescence varied and are discussed in Section 6-6, Exposure Studies Using Video Imaging Techniques. A standard curve was generated by plotting the mass per pixel, or per cm^2, of each standard pipetted on the skin versus the adjusted fluorescence intensity (gray level or l.s.b's). No conversions to relative optical density were made. Basically, images of exposed subjects were analyzed by measuring the mean fluorescence intensity of tracer deposits on the skin using the standard curve to convert the gray value or l.s.b's to a mass per pixel or per cm^2, which then was multiplied by the area of the deposit to obtain the mass of tracer per deposit. The masses of all deposits were summed to determine the total mass of tracer on the subjects' skin. Interpolation was used to calculate quantities lying between calibration reference values. It is important to generate a standard curve using a wide range of tracer masses spanning all gray levels that conceivably could be present in an image because tracer calibration curves on skin are not linear but sigmoidal, and in some cases approach a quadratic fit (Houghton 1997). Extrapolation, therefore, is not recommended. When estimating pesticide quantities on subjects who directly contacted the spray solution, such as spray applicators, pesticide exposure is estimated by adjusting mass using the tracer/pesticide ratio deposited on the target and not the ratio in the spray mixture. As reported by Fenske (1984), the ratio of pesticide/tracer is often higher when deposited than it is in the original spray mixture. This is due to low solubility of some tracers, such as Calcofluor and Uvitex OB, in aqueous solution and difficulties in providing adequate agitation of the spray mixture prior to, and during, spraying. Since the degradation rates and dislodgeability of applied pesticide and tracer may not be the same over time, it is important in studies measuring exposure due to contact with treated surfaces to adjust exposure estimates using the appropriate dislodgeability ratio for the postapplication time period being studied. Houghton (1997) developed regression models for estimating exposure to chlorpyrifos and bendiocarb that account for temporal differences in dislodgeability of the pesticide and tracer.

It is necessary to select a pesticide/tracer ratio for the spray solution that is appropriate for the anticipated extent of exposure. This involves trial and error. To be most effective, dermal tracer deposits must fall within the dynamic range of the system and should be bright enough to be detectable but not so bright as to saturate the system.

6-2-7 Image Transformation

Frequently images must be modified to distinguish target areas from the background. There are four basic types of transformations (Imaging Research Inc. 1996):

1. *Lookup tables*. This is a method of mapping gray level values onto new values. There are two types: input and output. Input lookup tables transform data after digitization but *before* they are stored in image memory, thus *changing* the original data. Input lookup tables are used in target discrimination applications. Output lookup tables take original image data from memory and assign display characteristics to data values. They do *not change* original data. They are used in contrast enhancement, pseudocoloring, and other visual effects of imaging systems.
2. *Multiple image transformations*. These transformations require that mathematical operations be performed on discrete pixels aligned in two buffers. An example is the pixel-by-pixel subtraction of preexposure images from postexposure images performed by Houghton, Solomon, and Harris (1996).
3. *Spatial transformations*. These transformations process pixels in groups. For example, registration functions alter image position and shape. Convolution functions alter images by basing the intensity of each pixel on its immediate neighbours. Spatial convolutions are used to enhance image appearance. Morphology refers to the transformations that alter shape and size (e.g., erosion and dilation). Examples of morphology are the fill-holes feature available on most image software and functions used to distinguish discrete units within a larger image.
4. *Frequency domain transformations*. An example is the Fourier transform, where the pixel spatial function $f(x,y)$ yields a two-dimensional frequency function $F(u,v)$, where u and v are the spatial frequency coordinates. High spatial frequencies of the Fourier data contain information such as granularity, whereas low spatial frequencies contain luminance and gross feature information. Image properties can be processed selectively in the spatial domain before the modified frequency function $G(u,v)$ is transformed back into a modified spatial function $g(x,y)$. This function is used for image manipulations such as image compression, when less data than the original are required.

6-2-8 Image Display

Digital images, originally captured using a monochrome camera, are converted to analog form by digital analog converters before being sent to the monitor. Almost all image software is capable of pseudocolor image display. At least three digital analog converters are used in pseudocolor systems, one for each of the RGB colors in the monitor. The image is digitized in monochrome and the computer assigns artificial colors to the gray levels displayed on the monitor. A monochrome image is produced when equal proportions of red, green, and blue are presented to the monitor. Colors are created using varying proportions of the three primary colors. The monitor must offer resolution appropriate for the specific imaging system.

Some image software is designed to work with color (RGB) cameras to show the true colors in the specimen. True-color systems have 8-bit sensitivity (256 levels) for each primary color. Therefore, each pixel contains 24 bits of information, unlike monochrome systems which possess 8 bits/pixel. Drawbacks to using color systems include limited spatial resolution, increased processing time for color versus monochrome images, the expense of good-quality color cameras, and the additional memory required to store 24-bit/pixel images. As a result, color systems tend to be expensive and complex. For fluorescence work, monochrome systems are adequate, as past history has indicated.

Most commercial image software allows direct image data exchange via the Windows clipboard to other programs such as Microsoft Excel. This is a necessity for improving the speed and efficiency with which data can be reviewed and analyzed.

6-3 THEORY OF FLUORESCENCE

During fluorescence, a photon from the incident radiation is absorbed and its energy is transferred to the tracer molecules (Fig. 6-1). This energy raises an electron from the ground state (S_0) to an excited state (S_1) of energy. $S_0, S_1, \ldots S_n$ are called singlet states. Only photons with energy equivalent to the difference in energy between S_1 and

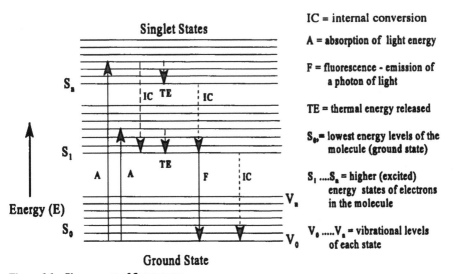

Figure 6-1 Phenomenon of fluorescence.
Source: Adapted from Vo-Dinh (1983).

S_0 can cause the excitation of an electron to S_1. Each singlet state is divided into different vibrational energy levels ($V_0 \ldots V_n$). Electron transition to an excited state always starts at the lowest vibrational level of S_0 but the upper level of the molecule that is reached depends on the energy of the absorbed photon or the wavelength (λ) or frequency (ν) of the absorbed radiation.

$$E = \frac{hc}{\lambda} = h \times \nu$$

where h = Planck's constant, E = energy, c = speed of light.

Unlike the ground state, the excited molecular state is unstable and has a very short lifetime. The excited molecule returns to the ground state (S_0) by losing or radiating its additional energy via several different phenomena, one being fluorescence whereby the molecule emits a photon. Chemical changes do *not* occur during fluorescence. If the lifetime of the singlet state (S_1) is 10^{-9} to 10^{-8} seconds and the electron returns directly to one of the S_0 vibrational levels, the energy difference is emitted as fluorescence radiation.

FWAs absorb UV light and emit light of a longer wavelength in the visible range, according to Stokes' first law which states that the frequency (ν) of emitted light must always have a lower value than the frequency of absorbed light. Emission maxima must fall within a narrow band from 425 to 435 nm. FWAs contain at least one of the following groupings, known as fluorogens, necessary for the formation of states that can be excited by UV light: —CH=CH—, —CO—, and —CN—, —CH=CH—COOH (Zahradnik 1982).

6-4 TOXICOLOGY OF FWAs

FWAs, often referred to as optical brighteners, can be classified according to their method of application or their chemical structure (Zahradnik 1982). Optical brighteners are classified into two large groups according to their use:

1. *Direct (substantive) brighteners*, which are predominantly water soluble, are used for brightening natural fibers and occasionally for synthetics such as polyamides.
2. *Disperse brighteners*, which are water insoluble, form aqueous suspensions and are used for synthetic materials such as polyester, polyamides, polyvinyl chloride, polyacrylonitrile, and occasionally for natural substances such as paper.

Optical brighteners can also be classified according to the materials they are used to brighten (e.g., brighteners for wools or cotton).

Chemically, FWAs fall into six basic structural groups: stilbenes, quinolones and coumarins, naphthalamides, 1,3-diphenyl-2-pyrazolines, benzoxazoles and benzimidazoles, and a combination of this latter group with aromatic, olefinic, or heteroaromatic systems (Gloxhuber, Bloching, and Kastner 1975). The structures of these compounds give them the ability to absorb short-wave UV energy (340–400 nm) in daylight and reemit it as wavelengths of 400–440 nm, as previously discussed (Gold 1975).

6-4-1 Calcofluor

The fluorescent tracer 4-methyl-7-diethylaminocoumarin (Calcofluor) is a coumarin derivative (Gloxhuber and Bloching 1979). Calcofluor has been used as a brightener in laundry detergents, paper coatings, labels, book covers, plastics, resins, varnishes, and lacquers and has been used as an invisible marking agent. Calcofluor use has declined because newer compounds with different structures have better coloristic properties (Gold 1975). Calcofluor was the tracer selected by Fenske (1988a–d), Fenske, Leffingwell, and Spear (1985, 1986), Fenske, Horstmann, and Bentley (1987), Fenske, Birnbaum, and Cho (1990), Archibald, Solomon, and Stephenson (1994a,b, 1995), and Methner and Fenske (1994a,b) in their studies evaluating exposure using the Video Imaging Technique for Assessing Exposure (VITAE). The structure of Calcofluor (compound 16) is given in Table 6-1.

For video imaging techniques to be advantageous from a safety standpoint, it is important to use a fluorescent tracer that is innocuous. The oral LD_{50} of Calcofluor is 5,000 mg/kg in rats and 2,400 mg/kg in mice, indicating low acute oral toxicity. The LD_{50} values for Calcofluor administered subcutaneously and intravenously in the mouse were 1,780 mg/kg and 180 mg/kg, respectively. The acute dermal LD_{50} is greater than 1,000 mg/kg in the rabbit and it was also found to be nonirritating to the skin and eyes of this species.

A review of skin irritation studies on FWAs, conducted using rats and rabbits, indicated that technically pure 4-methyl-7-diethylaminocoumarin did not cause skin irritation (Thomann and Kruger 1975). Dermal toxicity studies, reviewed in the same paper, included twenty compounds, none of which showed any signs of systemic toxicity. The authors concluded that the compounds were not absorbed through the skin. Unfortunately, 4-methyl-7-diethylaminocoumarin was not included in the dermal toxicity review by Thomann and Kruger (1975). Even though the dermal LD_{50} of Calcofluor is greater than 1,000 mg/kg in the rabbit (BASF technical data sheet), which is greater than the dermal LD_{50} of two related coumarins included in the review (500 mg/kg), conclusions regarding dermal absorption and systemic toxicity of Calcofluor cannot be made.

To be used as a tracer for the VITAE technique, Calcofluor was mixed with water in various concentrations. In previous research by Fenske (1984, 1988a–d), Fenske, Leffingwell, and Spear (1985, 1986a), Fenske, Horstman, and Bentley (1987a), Fenske, Birnbaum, and Cho (1990), Archibald, Solomon, and Stephenson (1994a,b, 1995), and Methner and Fenske (1994a,b) there were no reported cases of irritation or illness attributed to the use of Calcofluor. An extensive library search, the product Material Safety Data Sheet (MSDS), and discussions with the manufacturer, BASF, indicated that carcinogenicity, teratogenicity, and mutagenicity tests with this compound have *not* been conducted (BASF. 1993. MSDS on Calcofluor White RWP Concentrate, telephone communication with BASF in August 1993). There is also no published information regarding the metabolism of Calcofluor in humans. There have been no documented cases of health-related effects despite the extensive use of Calcofluor in detergents for many years; however, this fact alone does not proof that the product is innocuous. An important difference between the use of Calcofluor for measuring exposure in agricultural situations and its use in an indoor environment is that outdoor occupational exposures are of brief duration (maximum 8 h), whereas indoors, exposure of bystanders, which

Table 6-1 Trade names, chemical names, and structures of FWAs

Compound	Trade name	Chemical name	Structure
1	Tinopal RBS	Sodium 2-(4-styryl-3-sulfophenyl)-2*H*-naphtho[1,2-*d*]triazole	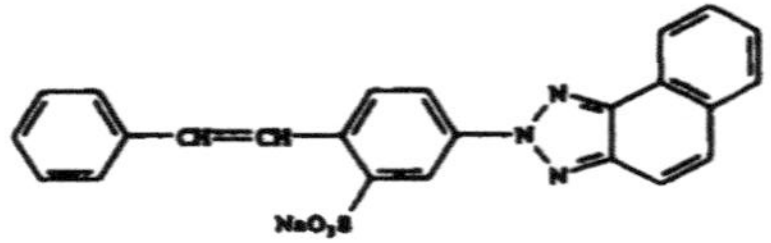
2	Tinopal AMS Blankophor DML	Disodium 4,4'-bis[(4-anilino-6-morpholino-1,3,5-triazin-2-yl)amino]stilbene-2,2'-disulfonate	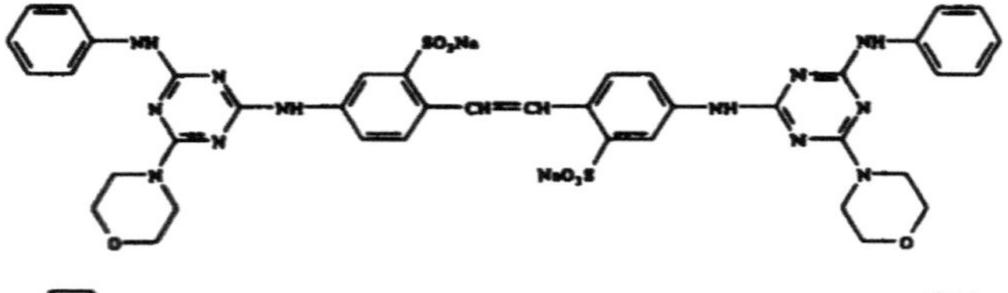
3	Tinopal 5BM-GX	Disodium 4,4'-bis[[4-anilino-6-(*N*-methyl-*N*-2-hydroxyethyl)-amino-1,3,5-triazin-2-yl]amino]stilbene-2,2'-disulfonate	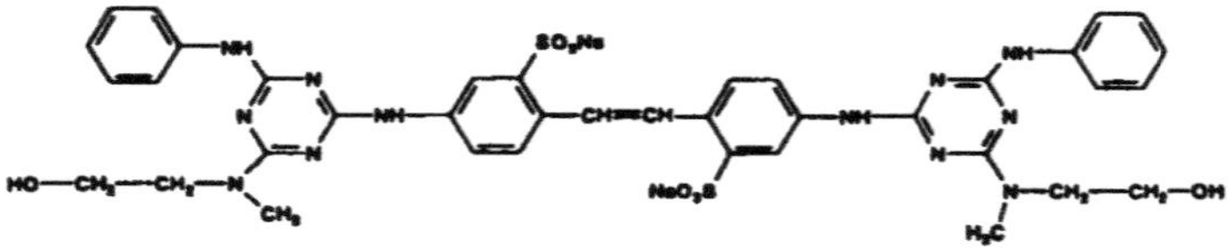
4[a]	Tinopal CBS-X	Disodium 4,4'-bis(2-sulfostyryl)biphenyl	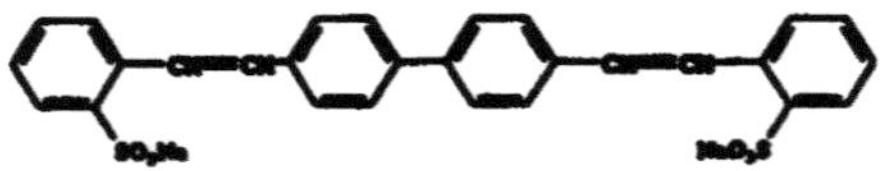
5[a]	Uvitex OB	2,2'-(2,5-thiophenediyl)bis(5-*tert*-butylbenzoxazole)	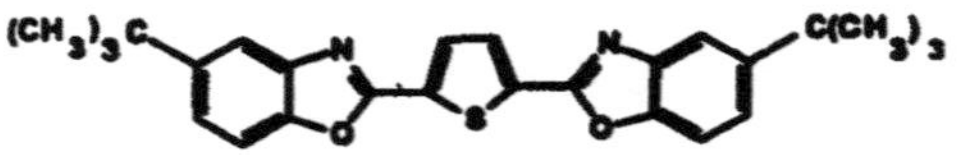
6	N/A[b]	1,3-Diphenyl-2-pyrazoline derivative	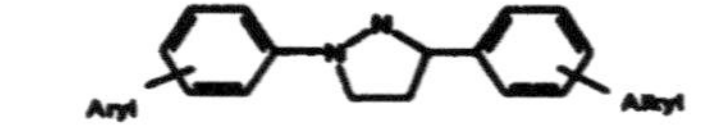

(*Contd.*)

148

Table 6-1 *Contd.*

Compound	Trade name	Chemical name	Structure
7	N/A	4,4'-bis[(4,6-disubst.-1,3,5-triazin-2-yl)amino]stilbene-2,2'-disulfonic acid derivative	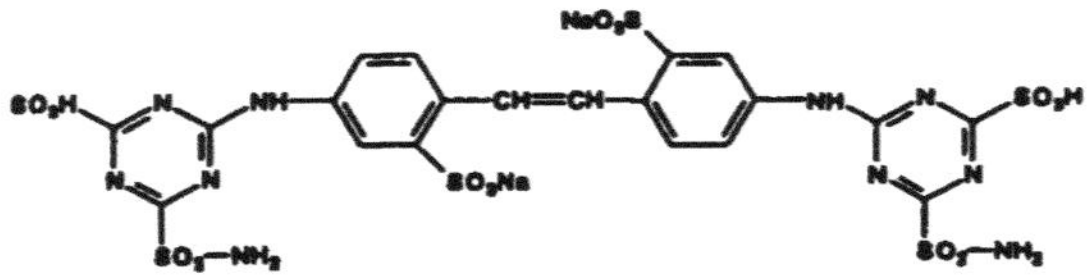
8	N/A	3-Phenyl-7-dimethylamino-2-quinolone derivative	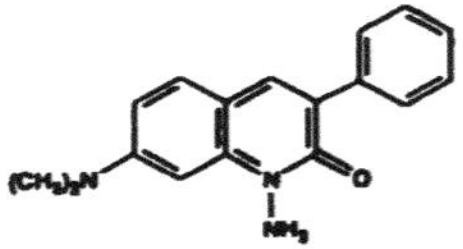
9	N/A	4,4'-Bis(styryl)biphenyl	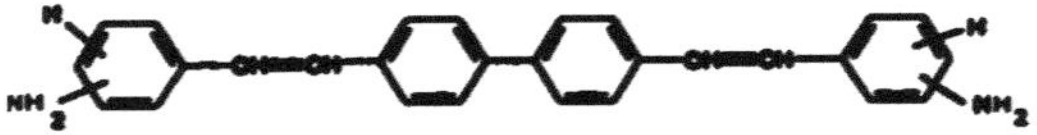
10	N/A	Disodium 4,4'-bis[(4,6-dianilino-1,3,5-triazin-2-yl)amino]stilbene-2,2'-disulfonate	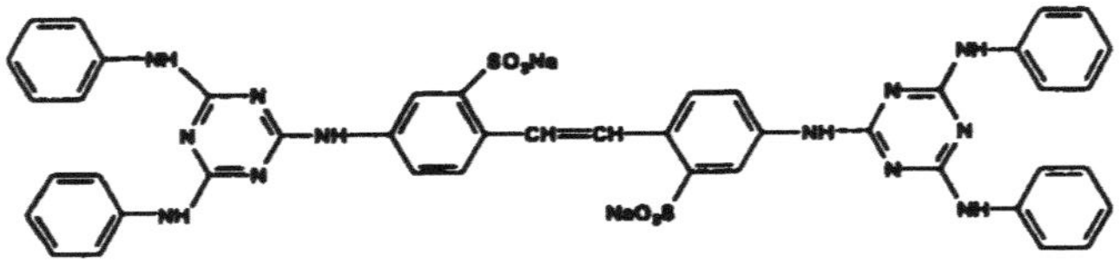
11	N/A	4,4'-Bis[(4-anilino-6-methylamino-1,3,5-triazin-2-yl)amino]stilbene-2,2'-disulfonic acid	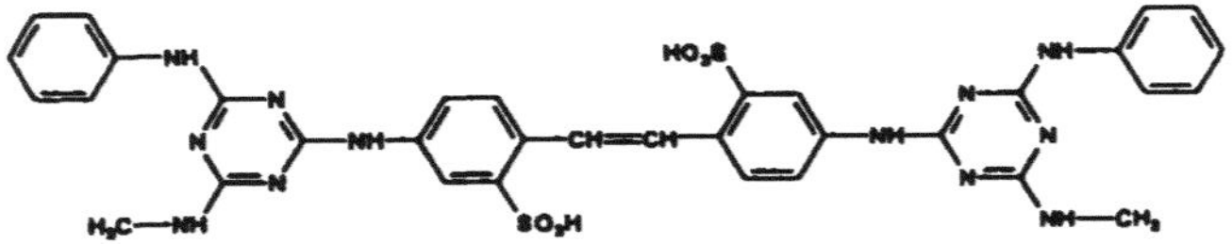

(Contd.)

149

150

Table 6-1 *Contd.*

Compound	Trade name	Chemical name	Structure
12	N/A	Dipotassium-4,4′-bis(4-phenyl-1,2,3-triazol-2-yl)stilbene-2,2′-disulfonate	
13	N/A	Disodium-4,4-bis{[4-anilino-6-bis(2-hydroxyethyl)amino-1,3,5-triazin-2-yl]amino} stilbene-2,2′-disulfonate	
14	N/A	1-(4-Amidosulfonylphenyl)-3-(4-chlorophenyl)-2-pyrazoline	
15	N/A	1-(4-Sulfonamidophenyl)-3-(4-chlorophenyl)-2-pyrazoline	
16[a]	Calcofluor	4-Methyl-7-diethylaminocoumarin	

[a]Used in human exposure studies.
[b]NA = Not available.

include children, occurs over prolonged periods (weeks or months) because the tracer is not easily removed from surfaces. Technically, this should not be a concern from a human health standpoint because studies indicate that FWAs in general are not dermally absorbed to an extent that is likely to cause a health hazard (Hess 1973; Gloxhuber, Bloching, and Kastner 1975; Buxtorf 1975). However, given the frequency of pica activity in children, the absence of carcinogenic, teratogenic, mutagenic, and metabolic data on this particular FWA, and general public concern regarding the use of chemicals for any purpose, the use of a tracer with a more complete toxicological package would be desirable from an ethical and legal standpoint.

6-4-2 Newer FWAs

A toxicology review of other FWAs is appropriate in the hope that a suitable compound with a complete toxicological data package will be found. Large quantities of FWAs are being used and the entire population is exposed regularly to these products. Since the early 1970s, toxicological tests have been conducted on the newer, more effective agents in anticipation of regulations being enacted.

The results of toxicological studies on fifteen FWAs tested at doses far in excess of the 4.4 μg FWA/kg to which humans would be exposed daily in a worst-case situation (Buxtorf 1975) are examined. The structures of these compounds are given in Table 6-1. Given the length of the products' chemical names, they are referred to in the text by the assigned numbers. The acute oral and dermal toxicities and results of dermal and eye irritation studies on these FWAs are given in Table 6-2.

6-4-3 Acute Toxicity Data

Dermal penetration, irritation, and sensitization in animals. A review of skin irritation and photoallergenic studies conducted on FWAs from 1954 to 1966 concluded that the DASC and NTS compounds tested were safe for use in laundry products (Stensby 1967).

In a review by Snyder, Opdyke, and Rubenkoenig (1963) that included compounds 1 and 10, daily topical application to rabbit skin of 1.5 mg of FWA/kg of body weight was without effect. Weight gains and hematologic findings were normal and there was no gross or microscopic evidence of alterations of internal organs or skin during the 13-week study. The authors calculated that, assuming a detergent contains a maximum of 0.2% FWA, the daily quantity of product required to give this dose would be 0.75 g/kg bwt. Since usage concentrations rarely exceed 0.5% for laundry and dishwashing, this is equivalent to application of 7.5 l/day for a 50 kg person and this would have to remain in contact with the skin. The test conditions were highly exaggerated and still no effect was noted.

A study of the effect of FWAs on the skin of hairless mice and shorn CF_1 mice, using fluorescence microscopy, indicated that compound 12, applied as a 0.1% or 0.01% aqueous solution over a 6- to 8-cm^2 area, did not penetrate into the subepithelial layers (dermis and subcutaneous tissue) after cutaneous application during a 21-day study (Luckhaus and Loser 1975).

Table 6-2 Acute oral and dermal toxicity of FWAs

FWA no.[a]	Trade name	Type[b]	Oral toxicity			Dermal toxicity			Skin and eye irritation		
			Oral LD50 (mg/kg)	Form tested[c]	Species	Dermal LD50 (mg/kg)	Form tested[c]	Species	Skin irritation, rabbit	Eye irritation, rabbit	Form tested[c]
1	Tinopal RBS	NTS	>10,250	Tech	rat	>2,000	Tech	rabbit	slight	slight	tech
			>10,250	60%	rat	—	—	—	—	—	—
			>15,000	60%	mouse	—	—	—	—	—	—
2	Tinopal AMS	DASC	>10,250	Tech	rat	>10,000	Tech	rabbit	≈none[d]	none	tech
3	Tinopal 5BM-GX	DASC	>10,250	60%	rat	>2,000	60%	rabbit	slight	≈none[d]	60%
4	Tinopal CBS-X	DSBP	5,580	Tech	rat	>2,500	Tech	rabbit	slight	extreme	tech
5	Uvitex OB	BEN	>10,000	Tech	rat	>1,000	Tech	rat	none	none	tech
6	—	DP	—	—	—	—	—	—	—	—	—
7	—	DASC	—	—	—	—	—	—	—	—	—
8	—	Quin	—	—	—	—	—	—	—	—	—
9	—	DSBP	—	—	—	—	—	—	—	—	—
10	Tinopal TAS, Uvitex SBA, Blankophor CA	DASC	>10,000	Tech	rat	—	—	—	—	—	—
11	—	DASC	>2,500	Tech	rat	>1,000	Tech	rat	≈none[d]	slight	tech
12	Blankophor	TS	>15,000	Tech	rat	>500	Tech	rat	none	moderate	—
			>15,000	Tech	mouse	—	—	—	—	—	—
			>5,000	Tech	rabbit	—	—	—	—	—	—
13	—	DASC	14,530	Tech	rat	>2,000	35%	rabbit	≈none[d]	moderate	tech
			14,530	35%	rat	—	—	—	≈none[e]	—	35%
14	—	DP	—	—	—	—	—	—	—	—	—
15	—	DP	—	—	—	—	—	—	—	—	—
16	Calcofluor RWP	COU	5,000	Tech	rat	>1,000	Tech	rabbit	none	none	tech
			2,500	Tech	mouse	—	—	—	—	—	—

Source: Adapted from Thomann and Kruger (1975).

[a] used in human exposure studies.

[b] Chemical family to which the compound belongs: DASC = diaminostilbenedisulfonic acid/cyanuric chloride; TS = triazolylstilbenedisulfonic acid; NTS = naphthotriazolylstilbene; COU = coumarin; DSBP = distyrylbiphenyl; BEN = benzoxazole; DP = diphenylpyrazoline; Quin = quinoline.

[c] Tech = technically pure product; a percentage refers to the concentration of active ingredient in the formulation tested;

[d] ≈ No means practically none.

Note: "—" Means not available.

A review of dermal irritation studies using pure technical compounds applied to rabbit skin and covered with occlusion bandages for 24 h indicated either no irritation or only slight irritation depending on the chemical (Gloxhuber and Bloching 1979). Compound 1, applied as a 0.01% solution in 60% aqueous ethylene glycol, did not cause irritation in mice and rats. Compound 2, applied at 0.5 g to rabbit skin for 24 h, did not cause irritation. Repeated applications of compounds 1, 3, and 4 did not cause irritation (Keplinger et al. 1974). Skin irritation studies conducted on compound 5 indicated that the chemical was not an irritant in rabbits (Ciba-Geigy Ltd. 1993).

Dermal irritation and sensitization in humans. Snyder, Opdyke, and Rubenkoenig (1963) reported that repeated patch tests in human volunteers showed no potential for dermal sensitization, despite the fact that doses applied were "10 times more concentrated with respect to brightness than detergent solutions to which the skin is normally exposed." Three weeks of continuous contact, followed by a challenge exposure 2 weeks later, was a very rigorous test, yet no sensitization occurred.

A study comparing skin irritation in infants wearing diapers washed with detergent containing compounds 13 and 15 to the incidence of irritation in those wearing diapers washed in a detergent that did not contain FWAs found no differences between the two groups (Weber 1968).

Skin sensitization studies conducted in human volunteers by Griffith (1973) tested 31 FWAs, at various concentrations, for skin sensitization potential in repeated insult patch tests. Materials were applied under occlusive patches during a series of nine applications, each of 24-h duration, over a period of 3 weeks. Additional applications were made 2 weeks later. Exposures were exaggerated compared to estimated consumer exposures. Only three compounds tested were sensitizers and they are not discussed in this review. Compounds 2, 4, 10, and 13 did not cause skin sensitization. A photosensitization test was conducted concurrently and was similar to the repeated insult test with the exception that twice weekly, after patch removal, the test areas were exposed to 30 min of sunlight. Compounds 2 and 10 did not cause irritation. Compound 5 did not cause skin sensitization in guinea pigs when applied as a 0.1% suspension in saline (Ciba-Geigy Ltd. 1993). Compound 5 showed no evidence of irritation or sensitization in 102 human subjects (Ciba-Geigy Ltd. 1993).

Patch tests in humans indicated that no skin reactions occurred with application of a 1% preparation of compound 3 and 2% preparations of compounds 2, 10, and 15 (with paraffin), during a 24-h exposure (Gloxhuber and Bloching 1979). There was one exception: One person reacted to a compound that is not discussed here.

Eye irritation studies. A rabbit eye test for mucous membrane irritation during which compounds 1–4 were applied *without dilution* and left in contact with the mucous membranes for either 2–4 s, or for an unlimited length of time, indicated that compound 2 caused low-level irritation; compounds 1 and 3 were moderate in effect; and compound 4 caused severe irritation (Thomann and Kruger 1975). Compound 5 was not an irritant in studies conducted in rabbits by the manufacturer (Ciba-Geigy Ltd. 1993).

Metabolism. Muecke, Dupuis, and Esser (1975) studied the metabolic behavior of ^{14}C-labeled compounds 2, 3, and 4 in rats. Oral doses of 5 mg/kg were excreted rapidly and completely, with an excretion half-life of 7–13 h. No more than 0.3% of the dose was excreted in the urine. Since virtually the only route of excretion was via feces (average of 95.4%) and the half-life of each compound was short, it was concluded that no significant amounts of FWAs were absorbed from the digestive tract. Ninety-six hours after dose administration, no radioactive residues were found in the liver, kidney, blood, brain, fat, or muscle. The limit of quantitation was 0.005–0.01 ppm.

6-4-4 Chronic Toxicity

A summary of chronic toxicity studies on FWAs is presented in Table 6-3.

Dermal studies. Solutions of compounds 1 and 10 in several vehicles, including water and acetone, applied to the skin of mice at doses of 0.05–0.2 mg and 0.02 mg per application, respectively, two to three times weekly, failed to produce malignant or benign tumors during the 63- to 92-week test period (Snyder, Opdyke, and Rubenkoenig 1963). The concentration of FWAs in this test were 50 to 10,000 times higher than the concentration in detergent solutions (0.001%), yet no tumors were induced.

Feeding studies. Keplinger, Lyman, and Calandra (1975a) studied the chronic toxicity and carcinogenicity of FWAs 1–4. None of these chemicals caused adverse toxic effects (reduced weight gain, decline in body weights, or decline in food consumption), or evidence of carcinogenicity (increase in tumor type or number) in 2-year feeding studies in albino rats at dietary levels of up to 1,000 ppm. Three of the four compounds (2, 3, and 4) were without effect at levels of 2,000 ppm in 2-year feeding studies in dogs. At this rate, however, compound 1 caused dose-related decreases in weight gains of some dogs (caused by decreased food consumption), development of inflammatory granulosa tissue on abdominal viscera, and increased hematopoietic activity of the spleen and liver. The authors concluded that this effect occurred because of failure to adjust the pH of the diet during the first 30 weeks of the study.

A 13-week oral toxicity study, conducted in rats, administered compound 5 at 1,000, 3,000, and 10,000 ppm in the diet (Bhatt et al. 1981). No reactions to treatment were noted, including effects on food and water consumption, feed conversion, and body-weight gain. Hematological assessment after 6 and 12 weeks did not reveal any changes attributable to treatment. At autopsy, no macroscopic or microscopic changes in tissues were noted in the rats fed 1,000 and 3,000 ppm in the diet. At 10,000 ppm, however, liver weights among both sexes were slightly higher.

Repetitive, daily administration of 0.025 and 0.25 mg of compound 5 per kilogram, given by oral intubation to male rats, revealed no toxic symptoms or mortality during the study. Body-weight gains were unaffected and no gross organ changes were observed on autopsy (Bathe 1979).

A 2-year chronic feeding study in Sprague-Dawley rats feeding 1,000 ppm of compound 5 continuously in the diet showed no treatment-related effects on mortality, morbidity, food consumption, food efficiencies, growth and development, hematological changes, renal function, or tumor incidence (Wheldon et al. 1970). Liver weights were slightly increased among males; however, there was no evidence of altered liver histology or altered liver enzyme activity. Some optic lens opacities were noted which may have been treatment related. Body fat of treated animals exhibited fluorescence under UV light. Histopathological examination of the respiratory tract, cardiovascular system, liver, kidney, and adrenal glands showed no treatment-related effects.

In mice, oral and simultaneous subcutaneous applications of compound 3 over a 2-year period caused no increased occurrence of tumors (Neukomm and De Trey 1961). Snyder, Opdyke, and Rubenkoenig (1963) applied FWAs topically to the skin of mice with the same noncarcinogenic results. The compounds tested included 1 and 10, among others not included in this review.

A study investigating the long-term administration of 1,000 ppm of compound 5 in the diet of mice over 52 weeks, followed by observation for a further 26 weeks, indicated no treatment-related effects on growth or feed consumption (Wheldon et al. 1969). Tumor incidence and liver and kidney weights at sacrifice were unaffected. Histopathological and microscopic examination of thirteen different tissues indicated that no treatment-related effects had occurred. Deposition of tracer in adipose tissue, which was visible during examination under UV light, was the sole treatment-related finding.

A 3-month subchronic dietary toxicity study in beagle dogs, designed to obtain a no-observable-effect level (NOEL) for compound 5, involved administering doses up to 50,000 ppm in the daily diet (Ciba-Geigy Ltd. 1991). No signs or symptoms of toxicological relevance occurred in the test animals. There was no effect on body-weight gain, food consumption, hematology, blood chemistry, renal systems, organ weights, or organ to body weight ratios, and pathological examination indicated no abnormalities that could be attributed to treatment. Ophthalmoscopic examination revealed no reaction to treatment, unlike the study using Sprague-Dawley rats (Wheldon et al. 1970). The authors concluded that, at 5% of the diet (50,000 ppm), compound 5 was well tolerated without effect. Compound 5 has been approved by the US Food and Drug Administration as an optical brightener in food wrappers (*Code of Federal Regulations*, 21 CFR§178.3297(e) [1992]) and has been exempted from tolerances by the US Environmental Protection Agency (US EPA) when applied to growing crops (40 CFR§180.1001(d) [1992]).

Phototoxicity and photocarcinogenicity studies. Forbes and Urbach (1975a–d) studied whether FWAs were phototoxic (producing an acute skin response following a single exposure to UV light) and whether they could enhance photocarcinogenesis (increased incidence or hastening of the appearance of skin tumors following several UV light exposures). Compounds 1, 2, 3, 4, and 10 were tested for phototoxicity using dorsal skin of hairless mice, miniature pigs, and healthy human adult males. Photocarcinogenicity studies involved treating hairless mouse skin with FWAs prior to daily exposure to UV light. In both studies, FWA response was compared to methoxypsoralin, a known phototoxic agent. UV-A, UV-B, and UV-C were used in the experiments. Compound 10 was included specifically to compare test results with those of Bingham and Falk (1970) who reported that the compound had photocarcinogenic properties. The Bingham and Falk (1970) study had several problems, including the use of dimethylsulfoxide (DMSO), which is known to affect skin metabolism and undergo photochemical alteration, and the selection of UV-C, which is not found in sunlight striking the Earth's surface or in artificial lighting. The Forbes and Urbach (1975a–d) studies used a different solvent and wavelength of light and most of the investigations were conducted using 10 μg/cm^2. This dose was selected as a compromise between too low a rate and an excessive one that would mask any biological effect due to the sunscreen properties of FWAs. The highest rate tested was 100 μg/cm^2. Results indicated that compounds tested, at rates up to 100 μg/cm^2, did not enhance tumor yield or development time and were not phototoxic.

Mutagenicity. A number of researchers have studied possible genetic effects of FWAs in yeast. This work was summarized by Kilbey and Zetterberg (1973) and by Lorke and Machemer (1975b) and indicated no evidence of mutagenicity; however, in a test on compounds 1–4 reported by Kilbey and Zetterberg (1973), selection of preexisting petite mutants that mimic a mutagenic effect were possible in certain situations. Since petite mutants are caused by alterations outside the nucleus, no conclusions can be drawn regarding mutations within the nucleus. Kilbey and Zetterberg (1975) conducted mutagenicity assays of compounds 1–4, a benzoxazole derivative of Uvitex OB, and a quinolone (not reviewed here) in microorganisms including yeast, *Salmonella typhimurium*, *Echerichia coli*, and *Neurospora crassa*. Fluorescent whitening agents tested did not produce any mutagenic changes or any other alterations of genetic material, including the development of petite mutants. This may be because microscopic examination suggests that FWAs concentrate in the cell wall and fail to reach the nucleus. Since mammalian cells do not contain a cell wall, a barrier effect may or may not exist in these cells. To the best of our knowledge, this has not been investigated in mammalian tissues.

Compound 5 was tested for mutagenic effects on several strains of *S. typhimurium* at 20, 80, 320, 1,280, and 5,120 μg/0.1 mL DMSO, with and without mammalian microsomal activation, to expose cultures to product metabolites from rat liver microsomes (Deparade, Durand, and Keller 1984). There was no evidence of induction of point mutations by compound 5 or its metabolites formed as a result of microsomal activation.

Lorke and Machemer (1975b) felt that tests on lower organisms were not appropriate when trying to determine the possible mutagenicity of FWAs, and so, they conducted mutagenicity studies using the dominant lethal test on male mice. This test involves mating treated males with untreated females, and studying the fetal mortality rate. If the mortality rate exceeds the spontaneous death rate, then it is assumed that genetic changes are responsible. The FWAs tested were compounds 2, 11, 12, 13, and 14. Oral administration of 5,000 mg/kg did not cause mutagenic effects during 8 weeks' mating. Effects studied included: ability of males to fertilize females; preimplantation losses, estimated using implantation rates; and postimplantation losses. Previous studies conducted by Keplinger et al. (1974) on compounds 1, 2, and 3 at doses as high as 50 mg/kg, and compound 4 at doses as high as 10 mg/kg, showed no evidence of a dominant lethal response. Keplinger also demonstrated that FWAs are not mutagenic when administered intraperitoneally. Keplinger et al. (1974) and Lorke and Machemer (1975b) both used several test parameters to assess mutagenic effects. There were no biologically significant differences with any of the compounds tested. Lorke and Machemer stated that, assuming absorption of 1 mg of "whitener X" per person and considering the oral dose of 5,000 mg/kg given in mice to be without an effect, the safety factor is approximately 1:250,000.

Further mutagenicity studies by Muller et al. (1975) confirmed previous results showing that compounds 1–4 were not mutagenic following oral administration of dosages up to one-third the LD_{50} in three different mammalian test systems (dominant lethal test in male mice, cytogenetic studies on metaphase chromosomes from Chinese hamster bone marrow, nucleus anomaly test in somatic interphase cells of Chinese hamsters).

Teratogenicity. In embryo toxicity studies of compounds 1, 2, 3, and 4, no teratogenic effects on rabbits were noted when the products were administered daily in oral doses as high as 30 mg/kg (Keplinger et al. 1974). Studies on compounds 11 and 12, reviewed by Lorke and Machemer (1975a), showed no teratogenic or embryotoxic effects in either rats or rabbits following daily oral doses of up to 1,000 mg/kg. At daily doses of 3,000 mg/kg, reductions in weight gain of the mothers, sporadic symptoms, and death were evident with compound 12. These doses were extremely high.

Reproduction. Three-generation reproductive studies conducted using albino rats (Keplinger, Lyman, and Calandra 1975b), evaluated compounds 1–4 for potential reproductive performance effects through three successive two-litter generations. At levels of 40, 200, and 1,000 ppm in the diet, no adverse effects were noted except for slight and random effects of compound 4 on pup survival that were not consistently related to dose or duration of exposure.

A three-generation reproductive study, conducted as part of a chronic toxicity study using CD-1 mice, evaluated compound 5 for effects on reproductive performance over 52 weeks of long-term dietary administration of 1,000 ppm (Wheldon et al. 1969). Treatment continued during mating, gestation, and lactation periods. Mating performance, conception rate, and gestation period of F_0 and F_1 generations were unaffected. Among the F_1 and F_2 generations, there were no significant differences between treated and control animals when assessed for litter size and mean pup weights at birth and weaning, external abnormalities, pup mortality, and litter weights. No morphological abnormalities were recorded at any stage of the study and mortality rates and causes throughout the experiment were comparable between treated and control animals. There was slightly decreased food consumption in the F_2 animals.

Estrogenicity. Since stilbene FWAs are chemically related to estrogenic stilbene derivatives, several of these compounds were tested using the Allen-Doisy test. In this test, sexually immature female mice are injected with test materials. The subsequent development of keratinized squamous cells in a vaginal smear indicates that the material has estrogenic activity (Critchley 1978; Landau 1986). A review of two German studies showed no evidence of estrogenic activity (Gloxhuber and Bloching 1979). Given the current interest in endocrine-disrupting chemicals, their possible effect on human fertility, and the development of such hormonally mediated diseases as breast cancer, this information is pertinent and reassuring when considering our daily dermal exposure to FWAs.

6-4-5 Conclusions

Although it is recognized that oral studies may not yield results applicable to dermal exposure scenarios because of possible differential toxicity between these exposure routes,

Table 6-3 Chronic toxicity of FWAs

FWA	Dermal	Oral	Reproductive	Teratogenic	Mutagenic	Carcinogenic
1	neg[1]	—	neg[2]	neg[6]	neg[1,7,8]	neg[1,2,4,5]
2	—	—	neg[2]	neg[6]	neg[1,7,8]	neg[1,2,3,4,5]
3	—	neg[1]	neg[2]	neg[6]	neg[1,7,8]	neg[1,2,3,4,5]
4[a]	—	—	neg[2]	neg[6]	neg[1,7,8]	neg[1,2,3,4,5]
5[a]	—	neg[1,2,]	neg[1]	—	neg[7]	—
6	—	—	—	—	—	—
7	—	—	—	—	—	—
8	—	—	—	—	—	—
9	—	—	—	—	—	—
10	neg[1]	—	—	—	—	neg[1,4,5]
11	—	—	—	neg[2,6]	neg[1]	—
12	—	—	—	neg[2,6]	neg[1]	—
13	—	—	—	—	neg[1]	—
14	—	—	—	—	neg[1]	—
15	—	—	—	—	—	—
16[a]	—	—	—	—	—	—

Note: Superscript numbers refer to the species in which the study was conducted, as follows: mice, 1; rats, 2; dogs, 3; miniature pigs, 4; healthy human adult males, 5; rabbits, 6; microorganisms, 7; Chinese hamsters, 8. Other symbols and abbreviations include (—) data not available, neg = no effect.
[a] Used in human exposure studies.

the data are included as support information. This toxicological review indicates that acute and chronic toxicity, carcinogenicity, teratogenicity, and mutagenicity data exist for compounds 1 to 5. These FWAs were negative for effects in all of these end points of toxicity. None of these five chemicals caused reproductive effects and four of the five are water soluble. This last point is important given that the chosen tracer must be dissolved in a water–pesticide emulsion for the video imaging technique to be successful. Use of Uvitex OB (compound 5) and Calcofluor (compound 16), which are not water soluble, requires the addition of a surfactant to create a pesticide–water emulsion (Black 1993; Archibald, Solomon, and Stephenson 1994a,b, 1995; Methner and Fenske, 1994a,b, 1996; Houghton, Solomon, and Harris 1996; Houghton 1997).

Since indoor residential exposure studies using video imaging techniques necessitate that tracers be sprayed in homes, it is critical from ethical and legal standpoints to demonstrate safety with a sound and extensive database. For this reason, any of tracers 1 to 5 would be appropriate for this particular use.

6-5 EXPOSURE TO FWAs

Gloxhuber and Bloching (1979) conducted an extensive toxicological review of FWAs and concluded that a wide margin of safety exists for these chemicals. This was supported by earlier reviews by Buxtorf (1975) and Burg, Rohovsky, and Kensler (1977). Calcofluor was not included in these two earlier reviews and was mentioned only briefly in Gloxhuber and Bloching's report because of lack of information on it. The average consumer may ingest FWAs in food and also may be exposed to them by dermal contact. Buxtorf (1975) calculated the average daily human exposure to FWAs, assuming they penetrate the dermis and are absorbed into the blood. Oral and dermal exposure values were summed to obtain a total exposure estimate. Buxtorf estimated that, in an extreme case, an individual might consume 0.001–0.03 mg of FWAs daily from food packaging materials. Detergent residues on badly rinsed cutlery and dishes may contribute another 0.02 mg per person per day. If an individual's diet contained 100 g of fish (which frequently contains FWAs) each day, daily intake would rise by 0.03 mg. Therefore, the maximum total amount consumed daily was estimated to be 0.08 mg or approximately 0.001 mg/kg body weight for an 80-kg person. In reality, daily oral ingestion of FWAs is probably closer to one-tenth to one-hundredth as much, or 10–100 ng/kg body weight.

Hand-washing tests with detergents containing 0.05–0.25% FWAs (Gloxhuber, Bloching, and Kastner 1975) indicated that the quantity of FWAs left on both hands after soaking for 15 min in a liter of lukewarm, 0.5% detergent solution containing FWAs varied from 0.07 to 0.17 mg. (Standard curves for quantifying the FWAs spectrophotometrically were created using the skin of hairless mice.) The whiteners tested were compounds 6, 7, 8, and 9. Even though compounds with different structures and fluorescence intensity on skin were selected, the skin was free of fluorescence 24 h post-exposure. This was attributed to shedding the outermost layers of the stratum corneum. Previous research supported this idea. After application to skin, FWAs are bound in the outermost keratin layers. If these layers are stripped using adhesive tape, the skin's fluorescence disappears (Jadassohn and Schaaf 1970; Gloxhuber 1972). Allergies to FWAs are rare, which supports the theory of shedding the outermost dermal layers. FWAs are also expected to photodegrade on skin. Even if all of the FWA bound to skin was absorbed, it still would not pose a health hazard based on oral LD_{50} values, which are several thousand times higher, and the results of chronic studies (excluding a remote risk of dermal metabolism producing a toxic metabolite). Since only compounds with high affinity for fabrics are suitable as FWAs, transfer of these chemicals to skin from washed fabrics would be slight; however, minute quantities have been transferred in situations of inadequate rinsing and excessive perspiration (Gloxhuber, Bloching, and Kastner 1975).

Dermal absorption studies of FWAs in laundry detergents indicated that the human hand adsorbs approximately 0.2 $\mu g/cm^2$ (Hess 1973; Gloxhuber, Bloching, and Kastner 1975). Taking the area of both hands to be 500 cm^2, Buxtorf (1975) calculated potential exposure from this route to be 0.1 mg. Adsorption of FWAs from clothing has been estimated at less than 0.01 to 0.17 $\mu g/cm^2$. Assuming 5,000–10,000 cm^2 of skin could be in contact with treated textiles, Buxtorf estimated that the total quantity absorbed would range from less than 0.05 mg to 1.7 mg per person. When data indicating that only 500 cm^2 of sweat-moistened material remains in contact with skin for any period of time were considered, dermal exposure was only 5–85 μg per person per day. Water-soluble FWAs are highly polar; therefore, adsorption and absorption probably are not equal; Buxtorf assumed equality to obtain a worst-case estimate. He calculated that absorption of FWAs from dermal contact could range from less than 0.015 to 0.185 mg per person (0.0017 to 0.003 mg/kg) per day. If all sources of exposure are summed, total intake of FWAs would amount to 265 μg per person or 4.4 $\mu g/kg$ daily in the most heavily exposed situation. An exposure estimate of 40 $\mu g/kg$ per day was calculated by Burg, Rohovsky, and Kensler (1977), who considered similar sources of exposure. Both researchers concluded that exposure would not be expected to cause adverse health effects.

6-6 EXPOSURE STUDIES USING VIDEO IMAGING TECHNIQUES

6-6-1 Historical Aspects

The first suggestion that fluorescent tracers could be used as a method for assessing dermal exposure to pesticides was made by Staniland (1958), who was studying the use of fluorescent dyes to assess spray distribution of malathion and DDT on plant foliage. Staniland commented that this method is a "valuable advisory tool of great educative value," which allows growers to visualize the type and degree of spray coverage obtained and effects of changing spray nozzles and pressures. An additional observation, complete with photos, was that skin and clothing of spray operators were contaminated and this exposure could be viewed easily during examination using UV lamps. The author noted that dermal contamination was clearly evident, "to an almost startling extent," particularly on hands. Facial contamination occurred, and when face shields were not worn, tracer was retained at the bases of the eyelashes, eyebrows and in hair. Staniland suggested that if tracers were periodically mixed with pesticide spray solutions, operators could be inspected for clothing and skin contamination using UV lamps. The value of this method for assessing equipment contamination and the efficacy of cleaning procedures also was noted.

The concept of using fluorescent tracers to evaluate spray coverage and droplet size was studied by several researchers including Edwards et al. (1961), Bullock, Brooks, and Whitney (1968), Himel (1969), and Carleton et al. (1981). Fluorescent tracers and fluorometric analysis were used to monitor pesticide drift by Yates and Akesson (1963), and by several pesticide manufacturers during the early 1980s (Cyanamid. 1985. Verbal communication with J. Snow). Ashbaugh (1977) reported results of a study designed specifically to assess dermal exposure of workers using a fluorescent tracer technique. A naturally fluorescent sulfur compound was applied to fruit trees prior to harvesting. Workers were examined and photographed, postexposure, under near-UV light. This study was qualitative in nature.

A portable, handheld, fluorometric monitor was developed by Schuresko (1980) to monitor polynuclear aromatic hydrocarbon (PAH) contamination of work surfaces in coal conversion facilities and on workers' skin. PAHs fluoresce naturally. The device contained a UV illuminator and a fluorescence detector, and was capable of separating background fluorescent illumination from the chemical's fluorescence signal. The device could be used at distances of ≤ 3 m in direct sunlight and indoors in artificial light, could quantify contamination, and could distinguish between organic and inorganic compounds on the basis of their fluorescence lifetimes.

Vo-Dinh and Gammage (1981) discussed the development of the lightpipe lumino-scope for monitoring dermal exposure to PAHs. This portable luminescence detector consisted of a bifurcated lightpipe that was capable of both emitting the excitation radiation and converging the emission toward a photon detector. The design of the lightpipe included a stethoscope-shaped end that could be pressed against the skin. This study, conducted on the skin of white mice, indicated that strong background fluorescence of skin interfered significantly with quantification of fluorescent coal compounds. The authors noted that brown hamster skin did not fluoresce as extensively. This skin-tone phenomenon was later noted by Fenske (1984), Archibald (1993), Archibald, Solomon, and Stephenson (1994a,b), and Black (1993). The lightpipe used a fixed excitation wavelength and was of limited use; however, it allowed rapid screening, monitoring of difficult-to-reach anatomical locations (e.g., armpits), and eliminated the need for anthropometric data correction and the use of UV protective eyewear. Increased fluorescence of callouses, fingertips, and dry skin was noted, and dirt was observed to obscure fluorescence. Background dermal fluorescence varied 10–35% between individuals, and 10–15% within individuals. The researchers suggested collection of preexposure data and stressed that background correction for each individual would be required to accurately quantify dermal fluorescence due to contaminants.

Franklin et al. (1981) used an unidentified tracer, in conjunction with dermal patches, to provide a qualitative assessment of dermal exposure to azinphos-methyl during low-volume airblast applications in orchards. Despite the use of gloves, respirators, boots, and long-sleeved coveralls, pesticide residues were detected on dermal patches placed beneath clothing. Patches indicated that the lower arm incurred the greatest exposure, followed by the chest. Video imaging, however, indicated that the most highly exposed bodily regions were the hands, neck, and face, which were not covered by dermal patches. This explained why patch estimates were not correlated with urinary metabolite excretion ($r = -.4$), and why wearing additional protective equipment on the legs, arms, and torso (i.e., rubberized coveralls) did not reduce exposure.

A review by Vo-Dinh (1983) explained that the foundations for quantitative luminescence spectrometry are based on the Lambert-Beer law of absorption which does not apply to surface luminescence of nontransparent, optically heterogeneous samples. Intensity equations for surface luminescence have been derived from Kubelka and Munk (1931). A unique relationship between the concentration of the analyte and luminescence allows quantification; luminescence intensity varies linearly with contaminant quantity *at low concentrations only* (Kubelka 1948). The relationship deviates from linearity above a certain concentration determined by the chemical's absorptivity. This was observed by several researchers (Black 1993; Archibald 1993; Groth 1994; Bierman, Brouwer, and van Hemmen 1995; Houghton, Solomon, and Harris 1996; Houghton 1997). Surface luminescence spectroscopy has advantages, however, because it allows quantification of contamination without chemical removal or extraction of the substance.

6-6-2 VITAE

Fenske (1984) described the development and preliminary testing of the VITAE in his PhD thesis. The technique was developed because of dissatisfaction with other dermal exposure techniques, primarily the patch technique. Problems with the dermal patch method are as follows:

1. Extrapolation of values unrealistically assumes a uniform distribution of dermal contamination which results in under- or overestimations of exposure.
2. Confusion exists over whether the mean patch value, or value nearest the region of interest, should be used for extrapolation.

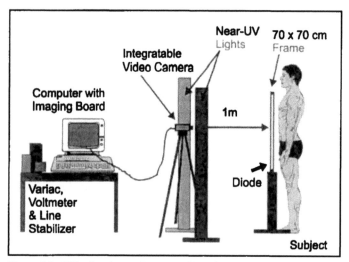

Figure 6-2 Video imaging equipment configuration.
Source: Adapted from Archibald, Solomon, and Stephenson (1994a).

3. Protocols place patches in locations that tend to overestimate exposure.
4. Placing only a few patches beneath clothing can result in inaccurate estimations of the extent of exposure that occurs as a result of penetration of fabrics or a "bellows effect" near garment openings. This assumption is a significant source of error.
5. Correlation is poor between exposure estimates obtained using dermal patches and those from urinary metabolite excretion.

Fenske's objective was to develop an accurate, affordable, highly sensitive system for detecting changes in fluorescence on skin, which provided reproducible results, was sturdy enough to be mobile, and would allow exposure assessment by anatomical region. The design and operation of VITAE was discussed in detail by Fenske, Leffingwell, and Spear (1986). An Ultricon tube video camera, equipped with a 16-mm lens, was used as a light detector and was coupled to a microcomputer with analog-to-digital capabilities. Software was developed in conjunction with colleague Dr. William Gibb. All images were captured in a darkened room. Two arrays of four 122-cm, F-40-watt long-wave UV (black light) bulbs were placed parallel to the subject plane to thoroughly illuminate the field of view. UV filter glass was placed over the bulbs to prevent rogue visible light from striking the subject. A Kodak 2E Wratten Gelatin filter, placed over the camera lens, prevented reflected UV light from entering the camera (detector). The equipment configuration was identical to that later used by Archibald (1993), Archibald, Solomon, and Stephenson (1994a,b), Black (1993), Groth (1994), and Methner and Fenske (1994a,b, 1996), and very similar to that of Bierman, Brouwer, and van Hemmen (1995), Houghton, Solomon, and Harris (1996), and Houghton (1997). The equipment configuration used by Houghton (1997) is shown in Fig. 6-2. Light intensity and spectrum of irradiance were determined to ensure subject safety. All subjects wore UV shielding goggles and minimal, nonfluorescent clothing.

The system used by Fenske (1984), Fenske, Leffingwell, and Spear (1986), and Fenske et al. (1986) operated at a resolution of 128 × 128 pixels. The gray-value scale ranged from 0 to 15. A touch screen was mounted on the monitor. When touched by the operator during image mapping, the screen's conductive layers contacted each other, creating a position-specific voltage that subsequently was converted to a binary number for each axis. Subjects were placed behind an imaging frame, 90 cm from the camera. Images were captured under long-wave UV light before and after exposure. Preexposure images were required because skin possesses a natural fluorescence that must be accounted for. Several adjustment factors were applied to the image data. Lens distortion, or vignetting, caused by light passing less efficiently at the edges of the lens than in the center, was measured and the images were corrected appropriately. Anthropometric adjustments, using mathematical correction factors, were necessary to adjust for foreshortening and diminution because body surfaces are curved rather than flat. VITAE assessed a pixel as "exposed" if its gray value exceeded that of the corresponding pixel in the preexposure image by a magnitude set by the operator.

A standard curve of fluorescence intensity versus mass/pixel was generated by pipetting 50 μl of various concentrations of Calcofluor in acetone on specific-sized areas on skin, such that the mass applied ranged from 10 to 1,400 ng/cm^2. Tracer deposits were mapped (outlined) using the touch screen and the median gray value and the area were calculated. Algorithms were used to adjust the final calculations of exposure according to the aperture setting used. The standard curve had two different slopes. One linear region was discernable from gray value 0–11, whereas another was seen from 11 to 15. The standard curve plateaued at high concentrations because of a buildup of tracer on the skin, masking material underneath, resulting in quenching. This effect also was noted by Vo-Dinh (1987), Black (1993), Houghton, Solomon, and Harris (1996), and Houghton (1997). Fenske, Leffingwell, and Spear (1986) and Groth (1994) stressed the importance of selecting an appropriate dose of tracer for each study based on the anticipated exposure because unexpectedly high exposures (e.g., spills or splashes) are underestimated due to quenching.

In preliminary deposition studies the recovered ratio of pesticide to tracer was always higher because of insolubility of Calcofluor in water and difficulties in providing sufficient agitation in the spray tank (Fenske 1984; Fenske et al. 1986). The deposition ratio (pesticide/tracer) on targets could not be predicted using the ratio in the spray solution. It is critical to measure deposition ratios on the target in order to adjust tracer estimates and obtain a more accurate estimate of pesticide exposure.

Clothing penetration studies indicated a high correlation between Calcofluor and malathion penetrating 50/50 cotton/polyester work shirt material ($r = .98$), but the correlation was lower for 65/35 cotton/polyester coverall material ($r = .84$) (Fenske et al. 1986). Ratios of pesticide/tracer were higher on gauze placed under the fabric than on fabrics themselves, suggesting differential penetration which must be accounted for when estimating field exposures. Subsequent work confirmed this (Ramaswamy and Boyd 1993).

Environmental stability studies with Calcofluor and a stilbene FWA involved the application of known concentrations of tracer on gauze pads and their exposure to midday sunlight (Fenske 1984). Loss of the stilbene was high (30–40%) after 1 h and 40–50%

after 2 h, compared to spiked pads placed in the dark. These values concur with those reported by Anders (1975), who discussed the *trans-cis* isomerization of the stilbene central double bond which results in excitation at shorter wavelengths. Calcofluor was more stable, sustaining 10–12% loss after 1 h, and 20–25% after 2 h. Spiked pads placed in sunlight during a study could be used to adjust for tracer photodegradation.

Fenske, Leffingwell, and Spear (1985) reported results of a study examining diazinon exposure among workers following airblast applications in pear orchards. Air samples, dermal patches, and VITAE were used. The ratio of pesticide to Calcofluor used was 3.8 to 1. All subjects wore gloves, boots, respirators, short-sleeved cotton T-shirts, long-sleeved cotton coveralls, and either baseball caps, hard hats, or helmets. Images of a standard target were taken before and after each image session to monitor changes in system stability (e.g., intensity of light). As previously discussed, appropriate adjustments were made to each image. A standard curve developed previously in the laboratory was used to quantify dermal deposition of tracer. Spiked pads placed in sunlight were used to correct for photodegradation of tracer. Deposition ratios, obtained from gauze pads placed on the tractor, were used to convert the image data to quantity of pesticide on skin. Four of six applicators had exposure under the coveralls that was visible but too low to quantify. One applicator, whose chest-pad exposure was high, was significantly exposed under the coveralls ($>1\mu g/cm^2$). A ring of fluorescence was noted on skin between glove cuffs and coverall sleeves, around face shields, down the back of the neck, and on the hands. For two subjects who experienced full-body exposure, protected areas accounted for 42% and 71% of the total exposure, indicating that pesticides can penetrate protective clothing. Dermal-patch estimates of exposure were 10–60 times greater than VITAE estimates. Exposure was nonuniform because of differences in protective equipment, work practices, and conditions.

These results prompted a review comparing the dermal patch technique and VITAE in studies of airblast applications in orchards and backpack applications in greenhouses (Fenske 1987, 1990). The comparison indicated that exposure estimates varied substantially depending on patch location chosen for extrapolation. The tracer technique indicated, visually, the extent of nonuniform dermal deposition of tracer and hence chemical. Splashes present on mixers, which would have been undetected using the patch technique, were noted. Substantial exposure occurred beneath clothing, which would have been missed by the interior patch. In situations where comparisons were possible, patch estimates were considerably higher than tracer estimates. For example, the tracer estimate for exposure to the head of mixers in the malathion orchard study was 35% of the patch estimate. The differences in exposure between subjects, using the two assessment techniques, were larger for applicators. For example, the tracer estimate was <2% of the patch value for the head. Fenske (1987) concluded that there is no standard patch location protocol that can overcome the nonuniformity of pesticide exposure. Exposure to the head was high in some cases and deposition may have exceeded the quantitative range of the VITAE system; therefore, exposure may have been underestimated. Calcofluor does not fluoresce on hair unless very high exposures occur and video imaging does not assess the top of the head. This may explain part of the underestimated value. Fenske (1987) commented that an accurate measure of dermal exposure lies between estimates provided by the two assessment techniques. Since deposition patterns were highly variable and activity and application dependent, Fenske (1990) stated that "nonuniformity of dermal exposure is the source of substantial uncertainty in exposure and risk assessments of agricultural workers."

VITAE was used to obtain a quantitative estimate of dermal exposure to chlorophenols among timber mill workers (Fenske, Horstman, and Bentley 1987). Wipe samples from treated lumber and residues on cotton-glove monitors were used to estimate masses of tetrachlorophenol (TCP) represented by Calcofluor on skin. Six of nine workers received measurable hand exposure, despite meticulous hygiene practices and use of chemical-resistant gloves. Most of this exposure (86%) occurred on the palms. This may have been partially due to glove damage, because careful examination of one subject's gloves revealed a razor-thin slice across the palm. This source of exposure would have gone undetected if not for the use of the tracer. Exposure to forearms accounted for 85% of the total dermal exposure. Due to abrasions and possible impairment of dermal barrier properties, percutaneous absorption of 100% of the compounds, and not 30% as estimated in studies using cadaver skin, may explain the 70% difference between VITAE-estimated doses and that predicted by excretion of urinary metabolites (total of 1,840 μg of TCP/day). If 100% absorption is assumed, the VITAE-estimated dermal dose was 1,420 μg of TCP/day. Together with inhalation exposure, this represents a total dose of 1,514 μg of TCP/day.

Vo-Dinh (1987) discussed development of two prototypes of the Luminoscope II which were improved versions of the lightpipe luminoscope. An electronic system for subtracting background fluorescence of skin was incorporated. Both lab and handheld field prototypes were designed with the stethoscope-like attachment that was placed directly on skin to obtain readings. An initial filter-paper study indicated that quantification of PAHs was possible only in the linear range of the standard curve of net fluorescence versus concentration (0.2–25 ng/cm^2). Fluorescence intensity decreased for masses greater than 600 ng/cm^2, because of quenching. On skin of hairless mice, quantification was possible up to 2 μg/cm^2. The mice were kept in the dark to eliminate the contribution of photodegradation. Fluorescence intensity of the PAHs on skin decreased over time, providing an indication of the compound's absorption through skin.

Fenske (1988a,b) used VITAE as a quantitative method to evaluate the effectiveness of protective equipment among twenty-five workers applying malathion and Calcofluor (5.7:1 ratio) during high-volume airblast applications in citrus groves. Workers wore baseball caps, neoprene gloves, respirators, cotton T-shirts, work boots, and 65/35 cotton/polyester work pants. Over this clothing, workers wore either Tyvek nonwoven coveralls, cotton/polyester (65/35) coveralls or work shirts made of cotton/polyester (50/50). Regardless of the type of clothing worn, little exposure occurred below the waist of mixers and applicators. All applicators incurred exposure on the face, the neck, and, despite the use of gloves, the hands. Hand exposure was highly variable, ranging from 52 to 263 μg (mean = 138.5 μg) for mixers and 4.9 to 221 μg (mean = 58 μg) for applicators. Observations indicated that hand exposure occurred because of either frequent glove removal, resulting in contact with pesticide concentrate or spray solution on the gloves, or glove failure. Hand exposure represented 41% and 13% of the total exposure for mixers and applicators, respectively.

Tracer penetrated the clothing and was evident on subjects' arms and torsos. Protected areas received 36% and 70% of the total exposure for subjects wearing Tyvek coveralls and work shirts, respectively. Mixers showed little difference in exposure regardless of coverall type, but applicator exposure to protected areas was 94 μg for the Tyvek coverall group, 205 μg (reported as 215.3 μg in Fenske [1988a]) for the cotton coverall group, and 432 μg for subjects wearing work shirts. The means of these groups were not significantly different because of high variability introduced by windy study conditions. To reduce variability within clothing type, all data for protected areas were normalized using unprotected head values. There were no significant differences between the two mixer groups; however, among applicators, those wearing either coverall type had significantly lower exposure than those wearing work shirts. Tyvek coveralls provided 80% exposure reduction relative to work shirts, whereas cotton coveralls reduced relative exposure by 50%.

Tracer patterns on protected areas suggested that exposure was not due primarily to fabric penetration but the result of material entering through openings at the collar and wrists. Of protected areas, forearms sustained the greatest exposure. Fenske (1988a,b) stated that this bellows effect, of material being drawn inside the clothing, had been observed during studies of air movement within protective garments. Among mixers, exposure to the forearms contributed 69% and 51% of the total exposure (after head exposure adjustment) for cotton coveralls and Tyvek coveralls, respectively. Among applicators, forearm exposure accounted for 48%, 43%, and 42% of the total exposure of work shirt, cotton coverall, and Tyvek coverall groups, respectively. Applicators wearing Tyvek coveralls sustained the lowest exposure of both groups (66.3 μg compared to 286.5 μg for the work-shirt group and 156.6 μg for the cotton-coverall group). Their torsos received the greatest portion of the total exposure (56%). Mixers wearing Tyvek coveralls also were better protected, with total exposure measuring 144 μg compared to 203.6 μg for those wearing cotton coveralls. Fenske (1988a,b) stated that the fluorescent tracer technique is an excellent tool for evaluating exposure beneath protective clothing, whether through garment openings or penetration through fabric. In the latter instance, however, exposure data must be corrected using a fabric penetration ratio for the pesticide and tracer.

In many study scenarios, researchers need simple, rapid, qualitative methods to evaluate the relative performance of different types of protective clothing, or work practices. Fenske (1988c) developed a qualitative visual scoring system for exposure estimation using VITAE. Calcofluor was used as a surrogate for pesticides applied using airblast sprayers in orange groves. The scoring system was based on a matrix, using exposure intensity as the abscissa and exposed area as the ordinate. Exposure intensity was rated on a scale from 1 (low exposure) to 5 (high exposure). The area exposed, as a percentage, was categorized into five classes each representing twenty percentage points on a scale from 0 to 100%. Each class was assigned a value from 1 to 5. The product of the two ranks produced the final score, which could range from 1 to 25. A total score for a body part was derived by adding the scores of the images. Volunteers who ranked images were consistent in their rankings. Scores were highly correlated with quantitative VITAE estimates. Spearman rank correlation coefficients (r) ranged from 0.83 to 0.97 for different body regions and volunteers. Hand exposures ranged from 0 to 329 μg and head exposures from 39 to 281 μg. Exposure to palms was 2.5-fold greater than exposure to the backs of hands. Mean exposure of mixers was more than four times greater than mean applicator exposure. The scoring system consistently showed these same differences.

Dermal exposure of applicators applying malathion–Calcofluor mixtures using airblast sprayers in citrus groves was measured using VITAE. Dermal exposure estimates were highly correlated ($r = .91$) with urinary metabolite excretion (Fenske 1988d). Mixer exposures were not as highly correlated ($r = .73$). Applicator exposures were more than three times greater than mixer exposures, indicating the tremendous exposure potential when using this type of sprayer. Exposure beneath protective clothing was more than 75% of the total exposure in both groups.

A qualitative determination of skin deposition patterns among pest control operators (PCOs) applying chlorpyrifos (Dursban TC termiticide) was obtained by using VITAE and Calcofluor at a ratio of 1:24 (tracer:chlorpyrifos) (Fenske and Elkner 1990). Quantitative measurements were obtained using dermal patches, personal air sampling, and biological monitoring for urinary metabolites. Tracer deposition revealed substantial hand exposure. Only one of seven workers wore gloves and this subject did not incur this exposure. Workers who rolled up their sleeves had extensive deposition on their forearms. Deposition of tracer on head and neck areas was evident in three workers, and one individual who wore only a T-shirt had exposure to the chest. Deposition was evident on work pants but not on skin underneath. This was possibly due to low water solubility, low concentration, and low fabric penetrability of the tracer, because chlorpyrifos was detected on dermal pads next to the skin. Fluorescent tracer observations concurred with dermal patch values, indicating that forearms and legs received the greatest portion of the total exposure (34% and 51%, respectively). Hand exposure was surprisingly only 7% of the total exposure, whereas exposure to the neck and head represented 7.5%. More than 50% of the total dermal exposure was found beneath the clothing. The estimated total absorbed dose, using respiratory and dermal exposure values and absorption rates, averaged 9.5 μg/kg daily with the highest exposure being 16.4 μg/kg. The strongest correlation found ($r^2 = .86$) was between 24 and 48 h excretion and total absorbed dose, with dermal dose being the major contributor to this association. Mean total dermal exposure was 5.94 mg/h. The small contribution of hand, head, and neck exposure to the total was attributed to the inclusion of exposure values from beneath clothing, skewing percentages compared to previous studies that did not measure exposure beneath garments. The highest exposure incurred by subjects in this study was half the NOEL for chlorpyrifos of 0.03 mg/kg per day based on changes in plasma cholinesterase (FAO/WHO 1973).

Karr et al. (1992) measured exposure of forty-eight applicators applying organophosphate insecticides to apple orchards by monitoring plasma and red blood cell (RBC) cholinesterase activity. Dermal exposure of six applicators was assessed by mixing sodium fluorescein dye with the spray solution and observing subjects under black light. Fluorescent areas on skin were noted, swabbed with alcohol, and the swabs subjected to fluorometric analysis. This study did not attempt to quantify exposure using a video imaging technique but used the tracer as a qualitative tool to identify exposed areas and used analytical fluorometric dye assessments to quantify exposure. Hands and heads incurred the greatest exposure. Log transformation of measured changes in RBC cholinesterase were inversely proportional to dermal exposure ($r = -.95$). There was no correlation between plasma cholinesterase change and dermal exposure.

6-6-3 VITAE2-Second Generation

Archibald, Solomon, and Stephenson (1994a) published details of the second-generation VITAE method (VITAE2). The original VITAE software was revised by Richard Fenske (Fenske, Birnbaum, and Cho 1990), and tested further by Archibald (1993). The program was essentially a collection of subroutines written in Microsoft C and utilized software written specifically for Data Translation video imaging boards. The computer was upgraded to a 386, 40-MHz system, with a Data Translation (DT 2851) frame grabber video imaging board that provided 512×480 pixel resolution with 0–255 gray levels. A monochrome Hitachi KP1161/2 CCD camera equipped with a zoom lens and Kodak 2E Wratten Gelatin filter was employed. Equipment configuration was the same as that of Fenske, Leffingwell, and Spear (1986). Pre- and postexposure images were recorded and, instead of using a touch screen to outline exposed areas, a mouse was used for this purpose. Corrections for lens distortion, background camera noise, light stability, and curvature of body parts were made as previously discussed. Correction for background skin fluorescence, however, was modified from the original VITAE software. Image histograms of the number of pixels exhibiting each of the 256 gray values were incorporated. Subtraction of preexposure histograms from postexposure histograms, and not a pixel by pixel subtraction of images as was later used by Houghton (1997), was used to determine the areas and extent of exposure. Since it was observed that dermal fluorescence increased with lighter pigmentation, a series of standard curves measuring changes in fluorescence intensity with increasing deposition of Calcofluor for various groupings of preexposure skin tones was developed (Archibald 1993; Archibald, Solomon, and Stephenson 1994a). These nine standard curves were highly linear ($r^2 = .88-.96$). Two calibration curves, one plotting the natural log of the slopes of the standard curves versus the natural log of the preexposure gray levels, and the other plotting the transformed y-intercepts of the standard curves versus the transformed preexposure gray levels, were used to produce regressions for subjects with any given preexposure dermal gray tone. The accuracy of the curves was tested by pipetting six known concentrations of tracer on the skin of volunteers. The solutions were applied on the *same-sized areas* as were used to generate the standard curve. VITAE estimates of tracer deposition were highly correlated with mass of tracer applied ($r^2 = .98$). This technique greatly improved the accuracy of the VITAE method. Stability tests on Calcofluor indicated that constant amounts were dislodged from foliage over a 48-h period, indicating its stability in the absence of UV, making it an excellent dye for use in greenhouse studies.

Exposure of eight greenhouse chrysanthemum workers to pirimicarb was measured using air monitoring, video imaging, and excretion of urinary metabolites (Archibald 1993; Archibald, Solomon, and Stephenson 1994b). Calcofluor (0.5 g) was mixed with a pirimicarb 50 WP formulation (1 g) and Assist Oil, prior to forming an aqueous solution. The addition of surfactant helped to suspend the tracer in the spray solution. Workers were imaged prior to, and following, a 4-h period of debudding chrysanthemums. Tracer deposition on skin increased linearly over the 4-h period. Rates of tracer and pirimicarb degradation and dislodgeability, measured during the study by collecting and analyzing plant tissue and wipes of treated foliage, were stable over time. These values were used to adjust tracer exposure estimates to predict pirimicarb exposure. Significant variation in tracer deposition over different anatomical locations occurred. This was attributed to personal habits, perspiration, and debudding techniques. Hands and arms incurred the greatest exposure, 42% and 20% of the total, respectively. Mean rate of tracer deposition per hour on hands (86.2 μg/hr) was constant over the work period ($r^2 = .98$); however, variation between workers was large. Air monitoring revealed no detectable pirimicarb residues during the work period, indicating the reentry times of 36–48 h are sufficient. The study was unable to correlate urinary metabolite excretion and VI-TAE2 exposure estimates because of failure to detect metabolites in the urine. This may have been due to low sensitivity of the analytical equipment, removal of pirimicarb residues from hands prior to absorption, or use of unrealistic absorption values in the calculation. In-vivo dermal absorption values for hands were not available; however, hands may absorb less than the in-vitro estimate of 5% due to thickness of the stratum corneum.

Exposure of five greenhouse applicators applying either pirimicarb or deltamethrin, using high- or low-volume applications, was measured using air monitoring, dermal patches, VITAE2, and excretion of urinary metabolites (Archibald 1993; Archibald, Solomon, and Stephenson 1995). The final concentration of Calcofluor in the spray solution was 0.5 g/l. VITAE2 procedures for collecting and adjusting images were performed as reported by Archibald, Solomon, and Stephenson (1994a). Dermal patches (10cm $\times$ 10cm) were placed on the skin and the clothing exterior. Tracer deposition on skin was significantly higher (2,224 μg/h) when using low-volume application methods and normal operating procedures (NOPs) than when using high-volume methods (164 μg/h). This was attributed to the fact that NOPs used limited protective clothing. Of the total tracer deposition, 38% was on hands when high volume and NOP were used compared to 31% for low-volume applications and NOPs. Protective operating procedures (POPs), where full protective gear was worn, significantly reduced tracer deposition during low-volume applications. Analysis of dermal patches on clothing indicated that the deposition ratio of tracer:pirimicarb was the same as the spray tank ratio (1:1); however, inner-patch deposition was highly variable for both compounds. Low-volume applications using POPs resulted in tracer and pesticide residues on inner patches that were below the limit of quantification. Regression analysis indicated that total pirimicarb metabolite excretion and VITAE2 estimates of dermal exposure were highly correlated ($r^2 = .93$). The authors concluded that VITAE2 is a quantitative and powerful educational tool and that, because dermal deposition of pesticides was shown to be nonuniform, the use of the patch technique has limitations.

Children's exposure to chlorpyrifos-treated turf was studied using respiratory monitoring and dermal patches and VITAE2 to quantify dermal exposure (Black 1993). Equipment configuration and image adjustments remained the same as those of Archibald, Solomon, and Stephenson (1994a,b, 1995). Black (1993) and Archibald (1993) were working with the new system concurrently. Standard curves for five different skin-tone groupings were developed. Summary calibration curves to quantify tracer deposition on skin were generated as done by Archibald (1993) and Archibald, Solomon, and

Stephenson (1994a). Dermal exposure to Uvitex OB mixed with a chlorpyrifos formulation blank to improve tracer solubility was evaluated by allowing children to contact treated turf during 30 min of unstructured play following a reentry time of 2.5 h. Adults were used for comparison and were asked to perform an 11-min choreographed exercise routine on duplicate turf plots.

Adults were more highly exposed than children, probably because of the type of movements in the exercise routine compared to sedentary movements of the children playing with toys on the turf. Hands were the most highly exposed area for both adults and children. Exposure to remaining body parts varied. Variability among adults was reflected by how energetically they exercised. Deposition of tracer and pesticide was determined using gauze deposition pads. Dislodgeability of chemicals was determined using two methods: surface wipes and removal of all turf and thatch in a 10 cm × 10 cm area demarked by a metal template. Correlations for tracer and pesticide deposited on turf and dislodged using the two methods were high ($r = .79$–0.93). Uvitex OB was more available from turf than chlorpyrifos.

Mean tracer deposition on children's hands, as determined by video imaging, was 64 ng/cm^2 compared with 189 ng/cm^2 for adults. Exposure to the face, neck, and lower legs of children ranged from 30 to 35 ng/cm^2, whereas exposure to upper legs and arms ranged from 11 to 20 ng/cm^2. Exposure to these areas on adults ranged from 42 to 58 ng/cm^2. For calculation of chlorpyrifos exposure, the adult data were used as a worst-case scenario and were scaled by multiplying the exposure by the children's body surface area before calculating tracer exposure for the entire body. The children's data were used to represent a typical case. In both cases, dermal exposure was estimated to represent 88–92% of the total exposure.

Estimates of internal chlorpyrifos dose were calculated using dermal and oral estimates based on the pesticide/tracer ratio from the wipe samples, turf residue samples, and the measured respiratory exposure among adults. When calculating internal dose, Black (1993) assumed the following: a 4-h contact with the lawn (2 h of typical exposure and 2 h of worst case), 25% of the hand exposure was ingested per hour, absorption of particulates and vapor was 100%, and 3% dermal and 70% oral absorption of chlorpyrifos occurred. For a typical case, internal doses estimated from the surface-wipe and turf residue samples averaged 30 μg/kg and 46 μg/kg, respectively. Oral exposure was the greatest contributor to internal dose, representing 46–52% of the total, whereas dermal exposure accounted for 25–28% because of low absorption. Because oral exposure is a function of hand exposure relative to pica behavior, Black (1993) stated that dermal exposure should be considered a major contributor to internal dose. For the worst-case situation, internal doses of 93 μg/kg and 149 μg/kg were calculated using the surface-wipe and turf residue samples to obtain tracer/pesticide ratios. Again, oral exposure was the dominant route, contributing 50–55% of the internal dose whereas dermal exposure contributed 30–32%. Dermal exposure exceeded respiratory and oral exposures, but when absorption was considered, oral dose exceeded both dermal and respiratory doses. Internal doses ranged from one to five times the NOEL for chlorpyrifos (0.03 mg/kg per day).

Breon (1993) reported difficulties obtaining homogenous solutions of Uvitex OB in acetone–water solutions while testing VITAE2 for accuracy. Mixing tracer with

Chemsurf surfactant provided a homogenous solution. Spray deposition studies of aqueous solutions of Uvitex OB, Chemsurf, and atrazine on filter paper indicated that deposition of tracer and insecticide were highly correlated ($r = .91$). Regression analysis of these data provided a regression equation that could be used to predict atrazine deposition from measured Uvitex OB masses. Breon commented extensively on the painstaking work required to obtain a suitable standard curve for quantifying Uvitex OB on filter paper using VITAE2. Houghton (1997) later encountered similar problems when working with Uvitex OB on skin. Breon reported that the standard curve of the natural logarithm (ln) of the mass/pixel versus the ln of the subtracted median gray level was curvilinear. This also was noted by Houghton (1997). Since only one gray value, that of filter paper, was used by Breon (1993), summary calibration curves were not generated. Multiple linear regression analysis produced a polynomial equation that most accurately represented the relationship ($r = .98$). The VITAE calculation program, however, was not designed to accept anything but linear curves. As a result, a linear relationship that most closely approximated the curve ($r = .91$) was found and used to quantify Uvitex OB.

A study examining exposure of greenhouse handgunners applying Calcofluor–natural oil as a pesticide surrogate indicated that high-speed, unidirectional ventilation (>5 m/s) reduced exposure to experienced applicators but increased exposure to inexperienced applicators (Methner and Fenske 1994a). Among experienced applicators, exposure was reduced 63% and 83% for upper arms and 74% and 94% for upper legs as measured by VITAE2 and patches, respectively. Inexperienced applicators incurred a ten- to thirtyfold increase in exposure compared to experienced applicators. This was most likely because trained subjects positioned themselves upwind of the spray aerosol. Without ventilation, there were no differences between the two subject groups. All image adjustments were conducted according to Archibald (1993), Archibald, Solomon, and Stephenson (1994a), and Black (1993); however, details on standard-curve generation were not reported. It is unknown whether curves for one or several skin tones were used. The authors commented that the greatest disadvantage of the imaging technique was the potential for excessive exposure to cause quenching, resulting in inaccurate exposure estimates.

Calcofluor and Uvitex OB, in combination with vegetable oil, were used as surrogates in two high-pressure handspray greenhouse applications to assess protective clothing performance (Methner and Fenske 1994b). The VITAE2 system was employed and results indicated that Tyvek, Saranex 23P Tyvek, Comfort-gard II, and Kleenguard coveralls all exhibited chemical breakthrough within 1 h after application began. Tyvek and Kleenguard fabrics, which do not possess additional chemical protective barriers, sustained tracer penetration 5–15 min from the start of spraying. Patch values proved misleading. Video imaging indicated that skin deposition was nonuniform and deposition rates for tracer beneath garments ranged from 2 to 939 μg/h. Since garments had tight-fitting closures, skin deposition was assumed to result from penetration of fabric. The extent of tracer and pesticide penetration may not be identical; however, using a surrogate tracer still provides valuable data regarding exposure. A third study using Uvitex OB as a surrogate to assess the effect of spray pressure on exposure during greenhouse handgunning applications used only dermal patches. Patch extracts were analyzed for tracer using fluorometry. No video imaging results were reported. Patches indicated that the upper body received the greatest exposure because of spraying technique and that dermal exposure was greatest with multidirectional ventilation > no ventilation > unidirectional ventilation.

Bierman, Brouwer, and van Hemmen (1995) used VITAE2 software to capture images but used a Scil image analysis software (TNO Nutrition and Food Research Institute) to analyze images of spray applicators and harvest workers following application of propoxur and Tinopal CBS-X to carnations in greenhouses. The Scil software contained a live alignment function similar to that later used by Houghton, Solomon, and Harris (1996) and Houghton (1997) and employed the use of lookup tables to enhance images with low fluorescent intensities (Brouwer 1996). The equipment configuration used was similar to that of Fenske, Leffingwell, and Spear (1986a) with the exception that light-bank design was modified. Two banks of two TL D-light bulbs (Philips), 122 cm long, were mounted with 50 cm between them and the central portions were covered. Reflector arms were added to the top and bottom of the light boxes. Light intensity within the imaging frame was consistent (coefficient of variation <5%). A Philips LDH0703/AS camera equipped with a zoom lens, Kodak 2E Wratten Gelatin filter, and infrared filter was integrated with an AT 386 computer and DT 2853 frame grabber and was used to capture images. A white reference was included in every image to correct for changes in light intensity. A Silicon Graphics 4D/35 computer (UNIX operating system) was used to analyze images. Background noise was subtracted, dark current was corrected, and dividing images by the standard white reference corrected for nonuniform illumination. No subtractions of histograms, or anthropometric adjustments, were made. Preliminary studies indicated that a rate of 62.5 g of Tinopal CBS-X/1000 m^2 was nonphytotoxic to carnations, and when dislodged from foliage onto skin, provided a fluorescent signal that fell within the linear range of the system. High-volume applications of Tinopal and propoxur were made using a spray boom or handgun. Subjects were imaged after a 1-h work period. Standard-curve data generated on different subjects and on different anatomical regions were pooled to create one standard curve. This curve was deemed acceptable for use for all bodily areas except the palms. Standards were applied by pipetting the concentrations onto object glasses and wiping them on the skin. Chemical analysis of the object glasses indicated that 95% of the tracer was applied dermally. Although no comments regarding the uniformity of the tracer deposits were made, it is assumed that this technique improved uniformity compared to pipetting the tracer directly on skin.

Tracer deposition was nonuniform on hands, forearms, and sometimes upper arms and lower legs. Head exposure was not routinely monitored. Among applicators, the hand not holding the spray gun was more heavily exposed compared to the opposite hand, possibly because of dragging the contaminated hose. Among harvesters, the flower-cutting hand had the greatest exposure. In many cases, tracer was visible on the skin but the intensity was too low for detection by the camera. Nine harvesters wore gloves and coveralls, whereas nine others wore no gloves or clothing on the forearms. The gloves and coveralls acted as surrogate skin and were analyzed for propoxur, Tinopal CBS-X, and methiocarb. The latter compound had previously been sprayed in the greenhouse. Spearman rank coefficients for propoxur, methiocarb, and Tinopal CBS-X (analyzed chemically) versus the amount of Tinopal CBS-X on the harvesters were 0.47, 0.45, and 0.52, respectively.

The low correlations may have been due to the fact that different workers were used for the two techniques and probably varied in their exposure. Deposition of Tinopal CBS-X and propoxur on applicators' clothing (analyzed chemically) and dislodgeability ratios from foliage to harvesters' clothing, were highly correlated (0.92 and 0.85, respectively). The higher correlation for deposition is a reflection of the solubility of Tinopal CBS-X in water, indicating its suitability as a surrogate for propoxur and the fact that applicators contact the spray solution itself. The correlation for methiocarb was lower (0.66), most likely because of changes in transfer rates from foliage to clothing since the pesticide was not sprayed at the same time as Tinopal CBS-X. Among applicators, $\pm 12\%$ of Tinopal CBS-X on clothing penetrated and reached the skin. Ratios of Tinopal CBS-X on harvester's clothes and skin indicated that the transfer rate for tracer to skin was much smaller than to clothing. The authors commented that dermal exposure estimates from both chemical analysis and imaging exhibited large variability. This has been noted for methods used in other exposure studies including hand washing, patches, and biological monitoring. Bierman, Brouwer, and van Hemmen (1995) stated that variability in the technique is insignificant compared to variability within and between workers. They suggested that researchers should reduce variability by using larger numbers of workers in studies and by standardizing working procedures.

Groth (1994) tested the performance of the VITAE2 system and the use of summary calibration curves for different skin tones using Uvitex OB. He reported that histogram subtraction set negative values to zero, creating a net histogram that was not a true integrated difference between post- and preexposure histograms. This method artificially inflated brightness, moving the standard curve for Uvitex OB to the right, underestimating exposure. Groth stated that the relationship between net brightness and mass/pixel was better described by a nonlinear function. Large differences were found between actual and predicted tracer masses and variability was large. This also may have been caused by differences in fluorescent properties of Uvitex OB versus Calcofluor RWP. After simulating workplace exposures, Groth concluded that VITAE2 was "not robust with respect to variations in tracer surface distributions."

6-6-4 Newer Advances in Imaging Technology

Fluorescence Interactive Video Exposure System (FIVES). Roff (1994, 1996a) reported a new method of collecting fluorescent images that uses less complex algorithms for anthropometric image adjustments. This method later was referred to as FIVES (Roff 1995, 1996b,c). The VITAE system used anthropometric correction software that fitted geometric shapes to limbs. The calculated curvature was used to correct area and fluorescence intensity prior to quantifying exposure using the calibration curves. To the best of the authors' knowledge, the effectiveness of the VITAE anthropometric correction was never published. The FIVES method is simpler and involves taking images with two different lighting systems. Ideally, spherical lighting allows a subject to be illuminated equally at all points regardless of the direction it faces. This removes effects of the inverse square law and Lambert's cosine law (see Section 6-2-3, Image Acquisition and

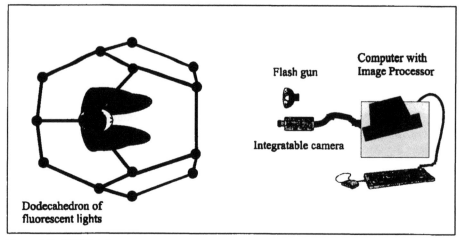

Figure 6-3 FIVES system.
Source: Adapted from Roff (1994).

Processing), making corrections for surface illumination unnecessary. Spherical shell lighting that does not interfere with image capture would be difficult to construct. Instead, Roff (1994) constructed a dodecahedron skeleton using twenty-four 0.61-m near-UV bulbs. The bulbs were covered with UV-A transmission filters that removed any light in the visible region. A Sony XC-77/CE monochrome integrating camera, fitted with a Kodak Wratten 2E filter over the lens, was placed 1 m outside the dodecahedron (Fig. 6-3). A near-infrared dichroic filter mounted behind the lens reduced internal reflections. Originally, an 800-J UV photographic flashgun fitted with filter glass to remove visible light acted as a point light source, and was placed adjacent to the camera. The lighting systems were controlled by a relay switching box (Roff 1994). This was later replaced by a 3,000-J flashgun. The Seecan Master image processor manipulated the 256 × 512 pixel image array and recorded two images in less than 0.15 s using the two lighting systems—one illuminated by the flashgun, the other by the dodecahedron. They were saved as an image pair. Color enhancement of the live image, prior to image capture, indicated under- or overexposure, which allowed the f-stop and number of integrations to be appropriately adjusted.

The rear of the dodecahedron had no lights, which prevented them from being captured in the images and causing bloom. As a result, the system can only correct foreshortening for angles up to 70 deg from perpendicular. Tests using two fluorescent standards, one perpendicular and one at an angle to the camera, confirmed that consistent illumination occurs with the dodecahedron over a wide range of angles and locations (variations <10%). The ratio of the light intensities from the point light source and dodecahedron provides an area correction factor that corrects for foreshortening. Foreshortening is a problem with the camera, not the lighting. A correction factor of a^2/b^2, where a and b are the distances of the subject and a fluorescent standard from the point light source, respectively, was used to correct for diminution. A standard gray card was placed in every image to obtain the correction factors and adjust data. No subtractions were done in Roff's original work (Roff 1994, 1996c). A larger dynamic range is obtained by using integration and different f-stops. Dodecahedral lighting helped to correct for anthropometric problems.

Roff (1994) used Tinopal CBS-X as the water-soluble tracer and found that a fifth-order linear regression function was required to fit the standard curve of tracer concentration in $\mu g/cm^2$ versus video response, or brightness (least significant bits, or l.s.b's) on skin. A custom software routine applied rejection criteria, the correction algorithm, and calibration curve to mapped image areas.

A study was conducted examining the exposure of subjects mixing and applying permethrin and Tinopal CBS-X in an enclosed booth. The two subjects wore hooded polypropylene coveralls, respirators, goggles, and neoprene gloves. Net exposure was 3.6 mg of tracer for the subject who liberally sprayed with a 1.0 g/l solution of Tinopal CBS-X, but nothing for the subject who carefully sprayed 0.5 g/l. Preexposure images were recorded to measure background skin fluorescence. Exposure was calculated by subtracting the preexposure values from the total of the postexposure values for the entire body. There was no pixel-by-pixel subtraction, or subtraction of histograms, with this method. A small degree of error was caused by poor performance of the algorithm on unexposed skin when its brightness did not provide a strong enough signal for accurate adjustment (darker skin tones). This situation was corrected by using a more powerful flashgun. Dermal exposure estimates, adjusted for skin absorption, metabolism, and other routes of exposure, correlated well with levels of permethrin metabolites in the urine. Since Tinopal CBS-X is water soluble, the deposition ratio of tracer:permethrin on the subject was assumed to be identical to that in the spray mixture (1:1); therefore no adjustment was made. In a later version of the software, two rejection criteria were applied to each pixel to remove unexposed pixels from the fluorescence and exposed surface-area measurements.

A second study (Roff 1994) was conducted by mounting the equipment in a truck and traveling to local farms. Exposure of workers wearing their usual work apparel while engaged in sheep dipping, using diazinon, was studied using Uvitex MST liquid. This tracer was shown to penetrate fabrics at the same concentration at which it was applied. Dermal assessments indicated exposure as high as 700 μg of diazinon. Hands were the most highly exposed region. These estimates underestimated actual exposure because mud and fecal matter in the baths obscured dermal fluorescence and because the system's dynamic range was not sufficient to cover the pesticide to water dilution ratio of 1:2,000. Excretion of urinary metabolites did not correlate with dermal estimates. Roff (1994) commented this may have been due to ingestion of splashed droplets or exposure from alternate routes, pesticide concentrate, or other pesticides used recently. Although the study demonstrated the technique's limitations, it also indicated its benefits by demonstrating the relative differences in exposure incurred by wearing different types of clothing (e.g., woolen sweaters vs. cotton coveralls). The technique is more accurate in situations in which dye remains where deposited on skin but may be less effective at measuring exposure from splashes where spray solution runs off skin, because the relative adsorption of tracer versus pesticide may not be known. Roff (1994) noted that background fluorescence of skin must be accounted for when quantifying tracer and stated that, although the technique's accuracy is limited, nonuniformity of exposure in the images indicated that the traditional patch technique would have been no more accurate.

Roff (1995, 1996a,b) described the generation of standard curves of tracer brightness (l.s.b's) versus concentration for FIVES and their correction for background skin fluorescence. No linear relationship was seen between mean pixel brightness and tracer concentration on skin, even at very low concentrations. Third- and fourth-order polynomials provided the best fit ($r > .995$) and linear polynomials and log-log polynomials gave acceptable fits. Tinopal SWN, which is soluble in oils, was mixed with a permethrin wood treatment formulation at concentrations of 10, 100, and 1,000 μg/ml. Tinopal CBS-X, which is water soluble, was mixed with the insecticide formulation at 10, 25, 100, and 250 μg/ml. Pre- and postexposure images were taken while the subject was standing adjacent to a fluorescent reference. Standard curves were generated by pipeting known amounts of the mixtures on forearms and palms of subjects. It was noted that palms naturally fluoresce much more intensely than other body regions. Higher concentrations were produced by pipeting on top of the previous application. Camera integration times were reduced for postexposure images because of increased brightness. A third image captured the subject holding reference standards to check the consistency of light intensity. Mean intensity of the reference was corrected to the preexposure reference by applying a multiplication factor, such that all mean reference intensities were identical. This accounted for changes in light intensity and images being recorded at different integrations.

Standard curves were corrected for preexposure brightness of the skin. One curve for each tracer was chosen as a calibration curve. Slopes of the other curves were aligned to the calibration curve using mathematical correction factors that approximated the ratios of the natural fluorescence of the skin in the anatomical areas where the curves were generated. Palms, however, required an additional correction factor of 0.6. Calibration curves were applied to each pixel separately and allowed direct conversions of pixel intensity into equivalent surface concentration of tracer. Mass was calculated by multiplying by the pixel area. The error was ±70% of the applied dose. At low exposure levels (< 100 ng/cm^2), the correction method was less accurate. Although this method of quantifying tracer on skin is not as accurate as the methods developed by Black (1993), Archibald (1993), Archibald, Solomon, and Stephenson (1994a), Houghton, Solomon, and Harris (1996), and Houghton (1997), Roff (1996a,b) commented that it is simple and the linear polynomials can be derived from unevenly coated surfaces. Given that fluorescence intensities of the standards on skin were not constant (coefficient of variation 20%), this represents a significant step forward. Both tracers degraded by 20% during the first 30 min of exposure to sunlight and then stabilized for an additional 30 min.

The calibration curves generated by Roff (1996b) were used to quantify tracer on skin in a study assessing exposure of nonoccupational users of wood preservative fluids (Roff 1996c). Two formulations, one spirit based and one water based, which contained 0.2% permethrin (w/w) were mixed with Tinopal tracer and painted on three-sided wooden trellis fences. Hands incurred the greatest exposure (21 mg out of a total of 28 mg). Subjects sustained twenty times more exposure when wearing shorts, T-shirts, and shoes than when wearing permeable gloves, trousers, shoes, and long-sleeved shirts. Variability between subjects was attributed to their behavior and painting technique. For every 1°C increase in temperature, exposure was reduced by a factor of 0.83. Examination of study variable interactions suggested that this was not a result of solution viscosity or evaporation but was caused by subject behavior. Subjects painted more rapidly in cooler temperatures, possibly increasing the extent of splattering from brush flicks. Subjects applying the spirit-based formulation sustained 3.7 times the exposure of those who applied the water-soluble formulation. Brush splattering occurred more readily with the spirit-based formulation. Surprisingly, the volume of formulation used was not correlated with exposure. Under worst-case conditions (1-h vs. 30-min exposure, minimal clothing, and spirit formulation), mean contamination was 2.3 ml at 20°C. At lower temperatures, the predicted exposure would exceed the US EPA assumption of up to 6 ml of fluid on unprotected skin (e.g., at 15°C, the predicted exposure was 14 ml).

The latest version of Roff's software (as this review is being written), uses a local background subtraction as measured in the preexposure images as opposed to a pixel-by-pixel subtraction (Roff, Martin 1997. Letter to author 10 October). This advance improves the accuracy of FIVES at lower exposures in the same way as the method of Archibald, Solomon, and Stephenson (1994a) does by changing the coefficients of the calibration curve.

Use of modified Northern Exposure software. Houghton, Solomon, and Harris (1996) and Houghton (1997) studied dislodgeability of chlorpyrifos and bendiocarb from household surfaces following broadcast applications using Uvitex OB. They began their study using VITAE2 but despite the generation of extremely linear standard curves for different skin tones ($r^2 = .95-.99$) and linear summary calibration curves (r^2 of .84 and .97 for the slope and y-intercept curves, respectively), these curves *did not accurately predict the tracer mass pipetted on skin when the area of the test deposits was different from that used to generate the standard curves*. Fitting third-order polynomial regressions to the curves improved the fit, but the same quantification difficulties were encountered. The difference between the accuracy test performed on curves in this study and that performed by Archibald, Solomon, and Stephenson (1994a), using Calcofluor RWP, was that spatial areas were different from those used to generate the standard curves. Under these conditions, VITAE2 was unable to accurately measure the mass of tracer deposited on skin. After Houghton, Solomon, and Harris (1996) reported their work, they discovered that this problem had been described by Groth (1994) in his M.Sc. thesis. It is most unfortunate that Groth's work was not published.

Houghton, Solomon, and Harris (1996) decided independently to modify a commercially available fluorescence microscopy software called Northern Exposure (Ver. 2.3b, Empix Imaging Inc., Mississauga, Ontario, Canada) to quantify dermal exposure to fluorescent compounds. This Windows-based software is very user friendly. Studies indicated that the modified software provided very accurate mean estimates of tracer mass present on skin when using average gray values to measure brightness, or when using total gray values, provided a grid was placed over the image to divide the area into sections equivalent to the area of the tracer deposits in the standard curve (Houghton 1997). It was noted, however, that large variability is inherent with fluorescent tracer techniques in general. This characteristic is due to variable area and thickness of tracer deposits on skin during real exposure episodes compared to those used to develop standard curves for quantification.

The method used by Houghton, Solomon, and Harris (1996) and Houghton (1997) to capture the images was modeled after that used in medical imaging analysis. The technique was used to estimate dislodgeability of bendiocarb and chlorpyrifos from various indoor surfaces including carpet and hardwood, following broadcast application for flea control. A standard curve was placed in every image by pipetting 2 μl of thirteen different concentrations of Uvitex OB, in acetone, over 2-cm^2 areas on one hand of the subjects, having them perform a hand press on the treated surface with the opposite hand and then imaging both hands in the same frame; thus, the standard curves were generated while the study progressed. Capturing a standard curve in every image ensured that the standards and target areas were subjected to the same light intensity. Standard curves were specific to each subject at that point in time, regardless of dermal pigmentation, perspiration, oiliness, or dryness, all of which may affect tracer fluorescence; therefore, summary calibration curves were not required.

The equipment setup was modeled after that of Fenske (1984); however, the addition of a photodiode to monitor incident light and a line stabilizer and Variac transformer to regulate current ensured that all images were captured at the same light intensity (Fig. 6-2). This was important in the event that the curves are used in the future to quantify tracer on subjects for which a standard curve does not exist. Provided that new images are acquired under identical light intensity and the subject's preexposure skin tone is matched to the appropriate standard curve, the curves generated by Houghton, Solomon, and Harris (1996) (also Houghton 1997) could be used. This was the basis for using standard curves for different preexposure skin tones in studies by Black (1993), Archibald (1993), and Archibald, Solomon, and Stephenson (1994a). A Sony XC-75 CCD black-and-white video camera was modified to perform image integration. This feature was extremely valuable because images could be integrated upon acquisition, allowing the operator an opportunity to brighten or darken an image to improve its quality. A Kodak 2B Wratten Gelatin filter that absorbs UV radiation below 390 nm was placed in front of the 16-mm macro camera lens. All images were captured at an f-stop of 5.6. A Mirage 486/DX-2, 66-MHz computer equipped with a super VGA color monitor and a 24-bit RGB Oculus TCX/MX frame grabber (512 × 512 pixel array) were employed. Northern Exposure uses *· tiff and *· bmp file formats, which are universal; therefore, images captured with this system can be analyzed using other image software packages. VITAE2 used a DT-Iris file format (Data Translation) which is not compatible with other systems.

A live alignment feature specially programmed into the Northern Exposure software allowed the operator to display a subject's preexposure image while the camera was live, so that subjects could place their hands in exactly the same position as in the preexposure image. A further image alignment was possible, after image capture, with the use of another added alignment feature. The preexposure image was then subtracted from the postexposure image by subtracting the corresponding pixels in the images using Boolean math methods. The resulting subtracted image represented the brightness and area exposed due to tracer deposition. The exposed areas were then highlighted by selecting a threshold brightness value that colored all of the tracer deposits red. These areas were mapped using the computer mouse and the intensity (average gray value) and area data were sent to an Excel spreadsheet via Dynamic Data Exchange. Images were corrected for background noise but no lens or anthropometric corrections were made because only hands, which are a planar surface and were placed in the center of the image frame, were imaged. Standard curve data were entered into the calibration portion of the program as nontransformed values. Standard curves were sigmoidal. Northern Exposure's Piecewise option was used to interpolate tracer masses (ng/cm^2) that fell between the calibration standard masses. The data output of $mass/cm^2$, for each tracer deposit, was then multiplied by the corresponding area of the deposit to solve for mass. The masses for all deposits on a hand were then summed to obtain the total mass.

Images of hands used to conduct hand presses of treated surfaces indicated that more tracer was dislodged from vinyl and hardwood than from carpet and upholstery fabric. A high Pearson's correlation coefficient was exhibited between bendiocarb mass dislodged using hand wipes and image estimates of Uvitex OB mass across all surfaces ($r = .94$), indicating that the tracer was an excellent surrogate for measuring exposure to bendiocarb. The correlation between chlorpyrifos and Uvitex OB masses dislodged was slightly lower ($r = .85$). Since dislodgeability ratios for different surfaces changed over time, it was recommended that regressions of mass dislodged over time be used to calculate exposure at any given time interval. Due to difficulties in quantification caused by quenching and effects related to the density of tracer deposits in the standard curve being different from those in the exposed image, the authors concluded that the technique is a semiquantitative one that has tremendous value as an extension education tool. They recommended the addition of lens correction factors, algorithms to correct for foreshortening and diminution, and the use of Roff's dodecahedral lighting to eliminate the need for anthropometric illumination adjustments when imaging curved body parts.

In addition to published information on video imaging techniques, Kross and Nicholson (1996) have been using VITAE2 both qualitatively and semiquantitatively to promote the use of personal protective equipment among applicators and to assist equipment manufacturers in designing application systems that reduce exposure. In one recent study among golf course workers, mixer/loaders sustained dermal deposits of Uvitex OB more than three orders of magnitude greater than tractor drivers. Dye deposition was greatest on subjects' hands (98% of the total), with face and torso exposure a distant second. Tracer was also visible on the back of the tractor cab window, near the air intake filter housing at the front of the cab, and on spray arms and nozzles. To assess exposure of the drivers, high concentrations of Uvitex OB were necessary. This caused clogging of spray nozzle and pump screens because of insolubility of tracer in aqueous solutions and quenching in mixer/loader images. The results of this study, which was sponsored by the National Institute of Occupational Safety and Health (NIOSH), prompted the equipment manufacturer and pesticide supplier to jointly design an enclosed pesticide loading system for sprayers.

A summary of human exposure studies using fluorescent tracer techniques is presented in Table 6-4.

Table 6-4 Summary of human exposure studies using fluorescent tracer techniques

Study	Method	Tracer	Contaminant studied	Exposure scenario	Quantitative/ qualitative
Staniland (1958)	black light	unidentified	malathion and DDT	strawberry crops	qualitative
Ashbaugh (1977)	black light	sulfur compound	none	orchards	qualitative
Schuresko (1980)	fluorometric monitor	none required	PAHs	coal plants	quantitative
Vo-Dinh and Gammage (1981)	lightpipe luminoscope	none required	PAHs	coal plants	quantitative
Franklin et al. (1981)	black light	unidentified	azinphos-methyl	orchards	qualitative
Fenske (1984), Fenske, Leffingwell, and Spear (1986)	VITAE	Calcofluor	none	method testing	quantitative
Fenske (1984), Fenske, et al. (1986)	VITAE	Calcofluor	malathion	orchards	quantitative
Fenske, Leffingwell, and Spear (1985)	VITAE	Calcofluor	diazinon	orchards	quantitative
Fenske, Hamburger, and Guyton (1987b), Fenske (1990)	VITAE	Calcofluor	—	orchards, greenhouse	quantitative
Fenske, Horstman, and Bentley (1987)	VITAE	Calcofluor	chlorophenol	lumber mills	quantitative
Vo-Dinh (1987)	Luminoscope II	none required	PAHs	coal plants	quantitative
Fenske (1988a,b,d)	VITAE	Calcofluor	malathion	citrus groves	quantitative
Fenske (1988c)	VITAE	Calcofluor	—	orange groves	qualitative
Fenske and Elkner (1990)	VITAE	Calcofluor	chlorpyrifos	structural/ homes	qualitative
Karr et al. (1992)	black light	sodium fluoroscein	organophosphate insecticides	orchards	qualitative
Archibald (1993), Archibald, Solomon, and Stephenson (1994a)	VITAE2	Calcofluor	pirimicarb	—	quantitative
Archibald, Solomon, and Stephenson (1994b)	VITAE2	Calcofluor	pirimicarb	greenhouse	quantitative
Archibald, Solomon, and Stephenson (1995)	VITAE2	Calcofluor	pirimicarb, deltamethrin	greenhouse	quantitative
Black (1993)	VITAE2	Uvitex OB	chlorpyrifos	turf	quantitative
Breon (1993)	VITAE2	Uvitex OB	atrazine	filter paper	quantitative
Methner and Fenske (1994a)	VITAE2	Calcofluor	none	greenhouse	quantitative
Methner and Fenske (1994b)	VITAE2	Uvitex OB	none	greenhouse	quantitative

(Contd.)

Table 6-4 *Contd.*

Study	Method	Tracer	Contaminant studied	Exposure scenario	Quantitative/ qualitative
Groth (1994)	VITAE2	Uvitex OB	none	standard curve testing	quantitative
Roff (1994)	FIVES	Tinopal CBS-X	permethrin	enclosed both,	quantitative
		Uvitex MST	diazinon	sheep dips	qualitative
Roff (1995, 1996b)	FIVES	Tinopal CBS-X, Tinopal SWN	permethrin	standard curve testing	quantitative
Roff (1996c)	FIVES	Tinopal CBS-X, Tinopal SWN	permethrin	fence painting	quantitative
Bierman, Brouwer, and van Hemmen (1995)	VITAE2 modified	Tinopal CBS-X	propoxur and methiocarb	greenhouse	quantitative
Houghton, Solomon, and Harris (1996), Houghton (1997)	Northern Exposure	Uvitex OB	chlorpyrifos bendiocarb	structural/ household	quantitative
Kross and Nicholson (1996)	VITAE2	Uvitex OB	none	golf courses	qualitative

6-7 CONCLUSIONS

The video imaging technique is rapid and allows examination of large numbers of workers. It is useful when determining relative exposures, and is an extremely valuable qualitative tool that indicates the anatomical distribution of exposure, allowing researchers to concentrate analytical techniques in these regions. The method's value from an educational standpoint should not be underestimated. The visual impact is tremendous. Video imaging techniques have generated extremely valuable information over the years. Data regarding the nonuniformity of dermal exposure and the potential for pesticide penetration through protective clothing and garment openings have helped to characterize dermal exposure. Video imaging techniques have also helped to clarify why excretion of urinary metabolites, after correction for absorption and other sources of exposure, often is not correlated with dermal patch estimates.

This review of research conducted using video imaging techniques indicates that obtaining quantitative estimates of dermal exposure is challenging. The introduction of dodecahedral lighting to reduce the complexity of algorithms required to correct foreshortening and diminution has been a significant advance. Adjusting standard curves for preexposure dermal gray tones has improved quantitative estimates dramatically. However, problems pertaining to the relative density differences between tracer calibration standards on skin and true dermal exposure deposits, in addition to variability inherent within and between individuals, suggest that it is a semiquantitative and qualitative technique.

REFERENCES

Anders, G. 1975. Limits of accuracy obtainable in the direct determination by fluorimetry of fluorescent whitening agents on thin layer chromatograms. In *Fluorescent whitening agents. Environmental quality and safety.* Edited by R. Anliker and G. Muller, 104–10. Supplement 4, edited by F. Coulston and F. Korte. Stuttgart: George Thieme.

Archibald, B. A. 1993. Video imaging as a technique for estimating pesticide exposure in greenhouse chrysanthemum production. Ph.D. thesis, University of Guelph.

Archibald, B. A., K. R. Solomon, and G. R. Stephenson. 1994a. A new procedure for calibrating the video imaging technique for assessing dermal exposure to pesticides. *Arch. Environ. Contam. Toxicol.* 26:398–402.

Archibald, B. A., K. R. Solomon, and G. R. Stephenson. 1994b. Estimating pirimicarb exposure to greenhouse workers using video imaging. *Arch. Environ. Contam. Toxicol.* 27:126–29.

Archibald, B. A., K. R. Solomon, and G. R. Stephenson. 1995. Estimation of pesticide exposure to greenhouse applicators using video imaging and other assessment techniques. *Am. Ind. Hyg. Assoc. J.* 56:226–35.

Ashbaugh, L. L. 1997. The use of fluorescent tracers as an index of dermal exposure to pesticides. M.Sc. thesis, University of California, Berkeley.

Bathe, R. 1979. Oral repetitive administration to male rats of TK 10326 (Uvitex OB). Ciba-Geigy Internal Report, 12 June.

Bhatt, A., H. West, R. Ashby, and J. C. Whitney. 1981. TK 10326: Toxicity in dietary administration to rats for 13 weeks. Confidential Report no. 81/CIA039/307, Life Science Research. Stock, Essex, England.

Bierman, E. P. B., D. H. Brouwer, and J. J. van Hemmen. 1995. Implementation, application and evaluation of the fluorescent tracer technique to assess dermal exposure. Report no. V 95-629, Netherlands Organisation for Applied Scientific Research Nutrition and Food Research Institute. Zeist.

Bingham, E., and H. L. Falk. 1970. Combined action of optical brighteners and ultraviolet light in the production of tumours. *Food Cosmet. Toxicol.* 8:173–76.

Black, K. G. 1993. An assessment of children's exposure to chlorpyrifos from contact with a treated lawn. Ph.D. thesis, Rutgers University.

Bonsall, J. L. 1985. Measurement of occupational exposure to pesticides. In *Occupational hazards of pesticide use.* ed. G. J. Turnbull, 13–34. London: Taylor and Francis.

Breon, T. A. 1993. Pesticide exposure assessment among agricultural workers: An application approach. M.Sc. thesis, University of Iowa.

Brouwer, D. 1996. Implementation and evaluation of the fluorescent tracer technique to assess dermal exposure. Presentation at the *Video Imaging Technique Assessing Exposure Workshop,* 20–21 June at University of Iowa.

Bullock, R. C., R. F. Brooks, and J. D. Whitney. 1968. A method of evaluating pesticide application equipment for Florida Citrus. *J. Econ. Entomol.* 61 (6):1511–14.

Burg, A. W., M. W. Rohovsky, and C. J. Kensler. 1977. Current status of human safety and environmental aspects of fluorescent whitening agents used in detergents in the United States. *Crit. Rev. Environ. Control* 7 (4):91–120.

Buxtorf, A. 1975. Toxicological investigations with fluorescent whitening agents. In *Fluorescent whitening agents. Environmental quality and safety.* Edited by R. Anliker and G. Muller, 191–98. Supplement 4, edited by F. Coulston, and F. Korte. Stuttgart: George Thieme.

Carleton, J. B., L. F. Bouse, H. P. O'Neal, and W. J. Walla. 1981. Measuring spray coverage on soybean leaves. *Trans. ASAE.* 24:1108–10.

Chester, G., and R. J. Ward. 1983. An accurate method for measuring dermal exposure to pesticides. *Human Toxicol.* 2:555–56.

Ciba-Geigy Ltd. 1991. 3-month sub-chronic dietary toxicity study in beagle dogs: TK 10326 (Uvitex OB). Final Report (Test no. 894086), 7 January.

Ciba-Geigy Ltd. 1993. Material safety data sheet: Uvitex OB. Revision 007, Ciba-Geigy Canada Ltd., 8 December.

Comer, S. W., D. C. Staiff, J. F. Armstrong, and H. R. Wolfe. 1975. Exposure of workers to carbaryl. *Bull. Environ. Contam. Toxicol.* 13:385–91.

Critchley, M. 1978. *Butterworth's medical dictionary.* 2d ed., 535 London: Butterworths.

Davis, J. E. 1980. Minimizing occupational exposure to pesticides. *Residue Rev.* 75:33–50.

Deparade, E., L. Durand, and A. Keller. 1984. Salmonella/Mammalian-microsome mutagenicity test: TK 10 326 (Uvitex OB). Confidential Report, Project no. 831044. Ciba-Geigy Ltd., 20 February.

Durham, W. F., and H. R. Wolfe. 1962. Measurement of the exposure of workers to pesticides. *Bull. WHO* 26:75–91, 279–81.

Edwards, G. J., W. L. Thompson, J. R. King, and P. J. Jutras. 1961. Optical determination of spray coverage. *Trans. ASAE* 4 (2):206–07.

Fenske, R. A. 1984. A fluorescent tracer technique for assessing dermal exposure to pesticides. Ph.D. thesis University of California, Berkeley.

Fenske, R. A. 1987. Assessment of dermal exposure to pesticides: A comparison of the patch technique and video imaging/fluorescent tracer technique. In Greenhalgh R, Roberts, TR (eds.) *Pesticide science and biotechnology: Sixth IUPAC Congress of Pesticide Chemistry*, 579–82. London: Blackwell Scientific.

Fenske, R. A. 1988a. Comparative assessment of protective clothing performance by measurement of dermal exposure during pesticide applications. *Appl. Ind. Hyg.* 3 (7):207–13.

Fenske, R. A. 1988b. Use of fluorescent tracers and video imaging to evaluate chemical protective clothing during pesticide applications. In *Performance of protective clothing: Second Symposium, American Society for Testing and Materials.* ed. S. Z. Mansdorf, R. Sager, and A. P. Nielsen, 630–39. Philadelphia: ASTM.

Fenske, R. A. 1988c. Visual scoring system for fluorescent tracer evaluation of dermal exposure to pesticides. *Bull. Environ. Contam. Toxicol.* 41:727–36.

Fenske, R. A. 1988d. Correlation of fluorescent tracer measurements of dermal exposure and urinary metabolite excretion during occupational exposure to malathion. *Am. Ind. Hyg. Assoc. J.* 49:438–44.

Fenske, R. A. 1989. Validation of environmental monitoring by biological monitoring: Fluorescent tracer technique and patch technique. In *Biological monitoring for pesticide exposure: measurement, estimation, and risk reduction.* ed. R. G. M. Wang, C. Franklin, R. C. Honeycutt, and J. C. Reinert, 70–84. *ACS Symposium Series*, no. 382. Washington, D.C.: American Chemical Society.

Fenske, R. A. 1990. Nonuniform dermal deposition patterns during occupational exposure to pesticides. *Arch. Environ. Contam. Toxicol.* 19:332–37.

Fenske, R. A. 1993. Dermal exposure assessment techniques. *Ann. Occup. Hyg.* 37 (6):687–706.

Fenske, R. A., S. G. Birnbaum, and K. Cho. 1990. Documentation for the second generation video imaging technique for assessing dermal exposure (VITAE system). Department of Environmental Sciences Report, Rutgers University.

Fenske, R. A., P. B. Curry, F. Wandelmaier, and L. Ritter. 1991. Development of dermal and respiratory sampling procedures for human exposure to pesticides in indoor environments. *J. Exposure Anal. Environ. Epidemiol.* 1:11–30.

Fenske, R. A., and K. P. Elkner. 1990. Multi-route exposure assessment and biological monitoring of urban pesticide applicators during structural control treatments with chlorpyrifos. *Toxicol. Ind. Health* 6:349–71.

Fenske, R. A., S. J. Hamburger, and C. L. Guyton. 1987. Occupational exposure to fosetyl-al fungicide during spraying of ornamentals in greenhouses. *Arch. Environ. Contam. Toxicol.* 16:615–21.

Fenske, R. A., S. W. Horstman, and R. K. Bentley. 1987. Assessment of dermal exposure to chlorophenols in timber mills. *Appl. Ind. Hyg.* 2 (4):143–47.

Fenske, R. A., J. T. Leffingwell, and R. C. Spear. 1985. Evaluation of fluorescent tracer methodology for dermal exposure assessment. In *Dermal exposure related to pesticide use.* ed. R. C. Honeycutt, G. Zweig, and N. N. Ragsdale, 377–93. *ACS Symposium Series*, no. 273. Washington, D.C.: American Chemical Society.

Fenske, R. A., J. T. Leffingwell, and R. C. Spear. 1986. A video imaging technique for assessing dermal exposure. 1. Instrument design and testing. *Am. Ind. Hyg. Assoc. J.* 47 (12):764–70.

Fenske, R. A., S. M. Wong, J. T. Leffingwell, and R. C. Spear. 1986. A video imaging technique for assessing dermal exposure. 2. Fluorescent tracer testing. *Am. Ind. Hyg. Assoc. J.* 47 (12):771–75.

Food and Agriculture Organization/World Health Organization (FAO/WHO) 1973. 1972 *Evaluations of some pesticide residues in foods.* WHO Pesticide Residue Series, no. 2. Geneva: World Health Organization.

Forbes, P. D., and F. Urbach. 1975a. Photocarcinogenesis: Lack of enhancement by fluorescent whitening agents. *Fluorescent whitening agents. Environmental quality and safety.* Edited by R. Anliker and G. Muller. Supplement 4, edited by F. Coulston and F. Korte, 212–22, Stuttgart: George Thieme.

Forbes, P. D., and F. Urbach. 1975b. Experimental modification of photocarcinogenesis. I. Fluorescent whitening agents and short-wave UVR. *Food Cosmet. Toxicol.* 33:335–37.

Forbes, P. D., and F. Urbach. 1975c. Experimental modification of photocarcinogenesis. II. Fluorescent whitening agents and simulated solar UVR. *Food Cosmet. Toxicol.* 33:339–42.

Forbes, P. D., and F. Urbach. 1975d. Experimental modification of photocarcinogenesis. III. Simulation of exposure to sunlight and fluorescent whitening agents. *Food Cosmet. Toxicol.* 33:343–45.

Franklin, C. A., R. A. Fenske, R. Greenhalgh, L. Mathieu, H. V. Denley, J. T. Leffingwell, and R. C. Spear. 1981. Correlation of urinary pesticide metabolite excretion with estimated dermal contact in the course of occupational exposure to guthion. *J. Toxicol. Environ. Health* 7:715–31.

Geno, P. W., D. E. Camann, H. J. Harding, K. Villalobos, and R. G. Lewis. 1996. Handwipe sampling and analysis procedure for the measurement of dermal contact with pesticides. *Arch. Environ. Contam. Toxicol.* 30:132–38.

Gloxhuber, C. 1972. Zur Toxikologie der Grundstoffe in Washund Reinigungsmitteln. *Fette, Seifen, Anstrichm.* 74:49. Summarized in C. Gloxhuber, and H. Bloching. Toxicologic properties of fluorescent whitening agents. *Toxicol. Annu.* 3 (1979):171–203.

Gloxhuber, C., and H. Bloching. 1979. Toxicologic properties of fluorescent whitening agents. *Toxicol. Annu.* 3:171–203.

Gloxhuber, C., H. Bloching, and W. Kastner. 1975. Studies on the reaction of skin when exposed to fluorescent whitening agents. In *Fluorescent whitening agents. Environmental quality and safety.* Edited by R. Anliker and G. Muller, 202–05. Supplement 4, edited by F. Coulston and F. Korte. Stuttgart: George Theime.

Gold, H. 1975. The chemistry of fluorescent whitening agents, major structural types. In *Fluorescent whitening agents. Environmental quality and safety.* Edited by R. Anliker and G. Muller, 25–46. Supplement 4, edited by F. Coulston and F. Korte Stuttgart: George Thieme.

Gold, R. E., T. Holcslaw, D. Tupy, and J. B. Ballard. 1984. Dermal and respiratory exposure to applicators and occupants of residences treated with dichlorvos (DDVP). *J. Econ. Entomol.* 77:430–36.

Gold, R. E., R. C. Leavitt, and J. Ballard. 1981. Effect of spray and paint-on applications of a slow-release formulation of chlorpyrifos on German cockroach control and human exposure. *J. Econ. Entomol.* 74 (5):552–54.

Griffith, J. F. 1973. Fluorescent whitening agents: Tests for skin-sensitizing potential. *Arch. Dermatol.* 107: 728–33.

Groth, K. M. 1994. Evaluation of calibration parameters and performance of the video imaging technique of assessing exposure (VITAE system). M.Sc. thesis, University of Washington.

Hess, R. 1973. Experimental toxicology of compounds used as optical brighteners. In *Fluorescent whitening agents, MVC Report 2, Proceedings of a Symposium at the Royal Institute of Technology,* Stockholm, 85–92.

Himel, C. M. 1969. The fluorescent particle spray droplet tracer method. *J. Econ. Entomol.* 62 (4):912–18.

Houghton, D. L. 1997. Development and validation of a fluorescent tracer method to estimate dermal exposure to pesticides used indoors. Ph.D. thesis, Department of Environmental Biology, University of Guelph.

Houghton, D. L., K. R. Solomon, and C. R. Harris. 1996. Development and validation of the fluorescent tracer method to estimate dermal exposure to pesticides used indoors. Final Report, to Health Canada, Canadian Network of Toxicology Centres, Guelph, Ontario Canada. 3 June.

Hsu, J. P., D. E. Camann, H. Schattenberg, B. Wheeler, K. Villolobos, M. Kyle, S. Quarderer, and R. G. Lewis. 1990. New dermal exposure sampling technique. In *Proceedings of 1990 EPA/A&WMA International Symposium on Measurement of Toxic and Related Air Pollutants,* 489–97. VIP-17. Pittsburgh: Air and Waste Management Association.

Imaging Research Inc. 1996. *Fundamentals of image analysis.* Imaging Research Inc., St. Catherines, Ontario.

Jadassohn, W., and F. Schaaf. 1970. Bemerkungen uber das Eindringen Gewisser Stoffe in Die Haut. *Dermatologica.* 140:275. As summarized in Gloxhuber, C., and H. Bloching. Toxicologic properties of fluorescent whitening agents. *Toxicol. Annu.* 3 (1979):171–203.

Karr, C., P. Demers, L. G. Costa, W. E. Daniell, S. Barnhart, M. Miller, G. Gallagher, S. W. Horstman, D. Eaton, and L. Rosenstock. 1992. Organophosphate pesticide exposure in a group of Washington State Orchard Applicators. *Environ. Res.* 59:229–37.

Kells, A. M., and K. R. Solomon. 1995. Dislodgeability of pesticides from products made with recycled pesticide container plastics. *Arch. Environ. Contam. Toxicol.* 28:134–38.

Keplinger, M. L., O. E. Fancher, F. L. Lyman, and J. C. Calandra. 1974. Toxicologic studies of four fluorescent whitening agents. *Toxicol. Appl. Pharmacol.* 27:494–506.

Keplinger, M. L., F. L. Lyman, and J. C. Calandra. 1975a. Chronic toxicity and carcinogenicity studies with FWAs. In *Fluorescent whitening agents. Environmental quality and safety.* Edited by R. Anliker and G. Muller, 205–12. Supplement 4, edited by F. Coulston and F. Korte. Stuttgart: George Thieme.

Keplinger, M. L., F. L. Lyman, and J. C. Calandra. 1975b. Three generation reproduction studies with FWAs.

In *Fluorescent whitening agents. Environmental quality and safety*. Edited by R. Anliker and G. Muller, 230–38. Supplement 4, edited by F. Coulston and F. Korte. Stuttgart: George Thieme.

Kilbey, B. J., and G. Zetterberg. 1973. A re-examination of the genetic effects of optical brighteners in yeast. *Mutat. Res.* 21:73–82.

Kilbey, B. J., and G. Zetterberg. 1975. Mutagenicity assays on fluorescent whitening agents using microorganisms. In *Fluorescent whitening agents. Environmental quality and safety*. Edited by R. Anliker and G. Muller, 264–77. Supplement 4, edited by F. Coulston and F. Korte. Stuttgart: George Thieme.

Kross, B., and H. Nicholson. 1996. Methods development study for measuring pesticide exposure to golf course workers using video imaging techniques. Unpublished report, University of Iowa.

Kubelka, P. 1948. New contributions to the optics of intensity light scattering materials. Part 1. *J. Opt. Soc. Am.* 38 (5):448–57.

Kubelka, P., and F. Munk. 1931. Ein Beitrag zur Opik der Farbanstriche (A contribution to the optics of colour coatings). *Z. Tech. Phys.* 12:593–601.

Landau, S. I. 1986. *International dictionary of medicine and biology*. Vol. 3, 2870. New York: Wiley.

Lewis, R. G., A. E. Bond, R. C. Fortmann, L. S. Sheldon, and D. E. Camann. 1991. Determination of routes of exposure of infants and toddlers to household pesticides: A pilot study. Report no. EPA/600/D-91/077. Research Triangle Institute, Research Triangle Park, N.C.

Lewis, R. G., R. C. Fortmann, and D. E. Camann. 1994. Evaluation of methods for monitoring the potential exposure of small children to pesticides in the residential environment. *Arch. Environ. Contam. Toxicol.* 26 (1):37–46.

Lorke, D., and L. Machemer. 1975a. Studies of embryo toxicity testing in rats and rabbits. In *Fluorescent whitening agents. Environmental quality and safety*. Edited by R. Anliker and G. Muller, 223–29. Supplement 4, edited by F. Coulston and F. Korte. Stuttgart: George Thieme.

Lorke, D., and L. Machemer. 1975b. Testing mutagenic properties with the dominant lethal test on the male mouse. In *Fluorescent whitening agents. Environmental quality and safety*. Edited by R. Anliker and G. Muller, 239–46. Supplement 4, edited by F. Coulston and F. Korte. Stuttgart: George Thieme.

Luckhaus, G., and E. Loser. 1975. Study by fluorescence microscopy of the effect of fluorescent whitening agents on the skin of mice. In *Fluorescent whitening agents. Environmental quality and safety*. Edited by R. Anliker and G. Muller, 198–202. Supplement 4, edited by F. Coulston and F. Korte. Stuttgart: George Thieme.

McArthur, B. 1992. Dermal measurement and wipe sampling methods: A review. *Appl. Occup. Environ. Hyg.* 7 (9):599–606.

Methner, M. M., and R. A. Fenske. 1994a. Pesticide exposure during greenhouse applications. Part I. Dermal exposure reduction due to directional ventilation and worker training. *Appl. Occup. Environ. Hyg.* 9 (8):560–66.

Methner, M. M., and R. A. Fenske. 1994b. Pesticide exposure during greenhouse applications. Part II. Chemical permeation through protective clothing in contact with treated foliage. *Appl. Occup. Environ. Hyg.* 9 (8):567–74.

Methner, M. M., and R. A. Fenske. 1996. Pesticide exposure during greenhouse applications. III. Variable exposure due to ventilation conditions and spray pressure. *Appl. Occup. Environ. Hyg.* 11 (3):174–80.

Muecke, W., G. Dupuis, and H. O. Esser. 1975. Metabolic behaviour of water soluble fluorescent whitening agents in the rat and bean plant. In *Fluorescent whitening agents. Environmental quality and safety*. Edited by R. Anliker and G. Muller, 174–79. Supplement 4, edited by F. Coulston and F. Korte. Stuttgart: George Thieme.

Muller, D., H. Fritz, M. Langauer, and F. F. Strasser. 1975. Nucleus anomaly test and chromosomal analysis of bone marrow cells of the Chinese hamster and dominant lethal test in male mice after treatment with fluorescent whitening agents. In *Fluorescent whitening agents. Environmental quality and safety*. Edited by R. Anliker and G. Muller, 247–63. Supplement 4, edited by F. Coulston and F. Korte. Stuttgart: George Thieme.

Naffziger, D., R. J. Sprenkel, and M. P. Mattler. 1985. Indoor environmental monitoring of Dursban L. O. following broadcast application. *Down to Earth*, no. 41. Dow Chemical Co. Midland, Mich.

Neukomm, S., and M. De Trey. 1961. Etude de Certains Azureurs Optiques du Point de Vue de Leur Pouvoir Cancerigene et Co-Cancerigene. *Med. Exp.* 4:298. Summarized in C. Gloxhuber and H. Bloching, Toxicologic properties of fluorescent whitening agents. *Toxicol. Annu.* 3 (1979):171–203.

Ramaswamy, G. N., and C. R. Boyd. 1993. Fluorescent dye and pesticide penetration tested in a computerized spray chamber. Part I. Nonwoven fabrics as barriers. *Bull. Environ. Contam. Toxicol.* 51:341–48.

Roberts, J. W., and D. E. Camann. 1989. Pilot study of a cotton glove press test for assessing exposure to pesticides in house dust. *Bull. Environ. Contam. Toxicol.* 43:717–24.

Roff, M. W. 1994. A novel lighting system for the measurement of dermal exposure using a fluorescent dye and an image processor. *Ann. Occup. Hyg.* 38 (6):903–919.

Roff, M. W. 1995. Fluorescence monitoring of dermal exposure: Suitability of calibration curves. Poster presented at *International Society for Environmental Epidemiology/ International Society for Exposure Analysis Conference*, 31st August, at Noordwijkerhout, The Netherlands.

Roff, M. W. 1996a. Image analysis with the FIVES system. Presentation at *Video Imaging Technique Assessing Exposure Workshop*, 20–21 June, at University of Iowa. 1996.

Roff, M. W. 1996b. Accuracy and reproducibility of calibrations on the skin using the FIVES fluorescence monitor. *Ann. Occup. Hyg.* 41 (3):313.

Roff, M. W. 1996c. Dermal exposure of amateur or non-occupational users to wood preservative fluids applied by brushing outdoors. *Ann. Occup. Hyg.* 41 (3):297.

Ross, J., H. R. Fong, T. Thongsinthusak, S. Margetich, and R. Krieger. 1991. Measuring potential dermal transfer of surface pesticide residue generated from indoor fogger use: Using the CDFA roller method. Interim report II. *Chemosphere* 22 (9,10):975–84.

Ross, J., T. Thongsinthusak, H. R. Fong, S. Margetich, and R. Krieger. 1990. Measuring potential dermal transfer of surface pesticide residue generated from indoor fogger use: An interim report. *Chemosphere* 20:349–60.

Russ, J. C. 1992. *The image processing handbook*. 1st ed., 1–52. Boca Raton, Fl: CRC Press.

Schuresko, D. 1980. Portable fluorometric monitor for detection of surface contamination by polynuclear aromatic compounds. *Anal. Chem.* 52:371–73.

Snyder, F. H., D. L. Opdyke, and H. L. Rubenkoenig. 1963. Toxicologic studies on brighteners. *Toxicol. Appl. Pharmacol.* 5:176–183.

Staniland, L. N. 1958. Fluorescent tracer techniques for the study of spray and dust deposits. *J. Agric. Eng. Res.* 3:110–25.

Stensby, P. 1967. Optical brighteners and their evaluation. Parts 1–5. *Soap Chem. Spec.* Vol? (April):41–102, (May):84–132, (July):80–88, (Aug.):97–103, (Sept.):96–138.

Thomann, P., and L. Kruger. 1975. Acute oral, dermal and inhalation studies. In *Fluorescent whitening agents. Environmental quality and safety*. Edited by R. Anliker and G. Muller, 93–198. Supplement 4, edited by F. Coulston and F. Korte. Stuttgart: George Thieme.

Vaccaro, J. R. 1990. Evaluation of dislodgeable residues and absorbed doses of chlorpyrifos following indoor broadcast applications of chlorpyrifos based emulsifiable concentrate (EC). Confidential unpublished report. Dow Chemical Co., Midland, Mich.

Vo-Dinh, T. 1983. Surface detection of contamination: principles, applications and recent developments. *J. Environ. Sci.* 23:40–43.

Vo-Dinh, T. 1987. Evaluation of an improved fiberoptics luminescence skin monitor with background correction. *Am. Ind. Hyg. Assoc. J.* 48:594–98.

Vo-Dinh, T., and R. B. Gammage. 1981. The lightpipe luminoscope for monitoring occupational skin contamination. *Am. Ind. Hyg. Assoc. J.* 42:112–20.

Weber, H. 1968. Uber die Frage der Einwirkung von WeiBtunern auf Die Haut Des Sauglings. *Monatssehr. Kinderheilk.* 116:69 Summarized in C. Gloxhuber and H. Bloching. Toxicologic properties of fluorescent whitening agents. *Toxicol. Annu.* 3 (1979):171–203.

Wheldon, G. H., D. D. Cozens, D. W. Jolly, A. E. Street, and L. E. Mawdesley-Thomas. 1970. Effects of BA-31'736 administered orally to rats over a two-year period. Confidential Report no. 2398/68/275, Ciba-Geigy, 18 November.

Wheldon, G. H., B. Hunter, L. E. Mawdesley-Thomas, and J. Tripod. 1969. Effects of Ba-31'736 (TAP-1159) administered orally to mice over a two-year period. Confidential Report no. 2445/68/322, Ciba-Geigy, 24 January.

Yates, W. E., and N. B. Akesson. 1963. Fluorescent tracers for quantitative micro residue analysis. *Trans. ASAE* 6 (2):104–07,114.

Zahradnik, M. 1982. *The production and application of fluorescent brightening agents*, 27. Chichester, Sussex: Wiley.

SEVEN

BIOMONITORING HUMAN PESTICIDE EXPOSURES

Robert I. Krieger

Department of Entomology, University of California Riverside, Riverside, CA 92521-0314

7-1 INTRODUCTION

Pesticide risk management requires multidisciplinary approaches to measure chemical exposures resulting from integrated pest management in agriculture, structural pest control, disease vector control in public health, and domestic use of household products (Fig. 7-1). Biological monitoring can make a substantial contribution to the process. Biomonitoring develops quantitative data about the magnitude of exposure and knowledge of the occurrence of chemicals in living things or the environment. The information is important to product development and stewardship as well as regulatory risk assessment and risk management. Contemporary risk management places a high premium on accurate exposure data (NRC/NAS 1983).

Human chemical exposure results from use of pesticides or other chemical technologies (Fig. 7-2). Intentional, unintentional, and unavoidable exposures occur as a result of dermal contact, inhalation, and/or ingestion. Exposure results in absorption of amounts ranging from negligible traces to toxic amounts (Table 7-1). Those substances or their biological derivatives in body fluids such as blood or their clearance via urine, sweat, saliva, or other biological matrices provide opportunity for biomonitoring (Nigg and Stamper 1989). Matrices other than blood and urine have received little attention, but some may have a place in the development of data concerning particular chemicals and use practices.

It is remarkable that biomonitoring is so underutilized in the study of pesticides. Analytical methods have been developed for fate and transport studies of these chemicals and some of their derivatives in amounts that are well below thresholds of apparent biological significance. Most current human exposure estimates are derived from environmental monitoring and include numerous uncertain default assumptions (Table 7-1). Underutilization of biomonitoring may result from factors including the low magnitude

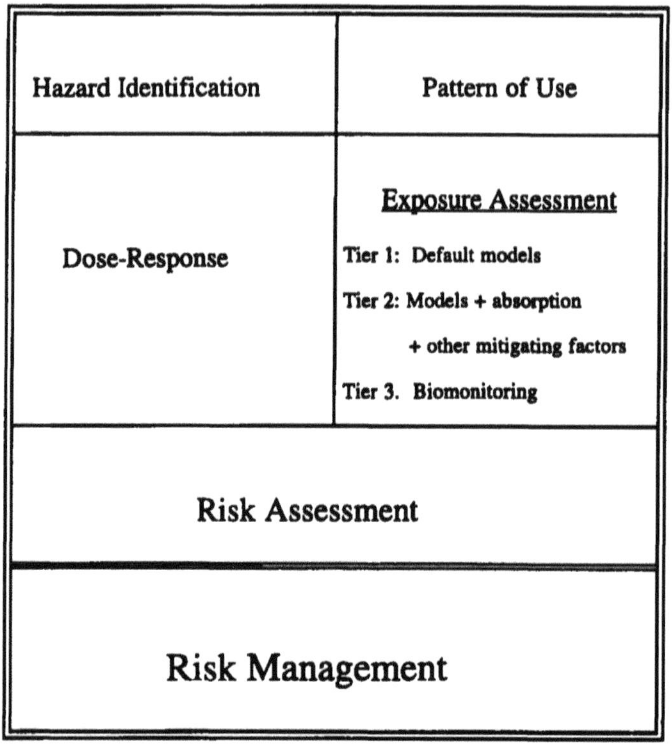

Hazard Identification	Pattern of Use
Dose-Response	**Exposure Assessment** Tier 1: Default models Tier 2: Models + absorption + other mitigating factors Tier 3. Biomonitoring
Risk Assessment	
Risk Management	

Figure 7-1 Contemporary risk assessment and risk management include four unit processes. Toxicology studies include hazard identification and dose-response studies. The no observed adverse effect level (NOEL, mg/kg) is usually derived from experimental studies in animals. Determinations of pattern of use and exposure assessment (mg/kg) represent dosage associated with particular human activities. Exposure assessments are based on default models founded upon passive dosimetry (Tier 1), models incorporating clothing penetration and measures of dermal absorption (Tier 2), or biomonitoring data (Tier 3). Each succeeding tier represents decreased number of default assumptions and increasingly specific exposure estimates. The margin of exposure (MOE, or the margin of safety [MOS]) is the ratio of the NOEL to human exposure. The MOE is the basic scientific element of risk assessment (or risk characterization) and a fundamental consideration during risk management (regulatory, product development, product stewardship, etc.).

of routine exposure, characterization of pesticides as a special threat to health because of their toxicity, and the viewpoint that exposure does not occur when persons wear protective equipment or reenter pesticide-treated fields or indoor environments after prescribed intervals following pesticide use. Regardless, pesticides are probably the best-characterized chemical technology with respect to knowledge of physical and chemical properties, formulation and pattern of use, environmental fate and transport, and toxicology. Widespread adoption of the risk assessment paradigm (Fig. 7-1) has increased the need for high-quality human exposure data used for risk management.

Pesticides may occur as solids, liquids, or gases (Table 7-2). Human exposures virtually always result from contact with a physically and chemically complex matrix in which pesticide is included in amounts ranging from undetectably low to high percentages. It is convenient to consider the environments in which residual pesticides and

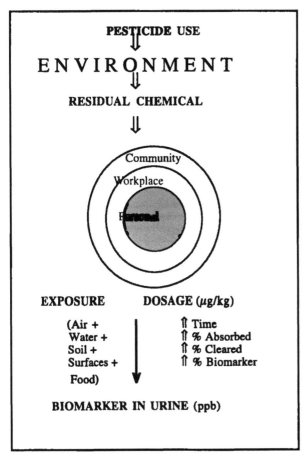

Figure 7-2 Pesticide use produces human exposures. Most exposures are unmeasurably low and only inferred from environmental levels reported in air, water, soil, food, and on treated surfaces. Opportunity for exposure occurs in the community environment (air, soil, and water), the workplace (air and treated surfaces), and personal environment (food plus all other sources). Biomarker in urine factored by stoichiometry of the reaction (formula weight (FW) parent/FW metabolite), percentage of dose cleared in urine, percentage absorbed, body weight, and time yields human exposure (μg/kg). When dermal exposure is under consideration, a clothing penetration factor (dermal exposure/potential dermal exposure) is included in the algorithm. Under usual conditions, only daily workplace and personal exposures directly resulting from pesticide use are monitored at levels of 0.1 μg equivalents/kg or more.

their degradation products are known to occur. Community, occupational, and personal environments each carry their own human pesticide exposure potential (Fig. 7-2).

In the vast majority of cases, investigators infer exposuresfrom environmental monitoring data derived from sensitive quantitative analysis. Biological half-life and route of clearance are important practical considerations that frequently determine feasibility of biomonitoring (Table 7-3). Such studies contribute to knowledge of environmental fate and transport of residual chemicals. The absorbed levels of residual chemicals from chemically complex media such as drinking water, air, meat, milk, and produce are usually

Table 7-1 Determination of absorbed dose for exposure assessment

Sample	Usual units	Common default assumptions	Impact on assessment
		Environmental matrix	
Air	mg/m^3, time-weighted average	$20\ m^3/day$, 100% retention and 100% absorption of chemical in airstream	Overestimates inhaled dose
		$20\ m^3/day$, 50% retention and 100% absorption of chemical in airstream	Usually overestimates inhaled dose
Water	ppb, $\mu g/l$	2 l/day, 100% absorption	Multimedia exposures difficult to evaluate and verify
Food	ppm, tolerances in fresh commodities	1 kg/day, 100% absorption, does not account for market penetration	Overestimates dose, very low incidence of foods at tolerance levels
	ppb, market-basket studies	100% absorption	Indirect estimate of consumption, over-estimates exposure of general population
	ppb or less, prepared foods	100% absorption	Indirect estimate, little data available
Treated surfaces	μg/unit area	100% absorption	Wash, rinse, and wipe samples overestimate absorbed dose; ignores protection of clothing and skin
		Biological matrix	
Blood	$<\mu g/l$	absorbed dose	Invasive procedure, human clinical database often limited
	cholinesterases	time and dose dependent	Not useful means to estimate dose
Urine	$\mu g/l$, creatinine correction (child 4–10 years old, 0.6 g/l per day; adult female, 1.0 g/l per day; male, 1.7 g/l per day	cleared dose or metabolites in spot or 24-h specimens	1. Stoichiometry-formula weight parent/FW metabolite, 2. Percentage of dose cleared in urine, 3. Steady state or percentage cleared per unit sampling time

Table 7-2 Biomonitoring and biological standards

Specimen	Analytical goal	Application	Reference points
Urine	Unchanged chemical	Lead, mercury, cadmium, substances excreted unchanged by kidney (e.g., 2,4-D)	Reference levels
	Pesticide metabolites	Organic pesticides of all major groups	Absorption, distribution, metabolism, and excretion; Biological Exposure Index
Blood	Organic and inorganic substances	Experimental studies and routine workplace monitoring	Absorption, distribution, metabolism, and excretion
			Biological Exposure Index
	Cholinesterase	Organophosphate (whole blood, red blood cells [RBC], plasma)	Exposure indicator
Breath	Volatile solvents Carbon disulfide Lower alcohols	Experimental studies and workplace monitoring	Reference levels (American College of Governmental Industrial Hygienists)

below existing analytical thresholds for meaningful, quantitative biomonitoring. Intermediate levels of exposure, detectable by biomonitoring following routine activities such as pesticide handling or reentry of treated fields or indoor environments, may be of analytical or regulatory relevance. Usually, such exposures are substantially less than amounts of health consequence. The extent of exposure and absorption usually are not investigated in cases of clinical morbidity and mortality when a history of chemical contact is available. Such data could contribute to knowledge of human dose-response and would doubtless decrease the number of cases attributable to excessive pesticide exposure.

Earliest measurements of human exposure were developed over 40 years ago to estimate pesticide mixer/loader/applicator exposure to DDT and parathion. Exposure

Table 7-3 Pesticide half-life in urine and blood and optimum sampling time

Half-life (h)	Optimum sampling time	Examples
<2	Too short for routine monitoring	Dichloromethane Carbon disulfide
2–10	End of shift, morning void Preshift	Organophosphate insecticides Phenoxy herbicides Pyrethroids 1,3-Dichloropropene (Brouwer et al. 1991)
10–100	End of shift, end of week	DDT
>100	Sampling time not critical	Pentachlorophenol Dieldrin Cadmium

to DDT resulted in deposition in tissues including fat and clearance of DDA in urine (Cranmer, Carroll, and Copeland 1972). Circulating amounts of persistent chlorinated hydrocarbons that are stored in tissues including body fat can be monitored in blood. Diet remains the source of low-level exposure to persistent chlorinated hydrocarbons. Although numerous measurements of persistent substances have been made as a measure of environmental quality and in epidemiological studies, these data are seldom used in contemporary exposure assessment.

During the same time interval, parathion exposures in citrus workers were studied using urinary p-nitrophenol and blood cholinesterases as biomarkers. Public health, medical, and analytical professionals established safe work conditions for use of organophosphate insecticides and developed the foundation for present-day human exposure assessment. Parathion exposures were measured by analysis of urinary p-nitrophenol in early studies by Batchelor and Walker (1954), Elliott et al. (1960), and Arterberry et al. (1961). These studies and methods used by pioneers of human and workplace (environmental) monitoring of organophosphates in agriculture include Gunther et al. (1977), Durham and Wolfe (1962), and Popendorf and Leffingwell (1982). The development and evaluation of safe use conditions were derived from these and similar other efforts to promote workplace hygiene.

The overall goal of current exposure assessment studies remains much the same but additionally includes efforts to forecast daily or short-term exposures in related work tasks and overall lifetime exposure. Extended forecasts of exposure characteristically include much uncertainty and numerous health-conservative default assumptions, and so, daily exposures must be defined accurately to avoid introduction of large cumulative errors. Biological monitoring is the method of choice to rank the exposure potential of particular work tasks and to develop estimates of human exposure driven by estimates of absorbed dose. When inflated and inaccurate exposure assessments are derived from unrepresentative exposure data, the resulting estimates promote unwarranted fear and anxiety among exposed persons and the general public. In addition, risk managers in industry, public health, and government regulatory agencies are forced to work from a flawed database.

This chapter is limited to discussion of some generally used methods and techniques for biomonitoring human chemical exposures in agricultural and domestic indoor settings. Several other sources of pesticide exposure are not considered because they are primarily of only regulatory importance and not amenable to biomonitoring, e.g., chemical residues in meat and produce of very low magnitude. Similarly, ambient air, drinking water, and soil are outside of current consideration because of their very low and/or negligible potential source strength and the limited bioavailability of the residues. Previous chapters include discussion of the occurrence of chemicals as environmental. Biomarkers of effects (responses) are considered in Chapter 8.

7-2 INTERNAL DOSE OR ABSORBED DAILY DOSE

Biological monitoring provides a quantitative measure of absorbed or internal dose based on knowledge of use, exposure, and the disposition of the chemical in humans and animals. Internal dose (ID) or absorbed dose (AD) may represent the amount recently absorbed (such as the amount of a volatile substance in exhaled air), during a workshift or day, or during an extended time period when long-lived substances are being sampled and analyzed. The most common measure of ID is absorbed daily dosage (ADD, $\mu g/kg$ body weight per day). ADD may result from measured plasma levels of parent compound or be backcalculated from active ingredient or metabolite levels in urine.

ADD doubtless better represents the amount of chemical associated or bound to sensitive, biological targets (biochemical lesions) than do environmental levels or measures of exposure obtained using passive dosimetry. Accessible receptors such as hemoglobin, albumin, and DNA may be associated with adverse effects of substances such as carbon monoxide, nitrate, and alkylating agents. Surrogate (nontarget) receptors such as plasma or erythrocyte cholinesterases also may be useful indicators of exposure and, as such, they may be valuable adjuncts for sampling and analysis of ID or ADD. In spite of positive attributes of measures of ID, substantial uncertainty still exists concerning critical biochemical lesions that produce an adverse effect.

7-3 RELATIONSHIP TO OTHER EXPOSURE MEASUREMENTS

Durham and Wolfe (1962) pioneered assessment of workplace pesticide exposures using dermal patches strategically attached to clothing, respirator pads, and hand rinses. These procedures were termed direct measures and similar methods have been used extensively. Results have been important in establishment of the protectiveness of clothing, improving equipment design, and in establishment of postapplication reentry intervals.

Potential dermal exposure (PDE) is represented in direct measurements by the total amount of pesticide retained on the patches and their relationship to the corresponding body area. Pesticide sprays, contaminated work surfaces, and residues on treated foliage can be transferred and retained by patches and exposed skin during agricultural activities. Sampling of these residues using patch dosimeters results in excessive estimates of exposure relative to AD resulting from patch placement, residue retention, and small size relative to corresponding body region. Data collected in this manner are commonly used to estimated AD. Default assumptions concerning clothing penetration and dermal absorption and estimates of exposure resulting from skin wipes or washes of surfaces not protected by clothing are factors that influence the magnitude of ADDs derived in this manner.

Such estimates have formed the foundation for current exposure modeling efforts in North America (Pesticide Handlers Exposure Database [PHED], Versar, Inc., Springfield, Va.) and the United Kingdom (Pesticide Operator Exposure Model (POEM), Pesticide Safety Directorate, York, UK). These databases are useful tools to rank the relative exposure potential of various application technologies or to gauge pesticide exposure potential. Their usefulness as a means to predict AD is limited by the methods used to collect data concerning the mosaic of exposure associated with use of chemical technologies including pesticides. Because the databases are based on PDE rather than ADD, the estimates of exposure may not be relevant without introduction of substantial defaults for the estimation of chemical exposure.

7-4 MONITORING CHEMICAL EXPOSURES

Biological monitoring, environmental (or ambient) monitoring, and health surveillance can each provide data concerning the magnitude and extent of chemical exposures (Hoet 1996). Substantial advances have been made in recent times, but even the most complete assessments are inadequate to describe the complete fate of a single substance. The same observation was made over 30 years ago by Durham and Wolfe (1962). The unknown will continue to promote innovative sampling strategies, increasingly lower analytical limits of detection, and expanded knowledge of the disposition of chemicals in humans, other living things, and the environment.

7-4-1 Biological Monitoring

The objective of biological monitoring is to provide data concerning ADD to health professionals for appraisal of potential hazards and risk associated with chemical technologies. These methods may utilize biological samples such as urine, plasma, and potentially a variety of other tissues to establish dose. Biological estimates of exposure rely upon knowledge of the time-dependent absorption and clearance of chemicals in humans so that residue levels in tissue or urine can be backcalculated to estimate dose (equivalents of parent compound). Biomonitoring studies may be augmented by other strategies to measure an adverse effect as a function of AD or to directly measure the amount of absorbed chemical reacting with biological active sites.

Biomonitoring data can contribute to the establishment of reference limits for unexposed persons, permissible exposure limits, biochemical exposure limits, biomonitoring action levels, or ADDs for calculation of MOEs or MOSs for risk management. These indices are calculated from NOELs (usually from toxicity testing in animals, mg/kg) and ADs (from human exposure assessment, mg/kg). Such data for pesticides are infrequently developed from biomonitoring, but increased utilization of exposure assessments in risk management should contribute to standards founded upon biomonitoring rather than environmental surveillance.

Biological monitoring can be used to determine the unintended extent of chemical exposure that results from pesticide use and environmental transport. A few products are used on humans or intentionally discharged into the air to protect stored products or to preserve structures including homes, office buildings, and institutions. The vast majority of other products are applied through the air and result in unintended or unrecognized exposure. Volatilization, photolysis, and hydrolysis are among numerous time-dependent processes that modify the nature and extent of human exposure to active ingredients. Chemicals in air may be absorbed following skin contact or by penetration through clothing. Biological factors such as behavior, routes of exposure, and differential rates of absorption influence AD. The biological disposition and fate of absorbed substances also will be determined by their physical, chemical, and biological properties which determine body burden, concentration at active and inactive sites, and clearance.

Biological monitoring data complement rather than replace environmental measures of potential exposure such as air levels or dislodgeable foliar residues or health effects (symptoms) reported by physicians participating in health surveillance activities. Some relationships between biological monitoring and other environmental measures of exposure are listed in Table 7-1.

7-4-2 Environmental or Ambient Monitoring

Environmental monitoring frequently supplies the foundation for estimates of ADD. These estimates usually rely upon a series of default assumptions which causes them to represent hypothetical worst-case conditions that may bear no relation to actual conditions of human exposure.

Table 7-1 lists environmental matrices—air, water, food, and treated surfaces—that may become part of aggregated human exposure. Analytical methods for determination of extremely low levels of many active ingredients (and occasionally their derivatives) can be coupled with default assumptions to form worst-case estimates of exposure. Resulting exposure estimates are likely to be two or more orders of magnitude greater than those determined by biomonitoring. The factors most easily recognized as responsible for the difference between an environmentally based measurement and a biologically based estimate are assumptions such as maximum ventilation rates (20 m^3 air/day) and ingestion of water and food, complete bioavailability of the active ingredient in all matrices, and complete absorption by persons of unspecified body weights.

An alternative is biological estimation of AD using a well-characterized substance under well-known conditions to develop an anticipated level of exposure. Exposures could be estimated initially from PHED or POEM factored by a clothing protection factor and dermal absorption. A more clear estimate of anticipated exposures would recognize the importance of the contact-transfer of pesticides from treated surfaces, dermal absorption, and body weight for establishment of biologically based ADD. The contributions from air, water, and food residues would be included in an estimate based upon biomonitoring, but the contributions are probably vanishingly small under usual conditions of indoor and outdoor use of pesticides.

Berteau et al. (1989) forecast mg/kg exposures for children based upon projected environmental levels and worst-case assumptions (Table 7-4). Subsequent situational exposure monitoring (Krieger et al. unpublished abstract 1991) using urinalysis has shown that usual internal chlorpyrifos dosages in adults range from 0.0001 to 0.030 mg chlorpyrifos equivalents/kg of body weight (bw) (based upon 3,5,6-trichloropyridinol). Estimates by Berteau et al. (1989) were based on a 4-h period for infants, and Krieger et al. (in review) has biomonitored full-day exposures of adults. This example is not fully satisfying because differences in the activities of children and adults will doubtless influence the ultimate magnitude of exposure. Table 7-5 includes experimental estimates of exposure derived from environmental and biological sampling. Foil deposition and cotton dosimetry (roller and whole body) must include a dermal absorption factor to adequately represent AD. The hypothetical estimates and biological monitoring differ by several orders of magnitude in this case, which illustrates the potential impact of cumulative health-conservative default assumptions on the accuracy of exposure assessments.

Table 7-4 Passive dosimetry and biomonitoring of indoor pesticide exposures

Assumptions for exposure assessment	Environmental monitoring	Biomonitoring
Contact	Removes total dislodgeable	Transfers available residue
Contacted area	25% of available	Actual area contacted
Dermal contact period	Occurs in 4 h	Actual exposure time
Dermal contact area	50% of body	Bare skin and clothed areas
Hand contact	Assume total ingestion	Actual ingestion from hands
Indoor air	24 h/day	24 h/day
Inhalation	100% absorbed	Actual retention
Ingestion	100% absorbed	Actual retention
Body weight	Assumed	Measured
Surface area	Calculated	Not a factor
Dosage	mgs/kg body weight	μgs/kg body weight

Source: Modified from Berteau et al. (1989).

7-5 RATING ENVIRONMENTAL AND BIOLOGICAL MONITORING METHODS

Sensitive and specific analytical methods for measurement of chemicals in diverse environmental samples provide exposure assessors with a rich armament for human studies. Environmental transport and fate studies are abundant for many important classes of chemicals used as pesticides. During the past 40 years, limits of detection of chemicals in environmental samples have decreased several orders of magnitude. This issue is discussed in detail in reviews of environmental pesticide chemistry (e.g., Plimmer [1982]). Considerable detail concerning the pattern of chemical use and the nature (route, rate, time, etc.) of exposure must be utilized in the development of sampling strategies. Lower limits of detection permit use of complex sample matrices. Knowledge of nondetectable levels is critical for development of biomonitoring strategies because each sample will be

Table 7-5 Estimates of human exposure derived from four sampling procedures

Procedure	Dermal dosage (μg/kg)	AD (μg/kg)
Foil deposition[a]	1,300	124
Surface roller[b]	77	7
Whole-body dosimeter[c]	192	19
Urine biomonitoring[d]	32 (calcd)	3

[a] Foil dosimeter in place to collect chlorpyrifos during fogging; dermal absorption = 9.6%.

[b] Dose dislodged by rolling carpet covered by cotton dosimeter (Ross et al., 1991).

[c] Dose dislodged by person wearing whole body suit (cotton gloves, socks, and union suits).

[d] Absorbed dose excreted during 3 days following two 20-min structured, high-contact activities; dosage is similar to that resulting from full day following fogging of residences.

assigned a finite level of analyte to reflect the sensitivity of the analytical procedure. On statistical grounds, samples are assigned half of the minimum detectable level when they are without measurable levels of environmental or human biological analytes. Lower apparent exposures are assigned when chemical analysis is performed on fewer numbers of samples or lower volumes using the most sensitive and specific method available. This issue must be addressed early in planning biomonitoring studies.

Planning also may be augmented by consideration of established workplace standards or measures of exposure such as PHED. Threshold limit values (TLVs) in the workplace can provide reference points for workplace exposures resulting from the occurrence of chemicals in air. The TLVs are published annually by the American College of Governmental Industrial Hygienists (ACGIH). They represent air levels of chemicals that have not been associated with disease for lifetime (40-year) work periods. Many chemicals carry a "skin notation" to indicate that exposure may result from both inhalation and dermal contact. Several dozen pesticide active ingredients are included among the standards (e.g., atrazine, borate, parathion, malathion). Many TLVs for inert ingredients also are listed and may be useful reference points for assessment of the consequences of workplace or indoor exposure. Biochemical exposure indices (BEIs) are lacking for active ingredients with established TLVs. BEIs could be a valuable reference point for pesticide biomonitoring in the workplace.

Biological monitoring affords exposure assessors several advantages over passive dosimetry or environmental monitoring for the determination of ADD and, ultimately, risk (Table 7-1). Routine ADDs in the μg/kg bw range (<1 to <100) are forecast for handlers and harvesters under conditions of use prescribed by pesticide labels. The importance of routes of exposure including dermal, inhalation, and ingestion is well known and fully accounted for by biological sampling. Distribution of dose, dermal absorption, and metabolic variables underlie interindividual variability. Personal hygiene and habits such as hand-to-mouth contact, smoking, hand washing, and eating or drinking may result in significant ingestion of chemicals. Personal protective clothing such as gloves also have the potential to promote ingestion and absorption. Biological monitoring provides the opportunity to assess the extent and significance of these personal factors in exposure assessment.

Table 7-5 lists human exposure estimates based on environmental and biological monitoring under experimental conditions. Foil coupon deposition, dislodgeable residue estimated by rolling a flat cotton dosimeter, and a cotton whole-body suit each retained significantly more analyte than did human skin, based on the ADD calculated from biomonitoring. The estimated 0.5-μg/kg dosage acquired during two 20-min exposure periods is similar to available daily exposures of 0.1 to 30 μg/kg that have been measured following indoor pesticide use (Krieger et al., in review). Protocols that include structured activity give investigators an opportunity to evaluate the extent of dermal contact in residences, which is the most important determinant of the ADD.

Ambient environmental monitoring is better suited to detection of acutely hazardous environments, and to the regulation and surveillance of exposures referenced to regulatory standards such as reentry intervals and TLVs (time-weighted average TLVs) (ACGIH 1997). When environmental monitoring is intended to contribute to human exposure assessment, data must include a concentration (mg chemical/m^3 air, μg/l water,

$\mu g/cm^2$ surface, etc.) and notation of time to permit calculation of exposure and AD. When hazardous conditions are apparent, the modification of the work environment can be guided by ambient monitoring without continuing human exposure.

Some important responses to chemicals may occur without significant systemic absorption of the stimulus. Effects related to odor, contact dermatitis, irritation of the eye and lung, and the common chemical-sense mediated by the trigeminal nerves of the nose are poorly, if at all, related to traditional estimates of AD. These responses also may be best related to ambient environmental conditions. Alarie et al. (1996) have shown that the reflex reduction in respiration rate in animals exposed to atmospheric irritants is correlated with the time-weighted TLV in the workplace. As understanding of the relationship between exposure and effects is clarified, it may be possible to more meaningfully link an agent in the environment with contact exposure and target organ concentration.

7-6 DEVELOPMENT OF BIOLOGICAL MONITORING APPROACHES

Biological monitoring requires a broad database concerning pesticide use, exposure, and physiologic disposition to allow reliable estimation of ADD ($\mu g/kg$ bw per day). For many active ingredients, data are too limited to propose a complete scheme for workplace or indoor monitoring. Characteristics of substances, analytes, and some general reference points are listed in Table 7-2. Careful measurements of parent pesticide or key metabolites may be used to establish relative measures of exposure related to types of work, productivity, reentry intervals, and protective clothing.

7-6-1 Kinetics

The relationship between absorption, distribution, metabolism, and clearance in urine, feces, or other biological media must be well understood and described in sufficient detail to allow for limiting estimates (Table 7-2). Variables including endogenous factors such as genetic constitution, health and disease, and anthropomorphic factors as well as exogenous factors such as workload, concurrent chemical exposures, tobacco smoking, alcohol consumption, and other personal habits may influence the determination of ADD.

The most common measure of biological clearance of chemicals is half-life, the time required to excrete half of the AD (Table 7-3). It is applied to chemicals that are excreted following first-order kinetics and represents a complex cascade of biological processes. The importance of half-life in blood or urine for biomonitoring is related to the determination of sampling time. Two to four half-lives of collection of plasma or urine are usually sufficient for estimation of ID. Twenty-four hour specimens are a great inconvenience to volunteers and unsuitable for most routine monitoring programs. As shown in Table 7-3, the half-life guides selection of sampling times, particularly when the monitoring goal is an ADD rather than a relative exposure indicator based on ID. Substances with extremely short biological half-lives, such as organic solvents and fumigants, cannot be biomonitored practically. The generic examples selected for the table are not exclusive, but they represent well-studied pesticides that have been successfully biomonitored. It is not critical to set sampling time for long-lived or persistent chemicals such as DDT, dieldrin, or pentachlorophenol that may be measured under apparent steady-state conditions.

7-6-2 Background Levels

The ubiquitous occurrence of trace levels of chemicals used as pesticides in food, air, water, and on surfaces and the ever-increasing sensitivity of analytical methods account for findings of trace levels of some substances in blood and urine. Background levels must be subtracted from analytical data to ensure that exposure has occurred. Background levels will vary depending on the magnitude of individual exposure associated with indoor or outdoor mixing/loading/application of pesticides, harvesting of treated crops, living in treated quarters, wearing treated clothing, or contacting treated lawn or turf.

A special background consideration in biomonitoring confronts investigators when both active ingredient and environmental transformation product are absorbed and excreted as a common metabolite. The occurrence in urine of phenolic leaving groups such as *p*-nitrophenol, nitrocresol, or trichloropyridinol (from parathion, fenitrothion, and chlorpyrifos, respectively) may inflate estimates of ID or ADD of organophosphate insecticides. A monitoring strategy involving a combination of passive dosimetry and biomonitoring may be required to determine the contribution of the active ingredient to urinary metabolite levels for risk assessment.

7-6-3 Biological Media

The principal media for sampling are blood and urine. Both fluids are free from environmental sources of contamination and absorbed chemicals or their metabolites appear in these media in measurable amounts after relatively short time intervals under routine conditions of use.

Blood. Blood constituents are the primary vehicle for the transport and distribution of chemicals. Parent chemicals and their metabolic derivatives can be detected in blood at short intervals after absorption. Determination of rates of biotransformation and clearance may represent the most sensitive and specific means of estimating AD owing to procedures developed for sensitive analysis of food and other environmental samples. Monitoring strategies dependent upon invasive sampling, usually blood, are less acceptable to volunteers than programs that utilize urine.

Urine. Urine is the most convenient body fluid for biomonitoring. Water-soluble metabolites may be collected in urine specimens by noninvasive procedures that are readily accepted by both males and females. As a result of water conservation mechanisms in the kidney, metabolites are more concentrated in urine specimens than in blood or plasma. As an additional consideration related to kidney concentration of urine, chemicals with relatively short half-lives are more effectively monitored in end-of-shift urine samples than in blood samples or exhaled air. Conditions that affect glomerular filtration, tubular secretion, and tubular reabsorption may alter rates of clearance and decrease the accuracy of determination of ADD.

Time is an additional important sampling consideration. Twenty-four hour specimens are preferred to spot samples, but they require considerable discipline on the part of volunteers and they are used infrequently in routine monitoring programs. End-of-shift spot samples may be too dilute (high liquid intake) or too concentrated (low liquid intake, excessive perspiration) for the analysis of short-lived substances such as inert ingredients or rapidly metabolized pesticides. This factor may be taken into account by measurement or urine creatinine or density (Elkins, Pagnotto, and Smith 1974) and used to estimate daily analyte clearance.

Incomplete daily urine collection may be corrected by measurement of creatinine concentrations that are both age and sex dependent, reflecting their dependence upon metabolic activity and muscle mass. Pesticide or metabolite concentrations ($\mu g/l$) must be divided by the creatinine concentration (g/l) to yield a measure that can be normalized by daily creatinine excretion to daily dose (Table 7-1).

Exhaled air. Analysis of alveolar air is limited to volatile substances of low solubility in blood. It is a noninvasive procedure that is not used currently to determine ID of pesticides or inert ingredients (Mikheev 1996). Solvents used as inert ingredients and volatile soil, commodity, and structural pesticides would be candidates for alveolar air analysis.

Other biological media. Other biological media including mother's milk, body fat, hair, fingernails, sweat, sputum, and saliva have been used to detect the occurrence of persistent chemical residues in research, but these media have not been adopted for routine biomonitoring (Nigg and Stamper 1989). Examples of use of other media are included in Section 7-7, Selected Biomonitoring Strategies and Results.

7-6-4 Analysis

Analytical methods must be accurate, precise, repeatable, sensitive, and chemically specific. Careful descriptions of analytical procedures allows scientific evaluation of their possible influence on exposure measurements. *Accuracy* represents the relationship between the analytical result and the true value determined by analysis of a large number of samples of known concentration. *Precision* is the closeness of separate analytical results. *Repeatability* is an expression of minimum variability, and *reproducibility* reflects its maximum. *Limits of detection* relative to background and signal noise as well as minimum *limits of quantitation* relative to expected levels are critical characteristics of analytical procedures for biomonitoring. A detailed discussion of these factors relative to biomonitoring has been published by Aitio (1996).

Time and costs of sample preparation and analysis are additional factors that will influence the selection of analytical procedures. These considerations become especially important when large numbers of samples are to be analyzed on a routine basis.

7-7 SELECTED BIOMONITORING STRATEGIES AND RESULTS

These examples illustrate pesticide biomonitoring programs to evaluate potential health risks where a measure of ID has been established. Organophosphate and organochlorine insecticide exposures have been studied using biomonitoring of urine in a variety of settings. These examples are useful to illustrate magnitudes of exposure associated with direct and indirect pesticide contact. Because similar sampling and analysis have been used, it is possible to consider relative amounts of exposure associated with particular activities in agricultural and urban settings.

7-7-1 Organophosphate Insecticides

Organophosphate insecticides have been in widespread use since the late 1940s. Exposures in agriculture are associated with handling products, mixing/loading and applying pesticide sprays (Nutley and Cocker 1993), and reentry of treated fields to harvest or perform other tasks (Krieger, Ross, and Thongsinthusak 1992). Other activities, including indoor pest control and urban use on private properties, are also sources of human exposure amenable to biomonitoring (Krieger 1995).

Mortality and morbidity prompted interdisciplinary studies that included the first biomonitoring studies in agriculture in the early 1950s (Batchelor and Walker 1954). Earliest studies were based on the p-nitrophenol leaving group of ethyl parathion. About 15 years later, Shafik and Enos (1969) reported measurement of alkyl phosphates in urine. The majority of the organophosphate insecticides are metabolized to six simple dialkyl esters which are stable in urine for at least 20 weeks (Ito et al. 1979). These include the following: O,O-dimethylphosphate (DMP), O,O-diethylphosphate (DEP), O,O-dimethylphosphorothioate (DMTP), O,O-diethylphosphorothioate (DETP), O,O-dimethylphosphorodithioate (DMDTP), and O,O-diethylphosphorodithioate (DEDTP). In general, 90% or more of a low oral dose is eliminated in urine within 6 to 24 h of administration. Accidental exposures may result in clearance of detectable levels in urine over substantially longer periods.

Leaving-group analysis of p-nitrophenol from methyl and ethyl parathion, trichloropyridinol from chlorpyrifos, and nitrocresol from fenitrothion is generally more sensitive and specific than analysis of dialkyl phosphates. Malathion mono- and diacids can be measured in urine at 3 to 5 ppb, at least five-times less than the detection limit (25–50 ppb) for DMP, DMTP, and DMDTP in urine.

Biomonitoring is well suited to evaluate exposure mitigation measures such as protective clothing, hand protection, or alternative work practices. Recent field studies to evaluate the protectiveness of rubber latex gloves demonstrate the importance of biological monitoring (Krieger 1995). Strawberry fields treated with malathion were harvested for three consecutive days by a crew of twenty-four harvesters. Dislodgeable foliar residues ranged from 0.005 $\mu g/cm^2$ to 0.15 $\mu g/cm^2$. Workers who wore gloves absorbed about 50% of the dose absorbed by workers who harvested strawberries bare-handed. If glove rinses had been used to estimate hand exposure, the ADD would have been set three to ten times higher than the ADD determined by biomonitoring. Environmental measures of exposure would be inadequate to gauge the protectiveness of gloves.

Mixer/loader/applicators used ultralow-volume equipment to treat apple orchards in British Columbia, where biomonitoring was used to show the importance of dermal organophosphate exposure resulting, in part, from clothing penetration (Franklin et al. 1981). In another study involving ninety-seven orchard workers, urine biomonitoring occurred during two seasons. Three insecticides were used, including methidathion, vamidothion, and azinphos-methyl. Only eight persons had background dialkylphosphate levels before spraying. In this study, azinphos-methyl exposure resulted in an ADD of less than 3 μg/kg bw. There was a lack of correlation between metabolite concentration and cholinesterase inhibition. Twelve persons had decreased serum cholinesterase (31–48%), but four of those had no urinary diakylphosphates. Retesting was not feasible. The study confirmed that dialkylphosphates in urine were more sensitive than cholinesterase inhibition in serum (Franklin et al. 1981). Similar studies by Franklin, Muir, and Moody (1986) correlated dermal exposure from patch dosimeters and urinary dimethylthiophosphate. Urine biomonitoring yielded more reliable and accurate estimates of exposure than patch dosimetry data.

Urine biomonitoring also was used to measure diazinon exposure of workers in a Medfly eradication program in California. Persons who applied granular diazinon were biomonitored by analysis of diethylthiophosphate in urine. Diazinon equivalents cleared in urine ranged from about 1–150 μg/kg bw per day for applicators and to about 0.4–4 μg/kg bw per day for supervisory personnel (Weisskopf et al. 1988; Weisskopf and Seiber 1989).

Persons in urban and rural indoor settings are exposed to a variety of products as a result of commercial or home-use products. Residents exposed to chlorpyrifos following indoor fogging have been biomonitored for periods of several days to more than a month (Krieger et al., in review). 3,5,6-Trichloro-2-pyridinol (TCP) may be measured in urine (limit of quantitation = 4 ppb) of persons monitored following day-to-day activities. Data in review indicate a range of exposure from 0.1 to 30 μg/kg bw. Results of biomonitoring reveal longer-term exposure than predicted by surface wipes and measurements of dislodgeable residues and air levels of chlorpyrifos in indoor environments. These studies and similar additional work represent results of situational exposure monitoring (Krieger et al. 1991) characterized by routine analysis of samples opportunistically obtained from volunteers following routine pesticide use. Such studies are an important means to establish the magnitude of exposure as foundation for establishment of meaningful testing procedures and critical experimental studies of pathways of exposure.

When produce contains pesticide residues, the amount is usually insufficient to yield urinary metabolites except under circumstances of excessive consumption (Krieger and Thongsinthusak 1993). Under experimental conditions, DMP, DMTP, and DMDTP were measurable within 2 h and detectable clearance continued over a period of about 2 days. The dosage was 124 μg dimethoate/kg resulting from ingestion of 7.3 g peas/kg containing violative levels (17 ppm) of the organophosphate. Clearance of an oral bolus dose and the residue from snow-pea ingestion were virtually identical. Neither plasma nor erythrocyte cholinesterase was inhibited at this level of dietary exposure. Urinary clearance of dialykylphosphate metabolites represents a more sensitive measure of acute exposure than acetylcholinesterase inhibition.

When initial concerns about organophosphate exposures were directed toward acute toxicity, there was widespread use of cholinesterase monitoring to detect exposure. The enzymes were inhibited by high or prolonged exposures, but contemporary exposure assessments for risk management require greater sensitivity and finite quantitative estimates of dose. Plasma pseudocholinesterase and erthrocyte acetylcholinesterase were used as surrogate indicators of the status of acetylcholinesterase in the nervous system. Cholinesterase inhibition cannot be quantitatively related to clinical status and ADD, but it does have the advantage of representing exposure experience over a period of time. Analytically significant inhibition may or may not be associated with illness; however, the measurements can be used to signal the need for evaluation of conditions of use and exposure. The most complete discussion of the relationship between cholinesterase inhibition and clinical signs has been published by Namba et al. (1971). As health concerns have focused upon quantitative measurement of low levels of exposure over longer periods of time, cholinesterase monitoring has become less useful as an exposure indicator.

Cholinesterase activity levels may be useful indicators of exposure and symptom interviews may contribute to exposure mitigation, but dialkylphosphate levels in urine are the more definitive measure of exposure for worker surveillance, investigations of illness, and general surveys of exposed persons (Richter et al. 1992). No changes in cholinesterase activities were found in workers with urine acephate levels of 1–2 mg/l following an 8-day period of use (Maroni et al. 1990). The importance of urine biomonitoring has been emphasized by Drevenkar et al. (1991), Richter et al. (1992), and Krieger and Thongsinthusak (1993).

Cholinesterase monitoring is of limited usefulness owing primarily to variable levels of response associated with particular organophosphate insecticides, insensitivity as an exposure indicator, intra- and interindividual variability, and analytical issues related to instrumental variables and sampling. Subclinical changes (less than 15%) in inhibition were correlated with dermal exposure by Barnhart et al. (1992). Greater inhibition was reported in surveys by Rama and Jaga (1992) and Forastiere, Papini, and Perucci (1992). A large cross-sectional monitoring study of Florida citrus field-workers was reported by Griffith and Duncan (1985). Applicators, mixers, loaders, tractor drivers, general workers, and harvesters in 436 Florida citrus groves were administered a health questionnaire. Some provided urine specimens which revealed higher levels of exposure during the spray season than during harvest. There was no apparent association between health symptoms and the relatively low levels of urinary dialkyl phosphate metabolites.

7-7-2 Organochlorine Insecticides

Persistent chemicals in the environment also are cleared slowly from the body. DDT and hexachlorocyclohexane (HCH) may be monitored in biota and human blood, fat, and breast milk (Bouwman, Becker, and Schutte 1994; Del Puerto et al. 1990; Kanja et al. 1992; Murphy and Harvey 1988; Nair and Pillai 1992; Sasaki et al. 1991). Reduced use in many regions is related to a general decline in tissue levels. Polychlorinated biphenyls (PCBs) are frequently reported in such surveys, but they should not be grouped with pesticides because of their different patterns of use. The comprehensive studies conducted by the US Public Health Service feature procedures for the detection and analysis of multiple organochlorines in sera (Burse et al. 1990). Continued use in other regions of the world will result in longer-term retention of residual pesticides. Active malaria control was associated with elevated total DDT in breast milk (12.2 μg/g before spraying to 19.5 $\mu g/g$ 6 months after spraying) (Bouwman, Becker, and Schutte 1994). These levels are substantially greater than levels resulting from less direct environmental exposures which result in the tissue levels reported in the following paragraph. They illustrate the range of tissue levels of persistent organochlorines that may be revealed by biomonitoring.

Other epidemiological biomonitoring studies have utilized maternal blood, milk, fat, and umbilical cord blood collected during cesarean births in Norway (Kanja et al. 1992). Samples were examined for organochlorines contained DDT (100%), p,p-DDE (100%), o,p-DDT (59%), dieldrin (27%), transnanochlor (15%), β-HCH (12%), and lindane (2%). Mean residue levels in fat ranged from 5.9 mg DDT (total)/kg to 0.03 mg β-HCH/kg and dieldrin (detected but amount not measurable). There were significant correlations between total DDT in subcutaneous fat and milk fat ($r = .963$), subcutaneous fat and maternal serum fat ($r = .843$), and maternal serum fat and maternal milk ($r = .868$). Similar results were obtained from analyses of breast milk and fat from samples from large numbers of nursing mothers in Australia (Stevens et al. 1993) and in a major, comprehensive study in the United States (Stehr-Green 1989). Liver, fat, blood cells, and serum were analyzed in Finland, where median serum levels were 1.8 mg DDT (total)/kg and 0.3 mg β-HCH and the investigator favored fat/serum ratios in the study of human exposure to organochlorine chemicals (Mussalo-Rauhamaa 1991). Results of this type of monitoring form an important link between the occurrence of persistent chemicals and their disposition, including deposition, in humans. They are particularly useful to reveal long-term trends of human exposure to persistent chemicals (Morris et al. 1984; Murphy and Harvey 1988) and in accidental contamination of the food supply (Stehr-Green et al. 1989).

Biomonitoring also may contribute to the clinical management of patients acutely or chronically exposed to persistent chemicals (Olanoff et al. 1983). Dose-response relationships in chlordecone studies were investigated by sensitive gas-liquid chromatographic analysis of body fluids and tissues (Guzelian 1992). Persons with toxic signs had blood:tissue ratios of 1:7. Data obtained from humans were more predictive of the significant toxic effects than were measurements made in rats. These findings underscore the importance of direct knowledge about the magnitude of pesticide exposure in humans.

7-7-3 Other Pesticides

Biomonitoring was used to assess the protectiveness of disposable coveralls worn by applicators and by mixer/loaders of chlorobenzilate in Florida (Nigg and Stamper 1983). Deposition of chlorobenzilate on clothing was four times greater on the applicators' than on the mixer/loaders' clothing, but AD was similar in both groups. Coveralls reduced AD by 24%. The study design permitted direct comparison of the protectiveness of coveralls. That work also demonstrated that clothing may intercept significant amounts of pesticide or other chemicals that are unavailable for dermal absorption. Excessive estimates of ADD result from use of passive dosimetry in human pesticide exposure studies.

Exposure of spraymen engaged in vector control expectedly vary depending on training, work practices, and general hygiene. In pesticide evaluation studies, Chester,

Sabapathy, and Woollen (1992) reported λ-cyhalothrin exposures of spraymen and villagers in dwellings treated for malaria. AD (54 μg/person) of spraymen was estimated from urine and serum. Trace levels were detected in villagers.

Nurseryworkers in a four-state study were biomonitored in a year-long study that included 2,4,5-T, 2,4-D, 2,4-DP, benomyl, bifenox, captan, carbaryl, chlorpyrifos, fenvalerate, picloram, and chlordimeform among fifteen products in use (Lavy et al. 1993). Low levels of exposure were inferred from the sparse number of positive urine specimens (42) of 3,134 samples analyzed. Those findings suggest that more sensitive procedures would be required to measure the pesticide exposure of persons engaged in conifer seedling production.

Dutch greenhouse workers who harvested carnations were monitored for propoxur exposure using urinary 2-isopropoxyphenol, the major metabolite. Total estimated dermal and respiratory exposure during harvesting ranged from 0.2 to 46 mg and from 3 to 278 μg, respectively. The Pearson correlation coefficient between dermal exposure and urinary phenol was .95%. Brouwer et al. (1993) estimated that dermal exposure could account for more than 80% of the AD. On the basis of the amount of phenol excreted, no increased health risks were predicted.

Homeowner use of pesticides outdoors has received very limited experimental study and even less evaluation by biological monitoring. Trace levels of 2,4-D were found in persons who were barefoot and wore shorts during a 1-h exposure to sprayed turf. All urines of persons exposed 24 h later contained nondetectable residues (Harris and Solomon 1992).

Epidemiological studies of chemicals in urine of the general population can reveal relatively persistent chemicals and their metabolites (Kutz et al. 1992). Pentachlorophenol was found in 71.6% (mean 6.3 μg/l) of the samples. Assuming 70-kg persons produce 1,500 ml of urine per day, ADD of 0.14 μg/kg may be estimated. Measurable levels of TCP (5.8% of samples analyzed), 2,4,5-trichlorophenol (3.4%), p-nitrophenol (2.4%), dicamba (1.4%), malathion mono- and dicarboxylic acids (1.1% and 0.5%, respectively), and 2,4-D (0.3%). These substances were detected in a 1976–1980 survey. Because the interval between exposure and sampling is unknown, it is difficult to reach meaningful conclusions about the occurrence of these substances or their parent pesticides in the environment.

7-7-4 Summary

Biological monitoring is a means to estimate ADD from a biological rather than an environmental database. It is axiomatic that there is an acceptable or safe level of exposure that can be predicted for chemicals including pesticides below which significant adverse effects, injury, illness, or discomfort will seldom occur. Exposure levels determined by biomonitoring rely upon fewer default assumptions than those developed by measuring chemical concentrations in environmental compartments. Biomonitoring allows investigators to avoid inflated and misleading estimates of AD resulting from defaults related to the magnitude and distribution of dose, the effectiveness of engineering controls, the protectiveness of clothing, and dermal absorption. Priority must be given to development of methods that can provide more accurate measures of dose. More accurate information about exposure should be helpful to clarify the potential impact of pesticide technologies on human health and well-being.

REFERENCES

American College of Governmental Industrial Hygienists (ACGIH) 1997. *Threshold limit values and biological exposure indices.* Cincinnati: ACGIH.

Aitio, A. 1996. Quality assurance. In *Biological monitoring of chemical exposure in the workplace.* ed. M. I. Mikheev, 20–51. Vol. 1. Geneva: World Health Organization.

Alarie, Y. 1981. Bioassay for evaluating the potency of airborne sensory irritants and predicting acceptable levels of exposure in man. *Food Cosmet. Toxicol.* 19:623–26.

Arterberry, J. D., W. F. Durham, and J. W. Elliott. 1961. Exposure to parathion: Measurement by blood cholinesterase level and urinary *p*-nitrophenol excretion. *Arch. Environ. Health* 3:476–85.

Barnhart, S., M. Miller, G. Gallagher, S. W. Horstman, D. Eaton, and L. Rosenstock. 1992. Organophosphate pesticide exposure in a group of Washington State orchard applicators. *Environ. Res.* 59:229–37.

Batchelor, G. S., K. C. Walker. 1954. Health hazards involved in the use of parathion in fruit orchards of North Central Washington. *AMA Arch. Ind. Hyg. Occup. Med.* 10:522–29.

Berteau, P. E., J. B. Knaak, D. C. Mengle, and J. B. Schreider. 1989. Insecticide absorption from indoor surfaces. In *Biological monitoring for pesticide exposure—measurement, estimation and risk reduction.* ed. R. G. M. Wang, C. A. Franklin, R. C. Honeycutt, and J. C. Reinert, 6–27. *ACS Symposium Series, no. 382,* Washington, D.C.: American Chemical Society.

Bouwman, H., P. J. Becker, and C. H. Schutte. 1994. Malaria control and longitudinal changes in levels of DDT and its metabolites in human serum from Kwa Zulu. *Bull. WHO* 72:921–30.

Brouwer, D. H., E. J. Brouwer, J. A. de Vreede, R. T. H. van Welie, N. P. E. Vermuelen, and J. J. van Hemmen. 1991. Inhalation exposure to 1,3-dichloropropene in the Dutch flower-bulb culture. *Arch. Environ. Contam. Toxicol.* 20:1–5.

Brouwer, R., K. van Maarleveld, L. Ravensberg, W. Meuling, W. de Kort, and J. J. van Hemmen. 1993. Skin contamination, airborne concentrations, and urinary metabolite excretion of propoxur during harvesting of flowers in greenhouses. *Am. J. Ind. Med.* 24:593–603.

Burse, V. W., S. L. Head, M. P. Korver, P. C. McClure, J. F. Donahue, and L. L. Needham. 1990. Determination of selected organochlorine pesticides and polychlorinated biphenyls in human serum. *J. Anal. Toxicol.* 14(3):137–42.

Chester, G., N. N. Sabapathy, and B. H. Woollen. 1992. Exposure and health assessment during application of lambda-cyhalothrin for malaria vector control in Pakistan. *Bull. WHO* 70(5):615–19.

Cranmer, M. F., J. J. Carroll, and M. F. Copeland. 1972. Determinations of DDT and metabolites including DDA, in human urine by gas chromatography. In *Epidemiology of DDT.* ed. J. E. Davis, and W. F. Edmundson, Appendix III. Mt. Kisco, N.Y.: Futura.

Del Puerto, K., A. M. Fernandez, M. Gimenez, L. Fernandez, C. Perez, and J. A. Verdes. 1990. Levels of DDT and its metabolites in human biological medium. *Gig. Sanit.* 1990:73–75.

Drevenkar, V., Z. Radic, Z. Vasilic, and E. Reiner. 1991. Dialkylphosphorus metabolites in the urine and activities of esterases in the serum as biochemical indices for human absorption of organophosphorus pesticides. *Arch. Environ. Contam. Toxicol.* 20 (3):417–22.

Durham, W. F., and H. R. Wolfe. 1962. Measurement of the exposure of workers to pesticides. *Bull. WHO* 26:75–91.

Elkins, H. B., L. D. Pagnotto, and H. L. Smith. 1974. Concentration adjustments in urinanalysis. *Am. Ind. Hyg. Assoc. J.* 35:559.

Elliott, J. W., K. C. Walker, A. E. Penick, and W. F. Durham. 1960. Insecticide exposure: A sensitive procedure for urinary *p*-nitrophelnol determination as a measure of exposure to parathion. *J. Agric. Food Chem.* 8:111–13.

Forastiere, F., P. Papini, and C. A. Perucci. 1992. The monitoring of cholinesterases in farm workers and tradesmen exposed to phosphoric esters and carbamates. *Med. Lav.* 83:135–45.

Franklin, C. A., R. A. Fenske, R. Greenhalgh, L. Mathieu, H. V. Denly, J. T. Leffingwell, and R. C. Spear.

1981. Correlation of urinary pesticide metabolite excretion with estimated dermal contact in the course of occupational exposure to guthion. *J. Toxicol. Environ. Health* 7:715–31.

Franklin, C. A., N. I. Muir, and R. P. Moody. 1986. The use of biological monitoring in the estimation of exposure during the application of pesticides. *Toxicol. Lett.* 33:127–33.

Griffith, J. G., and R. C. Duncan. 1985. Alkyl phosphate residue values in the urine of Florida citrus field-workers compared to the National Health and Nutrition Examination Survey (NHANES) sample. *Bull. Environ. Contam. Toxicol.* 34:210–15.

Gunther, F. A., Y. Iwata, G. E. Carman, and C. A. Smith. 1977. The citrus reentry problem: Research on its causes and effects, and approaches to its minimization. *Residue Rev.* 67:1–139.

Guzelian, P. S. 1992. The clinical toxicology of chlordecone as an example of toxicological risk assessment for man. *Toxicol. Lett.* 64–65:589–596.

Harris, S. A., and K. R. Soloman. 1992. Human exposure to 2,4-D following controlled activities on recently sprayed turf. *J. Environ. Sci. Health, Part B, Pestic. Food Contam. Agric. Wastes* 27 (1):9–22.

Hoet, P. 1996. General principles. In *Biological monitoring of chemical exposure in the workplace*. ed. M. I. Mikheev, 1–19. Vol. 1. Geneva: *World Health Organization*.

Ito, G., W. W. Kilgore, and J. J. Seabury. 1979. Effect of freezer storage on alkyl phosphate metabolites in urine. *Bull. Environ. Contam. Toxicol.* 22:530–35.

Kanja, L. W., J. U. Skaare, S. B. Ojwang, and C. K. Maitai. 1992. A comparison of organocholorine pesticide residues in maternal adipose tissue, maternal blood, cord blood, and human milk from mother/infant pairs. *Arch. Environ. Contam. Toxicol.* 22 (1):21–24.

Krieger, R. I., 1995. Pesticide exposure assessment. *Toxicol. Lett.* 82/83:65–72.

Krieger, R. I., and T. Thongsinthusak. 1993. Metabolism and excretion of dimethoate following ingestion of overtolerance peas and a bolus dose. *Food Chem. Toxicol.* 31:177–82.

Krieger, R. I., J. H. Ross, and T. Thongsinthusak. 1992. Assessing human exposures to pesticides. *Rev. Environ. Contam. Toxicol.* 128:1–16.

Krieger, R. I., T. Thongsinthusak, J. H. Ross, S. Taylor, S. Frederickson, S. Begum, and M. Dong. 1991. Situational chemical exposure studies provide human metabolism and urine clearance data for chloropyrifos (C), dimethoate (D), and malathion (M), (Abst. 570). *Toxicologist* 11:680.

Krieger, R. I., T. M. Dinoff, L. Fell, T. G. Osimitz, J. H. Ross, and T. Thongsinthusak. (In review) Biomonitoring and whole body cotton dosimetry to estimate the potential human dermal exposure to semivolatile chemicals used indoors. *J. Exposure Analysis and environ. Epidemiol.*

Kutz, F. W., B. T. Cook, O. D. Carter-Pokras, D. Brody, and R. S. Murphy. 1992. Selected pesticide residues and metabolites in urine from a survey of the US general population. *J. Toxicol. Environ. Health* 37 (2):277–91.

Lavy, T. L., J. D. Mattice, J. H. Massey, and B. M. Skulman. 1993. Measurements of year-long exposure to tree nursery workers using multiple pesticides. *Arch. Environ. Contam. Toxicol.* 24:123–44.

Maroni, M., G. Catanacci, D. Galli, D. Cavallo, and G. Ravazzani. 1990. Biological monitoring of human exposure to acephate. *Arch. Environ. Contam. Toxicol.* 19 (5):782–88.

Mikheev, M. I., ed. 1996. *Biological monitoring of chemical exposure in the workplace*, Vol. 1. *Geneva: World Health Organization*.

Morris, C. R., A. Gocmen, H. A. Peters, D. J. Cripps, and G. T. Bryan. 1984. Epidemiological study of populations previously exposed to hexachlorobenzene (HCB). *Proceeding of the Third NCI/EPA/NIOSH Collaborative Workshop: Progress on Joint Environmental and Occupational Cancer Studies* 463–79.

Murphy, R., and C. Harvey. 1988. Residues and metabolites of selected persistent halogenated hydrocarbons in blood specimens from a general population survey. *Environ. Health. Perspect.* 60:115–20.

Mussalo-Rauhamaa, H. 1991. Partitioning and levels of neutral organochlorine compounds in human serum, blood cells, and adipose and liver tissue. *Sci. Total Environ.* 103 (2–3):159–76.

Nair, A., M. K. K. Pillai. 1992. Trends in ambient levels of DDT and HCH residues in humans and the environment of Delhi, India. *Sci. Total Environ.* 121:145–57.

Namba, T., C. T. Nolte, J. Jackrel, and D. Grob. 1971. Poisoning due to organophosphate insecticides: Acute and chronic manifestations. *Am. J. Med.* 50:475–92.

National Research Council, National Academy of Sciences (NRC/NAS). 1983. *Risk assessment in the federal government: Managing the process*. Washington, D.C. *National Academy Press*. pp. 1–191.

Nigg, H. N., and J. H. Stamper. 1983. Exposure of spray applicators and mixer/loaders to chlorobenzilate miticide in Florida citrus groves. *Arch. Environ. Contam. Toxicol.* 12 (4):477–82.

Nigg, H. N., and J. H. Stamper. 1989. Biological monitoring for pesticide dose determination. In *Biological monitoring for pesticide exposure—measurement, estimation and risk reduction*. ed. R. G. M. Wang, C. A. Franklin, R. C. Honeycutt, and J. C. Reinert, 6–27. ACS Symposium Series, no. 382, Washington, D. C.: American Chemical Society.

Nutley, B., and J. Cocker. 1993. Biological monitoring of workers exposed to organophosphorous pesticides. *Pestic. Sci.* 38:315–22.

Olanoff, L. S., W. J. Bristow, J. Colcolough Jr., and J. R. Reigart. 1983. Acute chlordane intoxication. *J. Toxicol. Clin. Toxicol.* 20:291–306.

Plimmer, J. 1982. *Pesticide residues and exposure*. ACS Symposium Series, no. 182. Washington, D.C.: American Chemical Society.

Popendorf, W., and J. T. Leffingwell. 1982. Regulating OP pesticide residues for farm-worker protection. *Residue Rev.* 82:125–201.

Rama, D. B., and K. Jaga. 1992. Pesticide exposure and cholinesterase levels among farm workers in the Republic of South Africa. *Sci. Total Environ.* 122:315–19.

Richter, E. D., P. Chuwers, Y. Levy, M. Gordon, F. Grauer, J. Marzouk, Y. Levy, S. Levy, S. Barron, and N. Gruener. 1992. Health effects from exposure to organophosphate pesticides in workers and residents in Israel. *Isr. J. Med. Sci.* 28:584–98.

Ross, J. H., T. Thongsinthusak, H. R. Fong, S. Margetich, and R. I. Krieger, 1991. Measuring potential dermal transfer of surface pesticide residue generated from indoor fogger use: Interim report. *Chemosphere* 20, 349–360.

Sasaki, K., T. Ishizaka, T. Suzuki, M. Takeda, and M. Uchiyama. 1991. Accumulation levels of organochlorine pesticides in human adipose tissue and blood. *Bull. Environ. Contam. Toxicol.* 46 (5):662–69.

Shafik, M. T., and H. F. Enos. 1969. Determination of metabolic and hydrolytic products of organophosphorous pesticide chemicals in human blood and urine. *J. Agr. Food Chem.* 17:1186–89.

Stehr-Green, P. A. 1989. Demographic and seasonal influences on human serum pesticide residue levels. *J. Toxicol. Environ. Health* 27:405–21.

Stevens, M. F., G. F. Ebell, and P. Psaila-Savona. 1993. Organochlorine pesticides in Western Australian nursing mothers. *Med. J. Aust.* 158:238–41.

Weisskopf, C. P., and J. N. Seiber. 1989. New approaches to analysis of organophosphate metabolites in the urine of field workers. In Wang, Franklin, Honeycutt Reinert, ACS *Symposium Series*, no. 382, Washington, D.C.: American Chemical Society.

Weisskopf, C. P., J. N. Seiber, N. Maizlish, and M. Schenker. 1988. Personnel exposure to diazinon in a supervised pest eradication program. *Arch. Environ. Contam. Toxicol.* 17 (2):201–12.

EIGHT

BIOLOGICAL MONITORING: NEUROPHYSIOLOGICAL AND BEHAVIORAL ASSESSMENTS

Donald J. Ecobichon

Department of Pharmacology and Toxicology, Queen's University, Kingston, Ontario, Canada K7L 3N6

8-1 INTRODUCTION

Given the following three scenarios, which individual was poisoned by an organophosphorus ester insecticide, a phenoxyherbicide, and a dithiocarbamate fungicide? In one case, an unprotected farmworker was exposed to a pesticide during and after spraying in a cucumber field and, within hours, complained of dizziness, tiredness, muscle weakness, headache, and nausea, showed some disorientation, slurred speech, tonic and clonic convulsions, and lost consciousness. In a second incident, the afflicted individual reported flu-like symptoms, headache, nausea, vomiting, physical weakness, muscle spasms, slurred speech, disorientation, mental confusion, and memory loss. The third agricultural worker suffered headache, nausea, vomiting, dizziness, muscle pain, myotonia, weakness, and fatigue. Note that the postexposure signs and symptoms were similar in all three poisonings. In these incidents, the individuals were exposed to Manzidan (a fungicide mixture of maneb and zineb), diazinon (an organophosphorus ester insecticide), and 2,4-dichlorophenoxyacetic acid (the herbicide 2,4-D), respectively (Ecobichon and Joy 1994). Had the patients been monitored after recovery, similar signs and symptoms would still have been recorded, mostly indicative of neurological disorders. Of course, additional testing would have been required to identify the specific toxicants.

The above illustrates a general ramification of pesticide poisonings: The symptomatology can appear very similar. A second important point is the fact that acute, seasonal or annual (year-long) exposure to a variety of pesticides can result in some form of neurobehavioral, neuropsychological, or neuromuscular disorder (Ecobichon and

Joy 1994; Kaloyanova and El Batawi 1991). A third ramification is that there has been slow recognition that (1) victims of acute poisonings, after apparent recovery, may exhibit subsequent chronic, neurological effects; and (2) ancillary to this, chronic exposure to low-to-moderate levels of pesticides may elicit subtle, neurological sequelae at a late date, without any acute effects being observed. In such scenarios, the lesions may be difficult to detect and even more difficult to assess and quantify, the afflicted individuals either showing a slow recovery of normal functions or having persistent, even permanent, neurological deficits.

The preceding chapters have dealt with many technologies to monitor potentially exposed workers, to measure the activities of key target enzymes or the changes in affected neurotransmitters, to quantify the levels of surrogate chemicals and/or excreted pesticide residues and metabolites in biological fluids. This chapter addresses (1) acute signs/symptoms, which typically disappear within a few days, but which give valid clues to the nature of the poisoning; (2) the subtle, chronic, central and peripheral neurological and neuropsychological symptoms; and (3) the techniques used in assessing these debilitating conditions. Assessment of the subtle neuronal changes has been difficult because of the subjective nature of many of the techniques but advances have been made and we see that there is virtue in the old adage that "when you know what to look for, you can usually find it." Emphasis is placed on insecticide-related neurotoxicity because these agents are more hazardous and represent about 31% of the total pesticides used, but the herbicides (24% of total use) and fungicides (21% of total use) are not neglected (Mushak and Piver 1992).

8-2 ACUTE TOXICITY

The acute toxicity of pesticides has been the focus of many books, chapters, and papers over the years, the reader being referred to Ecobichon (1995, 1996), Ecobichon and Joy (1994), Kaloyanova and El Batawi (1991), Hayes and Laws (1991), and Marrs (1993) for both overviews and specific aspects of poisoning, mechanisms of action, and treatment. The general signs/symptoms of acute pesticide poisoning are described in Table 8-1, reflecting both the variety of specific mechanisms of action as well as the physicochemical properties of individual agents. Except for the insecticides and the bipyridyl herbicides, the human reactions, following exposure to other classes of pesticides, tend to be relatively nonspecific in nature, the agents generally eliciting ocular and dermal irritation and rashes, nausea, vomiting and diarrhea, muscle weakness, as well as dizziness, lethargy, fatigue, and other centrally related effects.

Insecticides, whether organophosphorus, carbamate, organochlorine, or pyrethrin in nature, exert actions on various target sites in both the peripheral and central nervous systems, giving rise to distinctive, agent- or class-specific signs and symptoms of toxicity (Table 8-1). The mechanisms by which these different insecticide classes elicit their actions have been described in detail, concise summaries being given by Ecobichon (1995, 1996) and Matsumura (1985). Of the bipyridyl, nonselective contact herbicides, paraquat is a unique mammalian poison, acting specifically in pulmonary tissue following active uptake by a diamine/polyamine transport system; it plays the role

Table 8-1 Acute pesticide toxicity, general signs and symptoms

Pesticide classification	Symptomatology
Insecticides	
Organophosphates Carbamates	*Mild:* Headache, dizziness, perspiration, lacrimation, salivation, blurred vision, tightness in chest, twitching of muscles in eyelids, lips, tongue, face.
	Moderate: Abdominal cramps, nausea, vomiting, diarrhea, bronchial hypersecretions, bradycardia or tachycardia, muscle (skeletal) spasms, tremors, general weakness.
	Severe: Pinpoint pupils, profuse sweating, urinary and/or fecal incontinence, mental confusion, pulmonary edema, respiratory difficulty, cyanosis, progressive cardiac and respiratory failure, unconsciousness leading to coma.
Organochlorines	*Mild:* Systemically, toxic action is confined to the central nervous system with stimulation resulting in dizziness, nausea, vomiting, headache, disorientation.
	Moderate to severe: Hyperexcitability, apprehension, weakness of skeletal muscles, incoordination, tremors, seizures, coma, and respiratory failure (progressive clinical findings related to severity of poisoning).
Pyrethroids	Generally of low toxicity but can cause irritation of oral and nasal mucosa. Some agents cause dermal tingling, stinging, or burning sensation followed by numbness (parathesia). Facial contamination results in lacrimation, pain, photophobia, congestion, and edema of eyelids and conjunctiva.
	Ingestion of large amounts may cause salivation, epigastric pain, nausea, vomiting, headache, dizziness, fatigue, coarse muscular twitching in limbs, convulsive seizures, loss of consciousness.
Herbicides	
Bipyridyls	Irritation of nose and throat, hemorrhage, eye irritation with conjunctivitis and corneal stripping, skin irritation, dermatitis.
	Ingestion results in initial signs of ulceration of tongue, throat, and esophagus; sternal and abdominal pain; general muscular pain; vomiting; and diarrhea. After 48 to 72 h, signs of renal and hepatic damage (oliguria, jaundice), respiratory difficulty (cough, dyspnea, tachypnea, pulmonary edema), and progressive respiratory failure.
Phenoxyacids Ureas Triazines Chloroaliphatics Arylcarbamates	Generally of low toxicity but, during handling, may cause irritation of eyes, nose, throat, and skin. Ingestion may cause gastroenteritis, nausea, vomiting, diarrhea. Respiratory signs include burning sensation, cough, chest pain. Muscle weakness and muscle twitching may be encountered, and central nervous system signs include dizziness, weakness, anorexia, and lethargy.
Fungicides	
Dithiocarbamates Phenolics Chlorobenzenes Benzimidazoles Thiophanates	Generally of low toxicity but local contamination of skin may cause itching, rash, and dermatitis. Ingestion of large amounts may cause nausea, vomiting, diarrhea, and muscle weakness.
	The dithiocarbamates exert a disulfiram-like effect in the presence of alcohol, causing flushing, sweating, dyspnea, hyperpnea, chest pains, hypotension.
Rodenticides	
Fluorinated agents Zinc phosphide Strychnine α-Naphthylthiourea Anticoagulants	Accidental rodenticide poisoning is difficult to achieve because agents are packaged as baits attractive only to the pest. When ingested, these agents may cause nausea, vomiting, intestinal cramps, diarrhea, excitation, abnormal cardiac rhythms, muscle spasms, seizures.
	The anticoagulants require laboratory assessment of coagulation status and treatment with vitamin K_1.

of a catalyst in the generation of superoxide anion (O_2^-) and hydrogen peroxide, both of which can initiate lipid peroxidation of cellular membranes with a critical loss of cellular integrity and function in gas exchange (Ecobichon 1996; Smith 1987). Technically, herbicides tend to affect biochemical pathways in plants that have no counterparts in mammals: For example, the phenoxy group acts as auxins (growth promoters); the triazines, ureas, and N-acylcarbamates inhibit photosynthesis; others act as inhibitors of protein or carotenoid synthesis or of nucleus and cell division (Ecobichon 1996). However, toxicity to mammalian systems has frequently been associated with trace-level contaminants or byproducts of synthesis. Fungicidal agents are derived from a variety of structures ranging from simple inorganic compounds (sulfur, copper sulfate) through the alkyl- and aryl-mercurials, the phthalimides and chlorinated phenols to metal-containing derivatives of thiocarbamic acid, all with cytotoxic properties related to the disruption of nucleic acid replication, with some showing capability of causing mammalian neurotoxicity and teratogenicity. Rodenticides arise from a number of chemical classes and most do not present a hazard to humans either because of species-selective mechanisms of action or their being dispensed in an unpalatable grain-based bait. However, poisonings occur through intentional suicide, attempted murder, or accidental ingestion as a food (Ecobichon 1996).

Diagnosis and treatment of pesticide poisonings are beyond the scope of this book. Diagnosis of poisoning by a particular pesticide relies on detailed attention to the observed signs and symptoms, measurement of suitable biological parameters, and analysis of residues obtained from regurgitated stomach contents, blood, urine, feces, exhaled breath, etc. Initial emergency treatment should involve:

1. *Decontamination* of the skin to prevent further absorption by removal of contaminated clothing and washing the skin thoroughly with warm water for 10–15 min without scrubbing. If the agent was ingested, administer chewable activated charcoal tablets or induce vomiting only if the patient is conscious and if the agent is neither very toxic nor ingested in a large volume nor in a petroleum distillate;
2. *Supportive Treatment.* Maintain adequate cardiac function by cardiopulmonary resuscitation or by keeping a patient's airway open with a tracheal tube. Treatment regimens are described by Ecobichon (1995, 1996), among others.

8-3 CHRONIC (DELAYED) TOXICITY

There is an early history of adverse health effects following chronic exposure to insecticides, although this information was largely ignored or dismissed as irrelevant and anecdotal. In 1961, about 14 years after the introduction of parathion (O,O-diethyl-O,4-nitrophenylphosphorothioate) as an agricultural insecticide, three landmark papers reported central nervous system effects following exposure to organophosphorus ester insecticides (Barnes 1961; Bidstrup 1961; Gershon and Shaw 1961). These papers reported an association between chronic occupational exposure, primarily to parathion, and evident psychiatric sequelae among pilots, mixer/loaders, applicators, and soforth. Gershon and Shaw (1961) described a range of signs including giddiness, tinnitus,

nystagmus, pyrexia, ataxia, tremor, paresthesia, polyneuritis, paralysis, speech difficulties including slurring, memory deficits with slow recall and difficulty in expressing
thoughts, insomnia, somnabulism, excessive dreaming, drowsiness, lassitude, generalized weakness, emotional lability, mental confusion, inability to concentrate, depression,
dissociation, and schizophrenic reactions. This detailed report was dismissed by many
because of the imprecise nature of the observations, based on the "impressions" of the
attending physicians and because of the absence of cholinesterase inhibition data. In an
earlier study, Holmes and Gaon (1956), examining 449 cases of insecticide poisoning,
listed headache, excessive dreaming, poor sleep patterns, irritability, fatigue, nervous-
ness, dizziness, muscle fasciculations and tremors, and paresthesia in the more severely
affected patients having plasma pseudocholinesterase (PChE) inhibition greater than
40%. Dille and Smith (1964) found similar effects, including memory impairment for
weeks or months following exposure. However, most of these early reports were considered to be anecdotal at best even though large numbers of subjects were involved in
some studies.

In 1964 and 1965, agricultural pilots were involved in 402 and 366 accidents, respectively, while engaged in aerial spraying operations, such accidents requiring investigation by committees of the U.S. Federal Aviation Agency (Smith, Stavinoha, and Ryan
1968; Wood et al. 1971). Despite confounding factors of long flying hours and possible exposure to a potpourri of chemicals, the accumulated evidence strongly suggested
an association between physical and mental state and accidents of the pilot-error type.
Erratic behavior, severe visual disturbances, impaired judgment, automatism, psychiatric symptoms (neuroses, obsessive-compulsive behavior, phobias, temper outbursts,
depression), marked personality changes, all have been reported to occur in pilots exposed to pesticides (summarized by Ecobichon [1994]). Once again, the evidence was
largely ignored, investigators experiencing considerable difficulty in convincing their
peers that these observations were valid and associated with exposure(s) (Clark 1971).
However, these adverse health effects have been confirmed repeatedly (Richter et al.
1992).

Supported by analyses of erythrocytic and plasma cholinesterase activity and the
subjective measurement of symptoms, recovery from poisoning appeared to require
approximately 6 months following exposure to a variety of organophosphorus ester insecticides (Tabershaw and Cooper 1966; Whorton and Obrinsky 1983; Midtling et al.
1985). Davignon et al. (1965) reported that insecticide-exposed apple growers experienced a higher degree of neurological manifestations (loss of reflexes, disturbances
of equilibrium, tremors) than did nonexposed individuals, a positive correlation being
found between the incidence of effects and number of years of exposure to insecticides.
A complication in this study was the "mixed" exposure to organochlorine insecticides
(DDT) as well as organophosphorus esters. Tabershaw and Cooper (1966), in an epidemiological study of 114 pesticide workers, found that complaints characterized as
optic, gastrointestinal, cephalic, cardiorespiratory, and neuropsychiatric persisted for up
to 3 years. Much of this early evidence has been summarized by Ecobichon (1994). With
the plethora of organophosphorus and carbamate insecticides in use today, the number of
reports of long-term, adverse health effects has increased and can no longer be ignored,
the same spectrum of neuromuscular and central nervous system (neuropsychological,

Table 8-2 WHO and UNDP core methods for health assessment following pesticide exposure

Type of assessment	Information collected
Exposure absorption	Exposure data, i.e., type of agent, amount Metabolites in urine Plasma and erythrocytic cholinesterase activities (Ellman method) for insecticide exposure
Health effects	Background i.e., health and occupational Symptoms and signs (based on VBC 82/1 and WHO/National Institute of Occupational Safety and Health questionnaire) Neurological examination (with semiquantitative assessment of parameters) Neurobehavioral examination Nerve conduction (sural, ulnar, and peroneal nerves) Neuromuscular junction testing Behavioral tests (WHO battery)

Source: Modified from Maroni, Jarvisalo, and La Ferla (1986).

behavioral) sequelae being relatively consistent between cases and studies. There are sufficient reports, anecdotal or otherwise, to point to the need for rigorous clinical assessment of the consequences of chronic effects associated with pesticide exposure.

The World Health Organization (WHO) and United Nations Development Program (UNDP) have developed a standardized protocol for the assessment of health effects of exposure to organophosphorus ester insecticides, the concept being to have a core battery of methods that could be implemented worldwide by any participants studying and reporting such poisonings. The core and supplementary (optional) tests are shown in Table 8-2, and their utility is discussed in detail by Maroni, Jarvisalo, and La Ferla (1986). Details concerning the WHO battery of behavioral tests are discussed later in this chapter. Although designed specifically for one class of insecticides, many of the health assessment tests could be applied to quantitation of adverse health effects caused by other classes of pesticides as well as by inerts (e.g., cosolvents, emulsifiers, dispersants, wetting agents) found in formulations.

8-3-1 Subjective Measurements

Many studies have developed questionnaires in association with surveillance/monitoring programs to ascertain whether or not exposed individuals experienced any signs or symptoms following exposure (Ames, Howd, and Doherty 1993; Kahn et al. 1992; Morgan, Lin, and Saikaly 1980). With overt poisonings, a range of distinctive effects will be reported, although not every person will describe every symptom. However, with suspect exposures, the litany of signs and symptoms may be long, varied, and somewhat irrelevant, even imaginary, reflecting individuals' concerns and anxieties (Kahn et al. 1992). If the respondents are knowledgeable about the toxicity of such chemicals, the list of signs and symptoms will be much more complete or focused, raising valid suspicions about the subjective and suggestive nature of questionnaires and responses to them. Personal experience with questionnaire responses, from individuals supposedly oversprayed during an aerial forest spraying program, was that they would claim a list of signs and symptoms, always in the same order of appearance as that presented in a well-circulated government booklet. In contrast, a similar survey carried out in Quebec elicited, in my opinion, a bona fida set of symptoms from the potentially exposed population who did not have access to the French version of the same booklet.

Questionnaires can be useful in cataloguing the signs and symptoms in poisonings (Summerford et al. 1953; Zwiener and Ginsburg 1988) or in differentiating between individuals showing lowered erythrocytic AChE activity and those involved in reported exposures to anticholinesterase insecticides (Ciesielski et al. 1994). In specific studies, such as a 4-month surveillance of nineteen farmworkers acutely poisoned by a combination of mevinphos and phosphamidon, the authors found that many of the subjective signs and symptoms (blurred vision, night sweats, headaches, nausea, muscular weakness) were persistent, though decreasing in intensity, but still quantifiable at the end of the study (Whorton and Obrinsky 1983).

8-3-2 Objective Measurements

Characteristic of many pesticides including the organophosphorus and carbamate ester insecticides, some herbicides and fungicides have common symptoms such as muscle fasciculations, tremors, generalized muscle weakness (especially with exercise), diminished tendon reflexes, flaccid or rigid muscle tone, and paralysis, all of these signs pointing toward a neuronal site of action or one at the neuromuscular junction of the skeletal muscles. With the anticholinesterase-type insecticides, a further complication is associated with the respiratory muscles (both diaphragmatic and intercostal), resulting in weakness, paralysis, and compromised respiratory function (Ecobichon and Joy 1994). The pyrethroid esters and some organochlorine insecticides specifically affect the sensory nervous system. Almost all pesticides, at high levels of exposure, appear to elicit toxicity in the central nervous system (Ecobichon and Joy 1994). This spectrum of adverse health effects suggests that monitoring various components of the central and peripheral nervous systems would prove advantageous to detecting and quantifying the toxicity. However, given the broad range of "normal" values, major concerns center on whether the developed assays detect changes substantial enough to permit their use as screening tests for the clinical effects of exposures in groups of exposed individuals (Richter et al. 1992).

8-3-3 Electromyography

Measurements of motor and sensory nerve conduction velocities have been standard practice in assessing the general well-being of individuals exposed to pesticides. With motor neurons, changes in velocities are examined in the fast-conducting fibers of the median, the ulnar, and the peroneal nerves (Misra et al. 1988; Vasilescu and Florescu 1980). Sensory nerve conduction is measured in the median and in the sural nerves. A number of noninvasive techniques have been described in the literature, those outlined by Stalberg et al. (1978), Ring et al. (1985), and Misra et al. (1988) being described briefly below. Interested readers should consult the original papers for specific details.

Motor neurons. Using the *ulnar nerve* for velocity studies, the active recording electrode is placed over the belly of the *abductor digiti minimi* muscle and the indifferent electrode is placed over the fifth metacarpophalangeal joint (Stalberg et al. 1978). The change in the amplitude of the compound nerve action potential is measured with pad electrodes on both sides of the *sulcus ulnaris* when the ulnar nerve is stimulated at the wrist.

Using the *median nerve* (elbow-to-wrist portion), the conduction velocity is measured by stimulating the nerve supramaximally at the wrist (between the tendon of *flexor carpi radialis* and *palmaris longus*) and at the elbow medial to the tendon of *biceps brachii* and the bracheal artery (Misra et al. 1988). The recording electrode is placed on the belly of the *abductor pollicis brevis*, the reference electrode being placed over the tendon of the muscle and the ground electrode on the back of the hand (Misra et al. 1988).

For the *peroneal nerve* (knee-to-ankle segment), a supramaximal square-wave stimulus (0.05–1 ms, gain of 5 mV/division, sweep speed of 5 ms/division, filter setting of 16 Hz to 16 kHz) ig given to the ankle between the tendons of *extensor digitorum longus* and *extensor hallucis longus*, the proximal stimulation being given proximal to the fibular head. Placement of the active recording electrode is on the belly of the *extensor digitorum brevis* and the reference electrode is positioned distal to it at the tendon (Misra et al. 1988).

In most of these experiments, the latency of the action potential is measured from the stimulus artifact to the first negative deflection; the distance between the two points is measured and divided by the latency difference to calculate the conduction velocity (Misra et al. 1988).

Sensory neurons. Using the *median nerve*, sensory conduction is measured orthodromically, using a ring electrode on the second digit to produce surface stimulation and placing the recording electrode at the proximal end and the reference electrode at the distal end of the interphalangeal joint (Misra et al. 1988).

Using the *sural nerve*, sensory nerve conduction is measured by antidromic stimulation between the two heads of the gastrocnemius muscle, recording the sural nerve action potential at the *lateral malleolus* with pad electrodes (Stalberg et al. 1978).

The use of sensory and motor neuron conduction velocities as indicators of exposure to pesticides has been disappointing because, in some studies, slight-to-significant reductions in velocities have been measured whereas, in other studies, the measured differences were not significant between the exposed and control individuals or were pronounced to be within "normal ranges." Stalberg et al. (1978) found a slight reduction in sensory conduction velocities. Ring et al. (1985) reported that a number of conduction parameters (motor distal latency of the median and ulnar nerves, motor conduction in the ulnar and popliteal nerves, sensory distal latency in the ulnar nerve) were significantly related to symptomatology in organophosphate-exposed workers, claiming that the measurements were different but that the values were within the normal range. Verberk and Salle (1977) found a 7% decrease in slow-fiber motor nerve conduction velocity in volunteers exposed to low-dose mevinphos for a month but saw no change in velocity in fast fibers or in the Achilles-heel tendon reflex. Misra et al. (1988), studying the effect of occupational exposure to the organophosphorus ester fenthion on neuromuscular function, found no differences between sensory and motor neuron conduction velocities in exposed and control populations. Similar observations have been reported by Jager et al. (1970), Jusic, Jurenic, and Milic (1980) and Ring et al. (1985), perhaps the key phrase being that all measured values were within the "normal" range. In contrast, Drenth et al. (1972) found significantly abnormal electromyogram (EMG) patterns in 40% of pesticide factory workers exposed to a mixture of organophosphorus and organochlorine compounds.

Such observed variability may be associated with the potency of the chemical, the technique(s) as conducted in the laboratory setting, or the time interval between exposure and clinical testing. Concerning discussions about EMG results, Roberts (1977) stated that there has been no evidence that the changes observed, or the underlying causes, posed any threat to workers' health, the individuals appearing and feeling healthy, frequently without any overt signs and symptoms. The equivocal nature of results from EMG studies suggests that these parameters are neither sensitive nor reliable for use in monitoring exposure-based effects of pesticides, particularly organophosphorus ester insecticides.

8-3-4 Vibrotactile Techniques

Sufficient evidence points to the fact that exposure to pesticides (active ingredients and/or formulations with cosolvents, emulsifiers, other "inerts") can result in transient or persistent motor and sensory polyneuropathies (Ecobichon and Joy 1994). Injury to the peripheral nervous system by chemicals generally manifests itself as a large-fiber axonopathy, the damage occurring first in the distal portions of large myelinated axons (Gerr, Hershman, and Letz 1990). Damage results in elevated sensory thresholds, affecting the perception of vibration and light touch. The stocking-and-glove neuropathy is characteristic of such injury, abnormal sensory function initially being detected in the distal sensory neurons of the limbs and then more proximally as the condition progresses. The pinprick stimulus test has been used extensively in sensory function assessment, requiring a yes/no response from the test subject to indicate awareness of the stimulus. Can sensory perception be quantified by other means?

Several vibrating instruments are presently available to assess changes in peripheral sensory function following occupational or environmental exposure to chemical and physical agents (Bove et al. 1986). These electromechanical devices deliver calibrated vibrating stimuli and permit noninvasive, nonaversive, quantitative and relatively rapid assessment of sensory perception at multiple anatomical sites by technicians with minimal training following simple but well-developed testing protocols (Gerr and Letz 1988; Gerr, Hershman, and Letz 1990). One such instrument is the Vibration II cutaneous threshold tester (Sensortek, Inc., Clifton, N.J.). These instruments usually consist of a voltage controller and one or two "slave" tranducers. The transducer has a plastic post, 1.5 cm diam, protruding from the surface, which provides a vibrating stimulus at 120 Hz. The amplitude of the vibration is determined by the applied voltage adjusted manually on the controller unit and is quantified as "vibration units" on a digital display.

Tests generally are conducted simultaneously on the index fingertip or the pad of the large toe of both the dominant and the nondominant hands and/or feet. With the method-of-limits mode, the stimulus is started at a high amplitude, and the subject must determine which of the two posts is vibrating. After each test, the intensity of vibration is reduced at a constant rate, say 10%, and the trial is repeated, the subject being asked to report verbally at the earliest time point when the vibration is no longer felt. A second test mode begins with the amplitude of vibration set below the threshold of the previous trial and, with the test subject's fingers on the stimulation posts of the two transducers, the technician raises the amplitude of vibration, the test subject verbally indicating when the vibration is first felt (Gerr and Letz 1988; Gerr, Hershman, and Letz 1990). The complete testing sequence consists of nine trials (five descending and four ascending). Other, more complex, testing paradigms can be developed to enhance the reproducibility and reliability of the experiments, for example, operating the transducers independently with a preselected, pseudorandom sequence of amplitudes being applied. The results must be corrected for age, height, and callus formation because, with all three parameters, there is a loss of sensory function and the observation of higher sensory thresholds (Gerr, Hershman, and Letz 1990; McConnell, Keifer, and Rosenstock 1994). Statistically significant differences in signs and symptoms of neurotoxicity also may be found between the dominant and nondominant limbs (Gerr, Hershman, and Letz 1990).

Vibrotactile thresholds have been measured in patients following poisoning by organophosphorus ester insecticides. In one study, an assessment of Nicaraguan workers about 10 to 34 months following acute organophosphate poisoning, abnormally high vibrotactile thresholds were measured in the lower extremities of these individuals when compared to control participants (McConnell, Keifer, and Rosenstock 1994). A complication of this study was that people poisoned by methamidophos were more severely affected than those poisoned by other organophosphorus esters (chlorpyrifos, acephate), suggesting that a different mechanism was involved, possibly related to classic organophosphate-induced delayed polyneuropathy (OPIDP) or some other effect of methamidophos. Neither clear indices of initial severity of poisoning nor dose-effect relationships could be established for these individuals. However, with the elapsed time between poisoning and evaluation, it appears that there was little improvement in the vibrotactile threshold, suggesting a persisting neurological deficit. In a second study, chronically exposed agricultural insecticide (guthion, phosphamidon, chlorpyrifos, phosmet, diazinon) sprayers showed poorer performance, with mean vibration threshold sensitivity scores that were significantly elevated for both the dominant and the nondominant hands compared to values reported for population-based controls (Stokes et al. 1995). However, the vibration threshold scores for paired dominant and nondominant feet were not statistically different between applicators and controls. These results suggest that previous exposure to organophosphorus esters was associated with a loss of peripheral nerve function. Questions that remain to be answered include whether the same test (1) would apply to other classes of pesticides, (2) would identify problems caused by organic cosolvents or inert ingredients in formulations, and (3) differentiate between the effects elicited by active agents and other organic components in formulations.

8-3-5 Electroencephalography

Many reports of acute and chronic pesticide poisonings and exposures have recorded transient or persistent alterations in memory, motor and cognitive functions, and neurobehavioral and neuropsychological parameters (Ecobichon and Joy 1994; Marrs 1993). A study, carried out in the early 1950s on workers from a plant manufacturing both organochlorine and organophosphorus ester insecticides, reported changes in specific electroencephalographic (EEG) patterns from deep, midbrain AChE-rich centers that were associated with sleep disorders, drowsiness, lethargy, and narcoleptic-like symptoms (Metcalf and Holmes 1969). There were observed increases in brief episodes (2- to 4-s duration) of low- to medium-voltage irregular activity in the theta range (2 to 6 Hz) which persisted up to 1965 when some of the workers were reevaluated (Metcalf and Holmes 1969). The results suggested that EEG data might be useful in assessing persistent effects of insecticide poisoning in the central nervous system (CNS).

Unfortunately, the complexity and variability of EEG patterns has not proved to be as useful as was originally thought. Korsak and Sato (1977) reported results similar to those of Metcalf and Holmes (1969), also noting a mixed pattern of increased slowing activity in the frontal ($F_{p1} - A_1$) region and increased fast activity bifrontally in subjects with high chronic exposure to organophosphorus ester insecticides. The neuropsychological and EEG data indicated that chronic exposure to organophosphorus esters may differentially affect only the left frontal hemisphere (Korsak and Sato 1977). The relationship between EEG abnormalities and pesticide poisoning may not be as straightforward as anticipated, given the multisynaptic neuronal pathways in which acetylcholine plays a neurotransmitter role.

Primate studies, in which rhesus monkeys were exposed to single large doses of the organophosphate sarin (5.0 μg/kg) or the organochlorine dieldrin (4.0 mg/kg), revealed that both agents caused significant increases in the relative amount of beta voltage (15 to 50 Hz) in EEGs, which persisted for 1 year (Burchfiel, Duffy, and Sim 1976). For the organophosphate sarin, the predominant effect was in the EEG derivation from the temporal cortex whereas, for the organochlorine dieldrin, it was from the frontal cortex. In a study of seventy-seven industrial workers having documented histories of one or more accidental exposures to sarin at least 1 year earlier than the examination, the findings suggested that exposure to organophosphorus esters can produce long-term changes in brain function (Duffy et al. 1979). The tests included standard clinical EEGs, computer-derived EEG spectral analysis, and standard overnight sleep EEGs. Univariate and multivariate statistical methods were required to identify between-group differences for the visually inspected and computer-derived data. Significant group differences included increased beta activity, slowing of delta and theta activity, decreased alpha activity, and increased levels of rapid-eye-movement sleep in the exposed workers.

However difficult the interpretation of neuroelectrophysiological results in terms of behavioral anomalies, the observed alterations provide evidence that exposures to organophosphorus esters and other pesticides can produce long-term changes in brain function (Duffy and Burchfiel 1980). As these authors state, the important questions—for example, the significance of the brain-wave differences, the persistence of the EEG effects, and the minimal dose necessary to precipitate long-term changes—cannot be answered without additional exhaustive study. Such answers will come only by performing concurrent EEG and detailed psychological testing on populations of control and exposed subjects (Duffy and Burchfiel 1980).

8-3-6 Neurobehavioral Evaluation

Changes in psychological function (mood, personality, memory, cognition) may occur at concentrations lower than those required for adverse effects on any other organ system (Stollery 1996a). As described by Gershon and Shaw (1961), Holmes and Gaon (1956), Levin and Rodnitzky (1976), Savage et al. (1988), and Stoller et al. (1965), among many others, behavioral changes following exposure to insecticides, particularly organophosphorus esters, have included (1) cognitive disturbances, (2) expressive language deficits, and (3) psychopathological sequelae, all appearing to be associated with EEG anomalies. A compilation of these centrally oriented sequelae is presented in Table 8-3. Earlier clinical studies suffered from (1) an absence of matched controls, (2) incomplete documentation of the exposures, (3) insufficient numbers of subjects, (4) insufficient quantitative measures of neurological and behavioral functions, and (5) incomplete statistical analysis (Savage et al. 1988). However, these same studies did demonstrate abnormalities in a wide range of functions, including problem solving, slowed digit-symbol speed and eye-hand coordination, and memory recall. How then does one objectively assess and quantify these biological effects, being aware that individuals exposed to other classes of pesticides, cosolvents and emulsifiers in formulations may show similar signs and symptoms of toxicity?

Table 8-3 Neurobehavioral sequelae related to pesticide poisonings

Features	Effects
Cognitive disturbances	Reduced vigilance and alertness
	Attentional deficits
	Slowed processing of information
	Psychomotor retardation
	Impaired memory functions
	Reduced comprehension
Expressive language defects	Speech difficulties
	Speech slurring, reduced enunciation
	Difficulties in saying what is intended
	Difficulty in formulating thoughts
	Difficulty in repetition
Psychopathological sequelae	Depression
	Restlessness
	Insomnia
	Excessive dreaming
	Emotional lability
	Weeping spells
	Schizophrenic reactions
	Irritability
	Poor sleep patterns
	Phobias
	Outbursts of temper (rage)
	Belligerent behavior
	Acute anxiety spells
	Obsessive-compulsive behavior

A large number of individual neurobehavioral or neuropsychological tests have been developed to quantify the CNS effects of solvents (Muijser et al. 1996; Stollery 1996b; Stollery and Flindt 1988), industrial chemicals (Tsai and Chen 1996), heavy metals (Stollery 1996c; Stollery et al. 1991) as well as pesticides, although the focus has been primarily on organophosphorus esters (Rodnitzky, Levin, and Mick 1975; Rosenstock et al. 1991; Savage et al. 1988; Stephens et al. 1995). Many of these tests could be considered to be generic, applicable to a wide range of agent-induced, central neurological damage. However, these tests are not used singly but as test batteries, concurrently measuring motor, cognitive, memory, and psychological parameters in adequate numbers of control and afflicted subjects in longitudinal, preferably double-blind, studies (Levin and Rodnitzky 1976). It is important to emphasize that such testing is by no means standardized, investigators picking and choosing or mixing and matching components from different established testing paradigms to produce a bewildering potpourri of testing strategies in the hope that they will be definitive. Although not all of the results from a test battery will identify abnormal behavior, a pattern of neurological sequelae may emerge to pinpoint the locus of neuronal damage. Feldman, Ricks, and Baker (1980) reviewed methods used in assessing neuropsychological effects of industrial toxicants, creating the basic list of parameters shown in Table 8-4. The WHO behavioral test battery is shown in Table 8-5 (Maroni, Jarvisalo, and La Ferla 1986).

Testing by paper and pencil and by questionnaires was conducted in the early days, revealing qualitative evidence of persistent psychological, cognitive, and behavioral anomalies associated with pesticide exposure. Metcalf and Holmes (1969) appear to be the first to use such assessments as the Wechsler Adult Intelligence Scale, Berton Visual

Table 8-4 Common tests used to detect behavioral effects of neurotoxins

Test type	Test functions
Memory	Wechsler Memory Scale
	Subtests: personal information, orientation, mental control, logical memory, digit span (forward and backward), visual memory, paired associate learning
Overall intelligence	Wechsler Adult Intelligence Scale (WAIS)
	Subtests: information comprehension, arithmetic, similarities, digit span, vocabulary, digit symbol, picture completion, block design, picture arrangement, object assembly
Sustained attention	Continuous Performance Test (COT)
	Bourdon-Wiersma Vigilance Test
	Neisser Letter Search
Dexterity and eye-hand coordination	Santa Ana Dexterity Test
	Flanagan Coordination Test
	Michigan Eye-Hand Coordination Test
	Finger-tapping test
Reaction time	Simple reaction-time test
	Choice reaction-time test
Psychomotor function	Mira Test
	Digit-Symbol Substitution Task
Personality or mood	Eysenck Personality Inventory
	Rorschach Test
	Feeling-Tone Checklist

Source: Modified from Feldman, Ricks, and Baker (1980).

Table 8-5 WHO Behavioral Test Battery

Test	Objectives or purposes
Reaction times	Auditory and visual
Santa Ana Test	Timed perceptual-motor coordination
Wechsler Digit-Symbol Test	Measuring perceptual and motor speed
Profile of Mood States Questionnaire	Measuring mood or affective states
Aiming Test	Measuring hand steadiness
Digit-span test	Immediate auditory memory
Benton Visual Retention test	Visual memory
Digit-symbol test	Measuring perceptual organization, motor dexterity, attention, and speed of performance
Helsinki Subjective Symptom Questionnaire	Investigates psychological, neurovegetative, gastrointestinal, and neurological symptoms

Source: Modified from Maroni, Jarvisalo, and La Ferla (1986).

Retention, and a story-recall test with workers exposed to organophosphorus esters in a manufacturing plant. No differences in the prevalence of overt signs were found between exposed workers and controls but difficulty in remaining alert and focused during testing, slowness of thinking and calculating, and poor memory were reported among the exposed individuals, suggesting problems in handling the test material.

The historical development of computerized neurobehavioral testing has been reviewed recently (Stollery 1996a; Williamson 1996). A summary of later, insecticide-related studies shown in Table 8-6 with brief descriptions of the test functions used, illustrates the extent of detail and complexity of testing schemes, especially the mix-and-match test approach in efforts to improve the quantification of the results.

Rodnitzky, Levin, and Mick (1975), using a sophisticated, computerized test battery, showed no exposure-related deficiencies in five parameters (memory, signal processing, vigilance, language, and proprioceptive feedback performance) in farmers and commercial pesticide applicators exposed to organophosphorus ester insecticides (Table 8-6). However, in a later paper, Levin and Rodnitzky (1976) suggested that mild-to-moderate changes in the same parameters could be measured in organophosphate-exposed workers.

In an extensive study of 100 matched pairs of previously poisoned individuals and nonpoisoned controls, Savage et al. (1988) found no significant differences in the poisoned subjects on audiometric or opthalmic tests, EEGs or clinical biochemistry (Table 8-6). However, anomalies were detected on measures of abstraction, memory, mood, and motor reflexes. Differences were much more apparent between the two cohorts in intellectual functions, academic skills, abstraction and flexibility of thinking, as well as in simple motor skills. Twice as many cases as controls (24 vs. 12) had Halstead-Reitan battery summary scores in the range characteristic of individuals suffering cerebral damage/dysfunction. The combination of clinical and neurophysiological assessment appeared to provide a more complete evaluation of brain function than use of either approach alone. Although some impairment of fine coordination and motor speed of the upper extremities was demonstrated, the major deficits were

Table 8-6 Test paradigms for neurobehavioral functions in insecticide-exposed individuals

Reference	Description of parameters studied
Rodnitzky, Levin, and Mick (1975)	*Memory.* Verbal recall of a consonant trigram presented 3 s after a ready signal and immediately followed by a 3-digit number, the subject repeating the number and beginning to count backwards in cadence with a metronome. At different intervals (3, 6, 9, 12, 15, 18 s), the subject was instructed to stop and repeat the initial trigram.
	Vigilance. Simple reaction-time test with the subject sitting in front of a display having a single centrally placed light stimulus. Subject is instructed to depress a single push-button with the (preferred) index finger. The light is activated at random intervals and the subject must release the push-button as soon after the onset of the stimulus as possible.
	Signal-processing time. Subject is seated before a display panel consisting of four lights, two in a vertical array and two in a horizontal array (top, bottom, left, right) and has access to two push-buttons, one for each index finger. Subject is instructed to release either the left or the right button in response to the appropriate stimulus. Subject responds to top light, ready signal, by releasing the left or right push-button in response to the left or right light flashing on at random intervals. See paper for other paradigms.
	Language. Tape-recorded administration of sentence repetition, repeating sentences of increasing length (progressively) upon hearing them.
	Proprioception. Blindfolded subject has to maintain position of a spring-loaded button within a 2-mm range of movement by applying a constant amount of pressure using the index finger, this being performed without visual feedback, with sensory cues only in the form of pressure of the spring-loaded button on the fingertip and a tone that is activated if the button strays beyond the 2-mm range. The force on the spring can be changed from 10 up to 300 g. Both hands are tested.
Savage et al. (1988)	*Wechsler Adult Intelligence Scale* (WAIS).
	Expanded Halstead–Reitan Battery. Eleven components measuring intelligence, attention, various cognitive functions, motor proficiency, sensory-perceptual functions, aphasia and related disorders, learning, and memory.
	Peabody Individual Achievement Test. Three parts measuring reading recognition, reading comprehension, spelling.
	Minnesota Multiphasic Personality Inventory. An objective personality test for the determination of increased tendencies toward psychiatric disturbances.
	Questionnaires. Subjective ratings of functioning with respect to memory, communication skills, academic skills, sensory and motor abilities, cognitive and intellectual abilities, emotional status.
Rosenstock et al. (1991)	*WHO Core Test Battery.* Subtests simple reaction time, pursuit aiming, the Santa Ana manual dexterity test, the Benton visual retention test.
	Wechsler Adult Intelligence Scale (WAIS-R). Including only subtests that measure digit span and digit symbol, vocabulary, block design, digit vigilance.
	Brief Symptom Inventory. Evaluates potential psychiatric and affective disturbances.
	Rey Auditory Verbal Learning Test.
	Finger-tapping test.
	Trials A.
	Scandanavian Questionnaire. Sixteen-item questionnaire asking about memory difficulties, concentration, headache, irritability, fatigue, depression.

(Contd.)

Table 8-6 *(Contd.)*

Reference	Description of parameters studied
Steenland et al. (1994)	*Neurobehavioral Evaluation System* (NES 2). Uses the following eight tests: Mood scale—(affect test)—subjects' self-reported transient state of tension, depression, fatigue, confusion, anxiety. Finger-tapping (motor-speed test)—measures how many times a key can be struck repeatedly in 30 s, using the preferred hand. Sustained visual attention (continuous performance test)—requires pressing a key quickly when a certain letter appears amid a temporal sequence of various letters (60 letters appear in 5 min). Hand-eye coordination (visuomotor-accuracy test)—measures the degree of error in tracing a moving sine curve with a cursor. Simple reaction time (visuomotor-speed test)—measures how fast one can respond to a visual stimulus by pushing a button. Symbol digit (coding-speed test)—requires matching digits to symbols as fast as possible following an exhibited matched pattern. Pattern memory (visual memory test)—requires selecting a previously seen pattern out of three similar patterns. Serial digit learning (learning memory test)—requires memorizing and replicating a series of eight digits as quickly as possible. *WHO Core Test Battery.* Noncomputerized tests of psychomotor function including the Santa Ana dexterity test (turning rows of successive pegs 180 deg) and the pursuit-aiming test (using a pencil to mark a point inside each of a series of circles as quickly as possible). *Postural Sway Test.* Quantitative analogue of the Romberg clinical examination, to measure CNS function.
Stephens et al. (1995); Stephens, Spurgeon, and Berry (1996)	*Questionnaires.* Demographic and lifestyle, including pesticide exposure history; general health (30-item version); subjective memory to measure subjects' own assessment of their memory; retrospective exposure (years and pesticide exposure). *Cognitive tests* (Automated Cognitive Test [ACT]). Short-term memory—digit-span test recalling the maximum number of digits (forward and backward) after presentation of a random single-digit sequence. Visual spatial memory—requires the subject to decide whether a question-mark probe appears at a location on the screen when one, two, four, or six circles have been presented previously. Sustained attention—simple reaction time, pressing a key in response to a stimulus appearing at random intervals over a period of approximately 5 min. Information processing speed—symbol-digit test in which subjects match symbols and numbers according to a coding key. Syntactic reasoning test—subject must judge truth or falsehood of a series of statements describing the order in which two following letters appear. Long-term memory function—category search consisting of a semantic classification task followed by recognition task involving the use of nouns that can be grouped into a category, say "birds" or "plants," with distractors, say animal nouns or unrelated nouns interspersed. The mean response latencies across item-types are recorded. Serial word learning test—three presentations of a 15-word list, each presentation being followed by an immediate written recall. The rate of learning is assessed by recording the mean number of words recalled after each presentation.

cognitive in nature (visuomotor, attention, language function) and were detectable by tests of ability.

Using a population of Nicaraguan agricultural workers, poisoned earlier by organophosphorus insecticides, Rosenstock et al. (1991) conducted a retrospective study to ascertain if single episodes of acute intoxication could lead to neuropsychological dysfunctions. The thirty-six poisoned males were examined some 24 months after poisoning and, with matched controls, were subjected to the test battery shown in Table 8-6. The poisoned group was significantly below the performance norms of the controls in auditory attention, visual memory, visuomotor speed, sequencing, problem solving, motor steadiness, reaction, and dexterity. The patients also reported symptoms consistent with CNS involvement, similar to those reported by Dille and Smith (1964), Midtling et al. (1985), and Savage et al. (1988). The study results suggest that single episodes of clinically significant organophosphorus ester intoxication can cause a persistent decline in neuropsychological functions that are detectable by an appropriate set of tests.

The extensive use of pesticides in California is well known, as is the requirement of reporting poisonings under the state worker health safety regulations (Brown, Ames, and Mengle 1989). With some 10,000 agricultural pesticide applicators in the state, it is not surprising that there are a number of reported poisonings annually, 162 between the years 1982 and 1985 (Brown, Ames, and Mengle 1989). Studies conducted by Ames, Howd, and Doherty (1995) and Steenland et al. (1994) have identified chronic neurological effects following acute intoxication by organophosphorus and carbamate esters using the NES 2 (Baker, Letz, and Fidler 1985; Steenland et al. 1994). On the battery of tests shown in Table 8-6, poisoned individuals ($n = 128$) performed significantly worse than the referent group ($n = 90$) on two neurobehavioral tests (sustained visual attention, mood scales). Among those having significantly lowered cholinesterase levels or who had been hospitalized, increased impairment in finger vibrotactile sensitivity was observed. Increased impairment in a wide spectrum of test functions was observed among workers with increased days of disability. However, in individuals with only a moderate cholinesterase inhibition, differences were more difficult to detect, only one of twenty-seven neurological tests—serial digit performance—being statistically significant from control values (Ames, Howd, and Doherty 1995). Once again, these studies have confirmed the presence of long-term neurological sequelae arising from acute insecticide poisoning, although pointing out the exposure dependence of the effects as well as the difficulty of detecting altered neuronal activity.

Recent reports have described adverse health effects among sheep farmers in the United Kingdom following exposure to commercial antiparasitic sheep dip, containing diazinon, propetamphos, or a mixture of diazinon and chlorfenvinphos (Stephens et al. 1995; Stephens, Spurgeon, and Berry 1996). Signs and symptoms included tension, anxiety, irritability, headaches, restlessness, poor mental health, and difficulties with memory and concentration. A total of 223 sheep farmers in two studies were assessed at least 2 months after exposure, using the battery of cognitive and neuropsychological tests shown in Table 8-6. The farmers performed significantly worse than controls (quarry workers) in tests measuring sustained attention and the speed of information processing and showed a greater vulnerability to psychiatric disorders (Stephens et al. 1995). Interestingly, there appeared to be no association between reported acute symptom levels and chronic neuropsychological effects, this observation suggesting that the chronic effects occurred independently of signs and symptoms that might follow immediately after exposure (Stephens, Spurgeon, and Berry 1996).

There are a number of computer-based, psychological assessment systems available for neurotoxicological investigation (Stollery 1996a). Although they have not been used to quantify pesticide-induced neurotoxicity, there is no reason why they should not be applicable because they are used in neuropsychology, neuropharmacology, and in assessing heavy metal, solvent, and plastic monomer-related neurotoxicity. Although many of these test systems originated as pencil-and-paper, task-related assessments, the use of the computer ensures that the testing is given in a standardized form with standardized feedback and a detailed recording of accuracy (Fray and Robbins 1996).The ACT system is an MS-DOS, key-press, menu-driven system for selecting and administering specially designed cognitive tasks, analyzing performance on twenty-six tasks in areas dealing with learning, memory, attention, reasoning, verbal and spatial abilities, using a keyboard as the sole response input device (Table 8-7) (Stollery 1996d). The ACT system has been used successfully in assessing neurological deficits in solvent- and lead-induced neurotoxicity (Stollery and Flindt 1988; Stollery et al. 1991).The cognitive-based tasks shown in Table 8-7 are sufficiently generic in nature to have wider application in neu-

Table 8-7 Components of ACT System

Parameters	Task name	Description
Reasoning	General reasoning	Verbal and numerical
	Vocabulary	Synonym recognition
	Syntactic reasoning	Linguistic transformations
	Semantic reasoning	Semantic relations
Verbal/spatial abilities	Category search	Words belonging to semantic category
	Acoustic search	Words rhyming with target word
	Word search	Lexical decisions
	Visual search	Single target letter among distractors
	Cognitive vigilance	Detection of rare target digit over time
Recognition	Paired associate recognition	Sequence of acquisition and cued recognition trials
	Word recognition	Immediate and delayed recognition of words
	Location recognition	Memory for location of items
	Pattern recognition	Memory for pattern of items
Memory	Memory search	Short-term memory scanning
	Paired associate recall	Delayed cued recall trial with relearning option
	Word recall	Immediate and delayed recall of words
	Repeated words	Memory for words
	Simple memory span	Forward and backward digit span
	Working memory span	Forward word span with concurrent processing
Learning	Paired associate learning	Sequence of acquisition and cued recall trials
	Serial word learning	Acquisition of supraspan list of words
Reaction	Simple four-choice	Continuous reaction time: simple mapping
	Complex four-choice	Continuous reaction time: six mapping rules
	Memory four-choice	Continuous working memory reaction time
	Serial four-choice	Continuous reaction time: pattern and random trials
	Delayed four-choice	Reaction-time task using RSI manipulations

Source: Modified from Stollery (1996d).

rotoxicity. The deficits identified among organophosphorus-ester-poisoned patients in earlier studies by other techniques should be detectable and quantifiable by the ACT system. A second computerized system is CANTAB, comprising three batteries of tests, each addressing a specific area of cognition: visual memory, attention, and working memory and planning (Fray and Robbins 1996; Sahakian and Owen 1992). Although it has been used in assessing neurodegenerative diseases, psychiatric disorders, neurosurgical cases, and acquired pathology, the battery of tests has not been applied to toxicant-induced neuropathies but appears to have many, if not all, of the attributes required for such assessments.

8-4 CONCLUSIONS

The original concerns persist regarding the validity of various tests for pesticide-exposed individuals, the potpourri of chemicals to which they may be exposed, possible interactions between chemicals, the anecdotal nature of reported poisonings, and the limited number of afflicted individuals. However, objective assessment of neurological and behavioral functions has developed far beyond that of 30 or even 20 years ago. Batteries of tests, including both simple and complex functions, now form a useful armamentarium to evaluate central and peripheral neurological deficits. Although applied almost exclusively to cases of organophosphorus ester poisonings, it is obvious from the published literature for other neurotoxicants that these test batteries could be employed to evaluate neurological functions in other pesticide poisonings. To date, the pattern of published results for exposed individuals has demonstrated that pesticide exposure did not cause distinctive differences in all tests when compared to suitable controls but that a spectrum of individual tests were positive, indicative of transient or persistent changes in neuronal function. In the future, greater use will be made of even more refined test batteries to evaluate the short- and long-term neurological effects of pesticide exposure.

REFERENCES

Ames, R. G., R. A. Howd, and L. Doherty. 1993. Community exposure to a paraquat drift. *Arch. Environ. Health* 48:47–52.

Ames, R. G., K. Steenland, B. Jenkins, D. Chrislip, and J. Russo. 1995. Chronic neurologic sequelae to cholinesterase inhibition among agricultural pesticide applicators. *Arch. Environ. Health* 50:440–44.

Baker, E. L., R. Letz, and A. Fidler. 1985. A computer-administered neurobehavioral evaluation system for occupational and environmental epidemiology. *J. Occup. Med.* 27:206–12.

Barnes, J. M. 1961. Psychiatric sequelae of chronic exposure to organophosphorus insecticides. *Lancet* 2: 102–03.

Bidstrup, P. L. 1961. Psychiatric sequelae of chronic exposure to organophosphorus insecticides (Letter to editor). *Lancet* 2:103.

Bove, F., M. S. Litwak, J. C. Arezzo, and E. L. Baker. 1986. Quantitative sensory testing in occupational medicine. *Semin. Occup. Med.* 1:185–89.

Brown, S. K., R. G. Ames, and D. C. Mengle. 1989. Occupational illness from cholinesterase-inhibiting pesticides among agricultural applicators in California 1982–1985. *Arch. Environ. Health* 44:34–39.

Burchfiel, J. L., F. H. Duffy, and V. M. Sim. 1976. Persistent effects of sarin and dieldrin upon the primate electroencephalogram. *Toxicol. Appl. Pharmacol.* 35:365–79.

Ciesielski, S., D. P. Loomis, S. R. Mims, and A. Auer. 1994. Pesticide exposures, cholinesterase depression, and symptoms among North Carolina migrant farmworkers. *Am. J. Public Health* 84:446–51.

Clark, G. 1971. Organophosphate insecticides and behavior, a review. *Aerosp. Med.* 42:735–40.

Davignon, L. F., J. St.-Pierre, G. Charest, and F. J. Tourangeau. 1965. A study of the chronic effects of insecticides in man. *Can. Med. Assoc. J.* 92:597–602.

Dille, J. R., and P. W. Smith. 1964. Central nervous system effects of chronic exposure to organophosphate insecticides. *Aerosp. Med.* 35:475–78.

Drenth, H. J., I. F. G. Ensberg, D. V. Roberts, and A. Wilson. 1972. Neuromuscular function in agricultural workers using pesticides. *Arch. Environ. Health* 25:395–98.

Duffy, F. H., J. L. Burchfiel, P. H. Bartels, M. Gaon, and V. M. Sim. 1979. Long-term effects of an organophosphate upon the human electroencephalogram. *Toxicol. Appl. Pharmacol.* 47:161–67.

Duffy, F. H., and J. L. Burchfiel. 1980. Long term effects of the organophosphate sarin on EEGs in monkeys and humans. *Neurotoxicology* 1:667–89.

Ecobichon, D. J. 1994. Organophosphorus ester insecticides. In *Pesticides and neurological diseases*. 2d ed. Edited by D. J. Ecobichon and R. M. Joy, 171–249. Boca Raton, Fla.: CRC Press.

Ecobichon, D. J. 1995. Pesticides. In *Principles of pharmacology. Basic concepts and clinical applications*. ed. P. L. Munson, R. A. Mueller, and G. R. Breese, 1563–79. New York: Chapman and Hall.

Ecobichon, D. J. 1996. Pesticides. In *Casarett and Doull's toxicology. The basic science of poisons*. ed. C. D. Klaassen, 643–89. New York: McGraw-Hill.

Ecobichon, D. J., and R. M. Joy. 1994. *Pesticides and neurological diseases*, 2d ed. Boca Raton, Fla.: CRC Press.

Feldman, R. G., N. L. Ricks, and E. L. Baker. 1980. Neuropsychological effects of industrial toxins: A review. *Am. J. Ind. Med.* 1:211–27.

Fray, P. J., and T. W. Robbins. 1996. CANTAB battery: Proposed utility in neurotoxicology. *Neurotoxicol. Teratol.* 18:499–504.

Gerr, F. E., D. Hershman, and R. Letz. 1990. Vibrotactile threshold measurement for detecting neurotoxicity: Reliability and determination of age- and height-standardized normative values. *Arch. Environ. Health* 45:148–54.

Gerr, F. E., and R. Letz. 1988. Reliability of a widely used test of peripheral cutaneous vibration sensitivity and a comparison of two testing protocols. *Br. J. Ind. Med.* 45:635–39.

Gershon, S., and F. B. Shaw. 1961. Psychiatric sequelae of chronic exposure to organophosphorus insecticides. *Lancet* 1:1371–374.

Hayes, W. J. Jr., and E. R. Laws, Jr. 1991. *Handbook of pesticide toxicology*. Vol. 1–3. San Diego, Calif.: Academic Press.

Holmes, J. H., and M. D. Gaon. 1956. Observations on acute and multiple exposures to antiChE agents. *Trans. Am. Clin. Climatol. Assoc.* 68:86–101.

Jager, K. W., D. V. Roberts, and A. Wilson. 1970. Neuromuscular function in pesticide workers. *Br. J. Ind. Med.* 27:273–78.

Jusic, A., D. Jurenic, and S. Milic. 1980. Electromyographical neuromuscular synapse testing and neurological findings in workers exposed to organophosphorus pesticides. *Arch. Environ. Health* 35:168–75.

Kahn, E., M. Berlin, M. Deane, R. J. Jackson, and J. W. Stratton. 1992. Assessment of acute health effects from the Medfly eradication project in Santa Clara County, California. *Arch. Environ. Health* 47:279–84.

Kaloyanova, F. P., and M. A. El Batawi. 1991. *Human toxicology of pesticides*. Boca Raton, Fla.: CRC Press.

Korsak, R. J., and M. M. Sato. 1977. Effects of chronic organophosphate pesticide exposure on the central nervous system. *Clin. Toxicol.* 11:83–95.

Levin, H. S., and R. L. Rodnitzky. 1976. Behavioral aspects of organophosphate pesticides in man. *Clin. Toxicol.* 9:391–405.

Maroni, M., J. Jarvisalo, and F. La Ferla. 1986. The WHO-UNDP epidemiological study on the health effects of exposure to organophosphorus pesticides. *Toxicol. Lett.* 33:115–23.

Marrs, T. C. 1993. Organophosphate poisoning. *Pharmacol. Ther.* 58:51–66.

Matsumura, F. 1985. *Toxicology of insecticides*. 2d ed. New York: Plenum Press.

McConnell, R., M. Keifer, and L. Rosenstock. 1994. Elevated quantitative vibrotactile threshold among workers previously poisoned with methamidophos and other organophosphate pesticides. *Am. J. Ind. Med.* 25:325–34.

Metcalf, D. R., and J. H. Holmes. 1969. EEG, psychological and neurological alterations in humans with organophosphorus exposure. *Ann. NY Acad. Sci.* 160:357–65.

Midtling, J. E., P. G. Barnett, M. J. Coye, A. R. Velasco, P. Romero, C. L. Clements, M. A. O'Malley, M. W. Tobin, T. G. Rose, and I. H. Monosson. 1985. Clinical management of field worker organophosphate poisoning. *West. J. of Med.* 142:514–18.

Misra, U. K., D. Nag, W. A. Khan, and P. K. Ray. 1988. A study of nerve conduction velocity, late responses and neuro-muscular synapse functions in organophosphate workers in India. *Arch. Toxicol.* 61:496–500.

Morgan, D. P., L. I. Lin, and H. H. Saikaly. 1980. Morbidity and mortality in workers occupationally exposed to pesticides. *Arch. Environ. Contam. Toxicol.* 9:349–82.

Muijser, H., R. B. M. Geuskens, J. Hooisma, H. H. Emmen, and B. M. Kulig. 1996. Behavioral effects of exposure to organic solvents in carpet layers. *Neurotoxicol. Teratol.* 18:455–62.

Mushak, E. W., and W. T. Piver. 1992. Agricultural chemical utilization and human health. *Environ. Health Perspect.* 97:269–74.

Richter, E. D., P. Chuwers, Y. Levy, M. Gordon, F. Grauer, J. Marzouk, S. Levy, S. Barron, and N. Gruener. 1992. Health effects from exposure to organophosphate pesticides in workers and residents in Israel. *Isr. J. Med. Sci.* 28:584–98.

Ring, H., S. Melamed, L. Heller, and P. Solzi. 1985. Evaluation of EMG examination as an indicator of worker susceptibility to organophosphate exposure. *Electromyogr. Clin. Neurophysiol.* 25:35–44.

Roberts, D. V. 1977. A longitudinal electromyographic study of six men occupationally exposed to organophosphorus compounds. *Int. Arch. Occup. Environ. Health* 38:221–29.

Rodnitzky, R. L., H. S. Levin, and D. L. Mick. 1975. Occupational exposure to organophosphate pesticides. *Arch. Environ. Health* 30:98–103.

Rosenstock, L., W. Daniell, S. Barnhart, D. Schwartz, and P. A. Demers. 1990. Chronic neuropsychological sequelae of occupational exposure to organophosphate insecticides. *Am. J. Ind. Med.* 18:321–25.

Rosenstock, L., M. Keifer, W. E. Daniell, R. McConnell, K. Claypoole, and The Pesticide Health Effects Study Group. 1991. Chronic central nervous system effects of acute organophosphate pesticide poisoning. *Lancet* 338:223–27.

Sahakian, B. J., and A. M. Owen. 1992. Computerized assessment in neuropsychiatry using CANTAB. *J. R. Soc. Med.* 85:399–402.

Savage, E. P., T. J. Keefe, L. M. Mounce, R. K. Heaton, J. A. Lewis, and P. J. Burcar. 1988. Chronic neurological sequelae of acute organophosphate pesticide poisoning. *Arch. Environ. Health* 43:38–45.

Smith, L. L. 1987. The mechanism of paraquat toxicity in the lung. *Rev. Biochem. Toxicol.* 8:37–71.

Smith, P. W., W. B. Stavinoha, and L. C. Ryan. 1968. Cholinesterase inhibition in relation to fitness to fly. *Aerosp. Med.* 39:754–58.

Stalberg, E., P. Hilton-Brown, B. Kolmodin-Hedman, B. Holmstedt, and K.-B. Augustinsson. 1978. Effect of occupational exposure to organophosphorus insecticides on neuromuscular function. *Scand. J. Work Environ. Health* 4:255–61.

Steenland, K., B. Jenkins, R. G. Ames, M. O'Malley, D. Chrislip, and J. Russo. 1994. Chronic neurological sequelae to organophosphate pesticide poisoning. *Am. J. Public Health* 84:731–36.

Stephens, R., A. Spurgeon, and H. Berry. 1996. Organophosphates: The relationship between chronic and acute exposure effects. *Neurotoxicol. Teratol.* 18:449–53.

Stephens, R., A. Spurgeon, I. A. Calvert, J. Beach, L. S. Levy, H. Berry, and J. M. Harrington. 1995. Neuropsychological effects of long-term exposure to organophosphates in sheep dip. *Lancet* 345:1135–39.

Stokes, L., A. Stark, E. Marshall, and A. Narang. 1995. Neurotoxicity among pesticide applicators exposed to organophosphates. *Occup. Environ. Med.* 52:648–53.

Stoller, A., J. Krupinski, A. J. Christophers, and A. K. Blanks. 1965. Organophosphorus insecticide and major mental illness. *Lancet* 1:1387–88.

Stollery, B. T. 1996a. Cognitive neurotoxicology: A luxury or necessity? *Neurotoxicol. Teratol.* 18:359–64.

Stollery, B. T. 1996b. Long-term cognitive sequelae of solvent intoxication. *Neurotoxicol. Teratol.* 18:471–76.

Stollery, B. T. 1996c. Reaction time changes in workers exposed to lead. *Neurotoxicol. Teratol.* 18:477–83.

Stollery, B. T. 1996d. The Automated Cognitive Test (ACT) system. *Neurotoxicol. Teratol.* 18:493–97.

Stollery, B. T., D. E. Broadbent, H. A. Banks, and W. R. Lee. 1991. Short-term prospective study of cognitive functioning in lead workers. *Br. J. Ind. Med.* 48:739–48.

Stollery, B. T., and M. L. H. Flindt. 1988. Memory sequelae of solvent intoxication. *Scand. J. Work Environ. Health* 14:45–48.

Summerford, W. T., W. J. Hayes, Jr., J. M. Johnson, K. Walker, and J. Spillane. 1953. Cholinesterase response and symptomatology from exposure to organic phosphorus insecticides. *Am. Med. Assoc. Arch. Ind. Hyg. Occup. Med.* 7:383–98.

Tabershaw, J. R., and W. C. Cooper. 1966. Sequelae of acute organic phosphate poisoning. *J. Occup. Med.* 8:5–20.

Tsai, S.-Y. and J.-D. Chen. 1996. Neurobehavioral effects of occupational exposure to low-level styrene. *Neurotoxicol. Teratol.* 18:463–69.

Vasilescu, C. and A. Florescu. 1980. Clinical and electro-physiological study of neuropathy after organophosphorus compounds poisoning. *Arch. Toxicol.* 43:305–15.

Verberk, M. M., and H. J. A. Salle. 1977. Effects on nervous function in volunteers ingesting mevinphos for one month. *Toxicol. Appl. Pharmacol.* 42:351–58.

Whorton, M. D., and D. L. Obrinsky. 1983. Persistence of symptoms after mild to moderate acute organophosphate poisoning among 19 farm field workers. *J. Toxicol. Environ. Health* 11:347–54.

Williamson, A. M. 1996. Historical overview of computerized behavioral testing of humans in neurotoxicology. *Neurotoxicol. Teratol.* 18:351–57.

Wood, W., J. Gavica, W. Brown, M. Watson, and W. W. Benson. 1971. Implications of organophosphate pesticide poisoning in the plane crash of a duster pilot. *Aerosp. Med.* 42:111–13.

Zweiner, R. J., and C. M. Ginsburg. 1988. Organophosphate and carbamate poisoning in infants and children. *Pediatrics* 81:121–26.

CONCLUSIONS AND HEALTH RISK
ASSESSMENT EXAMPLES

Donald J. Ecobichon

Department of Pharmacology and Toxicology, Queen's University, Kingston, Ontario, Canada K7L 3H6

9-1 INTRODUCTION

Having persevered through the preceding chapters and absorbed the appropriate lessons and techniques of obtaining relevant data, it is now time to look at potential exposure and health risks associated with residual pesticides in air and on surfaces. Risk is the probability of harm resulting from various activities or exposures and, because it is a probability, it carries numerical values between zero and one (Rodericks 1991). Risk assessment conjures up the concept of a specific methodological approach, extrapolating from sets of human and animal data under conditions of relatively intense exposure and developing quantitative estimates of risk at much less intense exposures experienced by human populations (Rodericks 1991). The conduct of risk assessment does not require any specific methodological approach but is best seen as a way of organizing our knowledge regarding potentially hazardous activities to use a systematic approach to question risks that may be posed under specific circumstances. Whatever methodology that is thought to be scientifically justifiable can be used, as long as the reasons for the choices have been explicitly set forth. As Rodericks (1991) concluded, risk assessment focuses explicit recognition of the various assumptions needed to reach conclusions about risk and the uncertainties their use introduces.

Scenarios can be developed, depending on the amount of information available about the specific circumstances of the exposure, to permit (1) crude exposure estimates to address issues following possible exposures, where no biological monitoring or chemical analysis has been done; or (2) more precise estimates, where the incident has permitted either limited or extensive monitoring of conditions. The scenarios are shown schematically in Fig. 9-1.

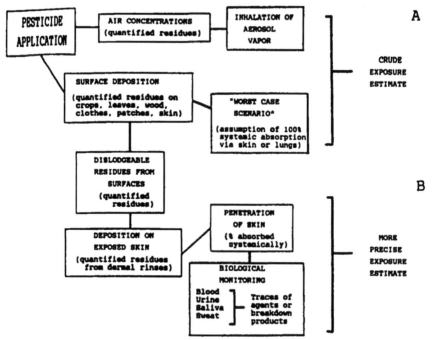

Figure 9-1 Exposure estimation of: A, a crude nature; or B, a precise nature. Estimation is based on the amount of quantifiable data available from the incident, including air and surface concentrations, dislodgeable residues, deposition on and systemic absorption through the skin, as well as the measurement of biological indices pertinent to the potentially toxic agent.

9-1-1 Crude Estimates

In the first situation, possibly a worst-case scenario, exposure may be suggested, e.g., aerial overspraying, downwind drift of spray onto bystanders, or an indoor residential spray (Fig. 9-1A). Aerial or surface samples may have been collected after the incident for analysis OR concentrations at the time of the incident may be estimated from experimental field trials with the same or similar agents or from simulation models. Certain assumptions have to be made concerning the potentially exposed individuals (Table 9-1). Exposure may include both inhalation and dermal routes. The lack of information on the toxicokinetics of the agent via either route may require an assumption of 100% absorption into the system.

Estimated absorbed dose via respiratory exposure is expressed as the product of time-weighted air concentrations (or an average concentration), respiratory volume, and percentage absorbed, all divided by body weight for the day of the incident and/or subsequent time intervals (Fenske et al. 1990).

$$\text{Resp. Exposure} = \frac{[\text{Air}] \times \text{Resp. Rate} \times \% \text{ Absorbed} \times \text{Time}}{\text{Body Weight of Individual}}$$

where [Air] is milligrams or micrograms of agent per liter of air, Resp. Rate is the liters of air inhaled/min, and % Absorbed is the amount known to cross the pulmonary blood

Table 9-1 Assumptions about the physiology of individuals exposed to pesticides, for exposure estimation

Parameter	Adult	Child
Body weight (kg)	70	9
Respiratory rate, at rest (l/min)	20	2.1
Respiratory rate, moderate activity (l/min)	40	6.3
Exposed skin surface (m^2)	0.3	?
Transfer of available residues by dermal and respiratory routes (%)	100	100
Absorption of contacted residues on skin and in lungs (%)	100	100

barrier into the systemic circulation. One could assume that, except in the unlikely case of very large particulate material, any agent entering the respiratory tract (nasopharyngeal, tracheobronchial, alveolar, and parenchymal regions) would be completely absorbed through the mucous membranes.

Estimated absorbed dose via dermal exposure is expressed as the product of surface area contacted by the agent, the available residues, the percentage absorbed, all divided by the body weight (Fenske et al. 1990).

$$\text{Dermal Exposure} = \frac{\text{Surface Area} \times [\text{Agent}] \times \% \text{ Absorbed}}{\text{Body Weight}}$$

where Surface Area represents the exposed skin, some fraction approximately 2.9 m^2 (adult); [Agent] is the quantified agent in milligrams or micrograms/m^2; % Absorbed is the amount known to penetrate the skin into the systemic circulation; and Body Weight is measured in kilograms. The percentage of agent absorbed systematically frequently is not available, is difficult to gauge, and usually ranges between 1.0 and 30.0%, depending on the physical and chemical properties of the agent in question. In a worst-case scenario, one could assume 100% absorption from the exposed skin but, in all likelihood, this would result in a gross overestimation of the exposure.

With estimations of inhalation and/or dermal absorption of the potential toxicant systemically, one can at least arrive at a body burden following exposure and at a conclusion as to whether one should be concerned about possible adverse human health effects. Although the potential body burden for a healthy young adult might be considered to exert little or no effect, the same body burden in children, the elderly, or individuals with a compromised health status might place these subjects at risk because of the inability to efficiently biotransform, detoxify, and eliminate the agent. This would necessitate the use of uncertainty or safety factors to adjust for a lower acceptable body burden for such individuals. Exposure scenarios, particularly duration and level of exposure, might be very different for these subjects than for the healthy young adult.

9-1-2 Precise Estimates

Human exposure can be extrapolated from extensive quantification of chemical residues in air and on surfaces as well as from transfer to, deposition on, and penetration of the skin. In addition to or in place of these parameters, biological monitoring of body fluids of potentially exposed person(s) for residues of the agent and/or breakdown products over a time period may provide a direct measure of exposure. The concomitant measurement of selected biochemical and physiological markers may confirm the body burden of the agent required to elicit specific adverse health effects. These, of course, have been the topics of the preceeding chapters.

In concluding this chapter and the book, it is my intention to examine three diverse scenarios as practical examples of strategies for calculating potential exposure, depending upon the amount of quantified data and information available.

9-2 RISK ASSESSMENT EXAMPLES

9-2-1 Scenario 1

In this example, no data were available at the time of the incident, and so, estimates of human exposure were derived from other studies and computer models.

The incident, observed by this author, arose from the aerial application of mancozeb (Dithane, Manzate), an ethylenebisdithiocarbamate fungicide used for potato blight, to a field separated by a community road from a number of homes (Ecobichon, unpublished).Concern was voiced by residents in a petition, requesting that the health department conduct a study of risk(s) to their health. Germane to this effort was an estimation of potential exposure. The application rate of the mancozeb was known (0.158 g/m^2) and the wind speeds, at $90°$ to the direction the aircraft flew, were estimated to be approximately 10 km/h at 2 and 10 m above the potato-crop canopy. Estimates of air and surface concentrations of the agent at different distances from the spray line were derived from a simulated field study by Crabbe and McCooeye (1985); the results extrapolated for mancozeb are shown in Fig. 9-2.

The estimates of dermal absorption were calculated for a casually dressed adult with an exposed skin surface of 0.3 m^2 (short-sleeved shirt, open "V" of shirt front, bare arms, exposed back of neck, face, and head). It was assumed that all residues deposited on the skin were absorbed. For the estimation of inhalation exposure in the 60-min period after spraying, it was assumed that the adult was breathing at a rate of 20 liters of air per minute and that all respired residues were absorbed systemically from the respiratory tract. It was also assumed that the potentially exposed individuals were some 30 to 75 m from the spray line and the field fencing (road allowance, width of road, and setback from road to fronts of homes).

The estimates of adult exposure are presented in Table 9-2. Surface deposition ranged from 0.0015 to 0.0003 g/m^2 over the 30- to 75-m distance and, assuming complete absorption of dermally acquired residues of mancozeb, the dermal exposure would be 0.02, 0.004, and 0.004 mg/kg of body weight at distances of 30, 50, and 75 m, respectively. From the derived air concentrations at the same distances, potential exposure for a 60-min period after spraying revealed low levels—1.23 to 1.02 μg/kg of body weight. Thus, total acquisition of mancozeb, under the conditions and assumptions described, were 21.3, 5.23, and 5.02 μg/kg of body weight at 30, 50, and 75 m, respectively, from

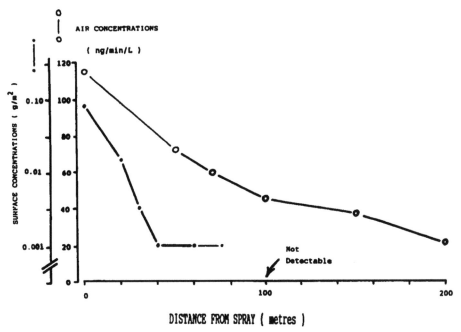

Figure 9-2 Aerial and surface concentrations of mancozeb, an ethylenebisdithiocarbamate fungicide, at varying distances from a sprayed field as estimated from a simulation model of off-target losses of pesticide. *Source:* Crabbe and McCooeye (1985).

the spraying operation. The actual exposures may be considerably lower than the values calculated because it is unlikely that more than 2.0 to 5.0% of the dermal burden of mancozeb would be absorbed systemically, although most if not all of the inhaled pesticide could be absorbed. It was concluded from this exercise that the mancozeb acquired would not elicit any adverse effects in adults.

Table 9-2 Estimated dermal and inhalation exposure to mancozeb sprayed aerially on potatoes

Distance from spray (m)	Dermal exposure		Inhalation exposure		Total exposure
	(g/m²)	(μg/kg)	(μg/h)	(μg/kg)	(μg/kg)
30	0.0015	20	90.0	1.28	21.03
50	0.0003	4	86.2	1.23	5.23
75	0.0003	4	71.3	1.02	5.02

9-2-2 Scenario 2

In this example, accurate measurements were made of residues in air and on surfaces at the time of application and the values were extrapolated to both children and adults.

The study conducted by Fenske et al. (1990) is used, and the details are summarized below.

Chlorpyrifos (0.5% aqueous spray) was applied in uninhabited, three-room apartments (one ventilated, one unventilated) using a CO_2-pressurized, handheld fan broadcast sprayer, the formulation being applied approximately 40 cm above the carpet, covering the entire floor and taking 5 to 7 min/room. Duplicate air samples were collected by battery-powered pump in each room at 0–0.5, 0.5–1.0, 1.0–1.5, 1.5–3.0, 3.0–5.0, and 5.0–7.0 h after application, with additional 2-h samples being collected 24 h postapplication. The air sampling devices were placed 25 and 100 cm above the carpet to represent breathing zones for a child playing on the carpet and a seated adult, respectively. Aluminum foil squares (10 × 10 cm) were placed in the rooms before treatment to measure residues deposited on the carpet. Five squares were collected from each room within 5 min of completing the application. Three replicate surface wipes (10 × 10 cm) were obtained at 20, 40, 80, 100, 170, 260, and 380 min and at 24 h, according to the procedure established by the U.S. Occupational Safety and Health Administration (OSHA 1984). The samples were analyzed by gas chromatography.

The results of the study are summarized in Table 9-3 and Fig. 9-3 modified from the data reported in the study. Higher chlorpyrifos concentrations were measured in the child's breathing zone (25 cm above the carpet) than in the adult's breathing zone (100 cm above the carpet). For available vapor and surface concentrations, the amounts were higher in the unventilated apartments. A risk assessment was based on the scenario of a 9- to 10-month-old infant playing in a carpeted room beginning at least 60 min after a 0.5% chlorpyrifos application. The assumptions made included (1) infant weight of 9.0 kg; (2) infant respiratory volume of 2.0 l/min resting (16 h/day), 6.3 l/min active (8 h/day); (3) 100% respiratory absorption of vapors; (4) total carpet contacted in one day, 2.3 m² (25 ft²); (5) 100% transfer of available surface (wipe sample) residues to skin; and (6) an estimated 3.0% dermal absorption of chlorpyrifos based on available literature (Nolan et al. 1984). The total chlorpyrifos estimated to be absorbed by infants on the day of application and on the following day, under ventilated and unventilated conditions, is shown in Table 9-3. Note that ingestion from deposits on hands was not considered. With a no observable effect level (NOEL) for chlorpyrifos in humans

Table 9-3 Estimated chlorpyrifos absorption via dermal and pulmonary routes by infants on day of application (day 1) and day following application (day 2)

Experimental conditions	Dose (mg/kg)		
	Dermal	Respiratory	Total
Day 1			
Ventilated	0.052	0.023	0.075
Nonventilated	0.120	0.038	0.158
Day 2			
Ventilated	0.022	0.016	0.038
Nonventilated	0.037	0.016	0.055

Source: Data from Fenske et al. (1990).

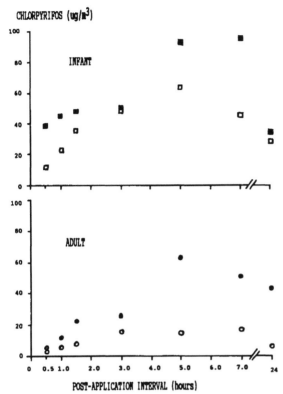

Figure 9-3 Chlorpyrifos concentrations in the breathing zones of an infant (25 cm above the carpet) and a seated adult (100 cm above the carpet) following application of the agent in ventilated (open symbols) and unventilated (solid symbols) quarters.
Source: Modified from Fenske et al. (1990).

established at 0.03 mg/kg per day, all of the absorbed dose estimates exceeded the NOEL by 2.5 to 5 times on day 1 and 1.2 to 1.8 times on day 2. The US Environmental Protection Agency (US EPA) derived human reference dose (RfD) for chlorpyrifos is 0.003 mg/kg per day, applying a tenfold uncertainty factor to the human NOEL. The exposure dose estimates of Fenske et al. (1990) exceeded the RfD by ten to fifty times.

9-2-3 Scenario 3

Exposure to and absorption of azinophos-methyl (Guthion) were estimated in orchard workers involved in routine mixing, loading, and applying of this agent in a wettable-powder formulation, using ultralow-volume airblast equipment. Air monitoring, patch techniques, alkyl phosphate (dimethylphosphorothioic acid, the sum of O,O-dimethyl-O-ethylphosphorothioate and O,O-dimethyl-S-ethylphosphorothioate) excretion in urine and erythrocyte and plasma cholinesterases were measured to estimate exposure and a fluorescent tracer technique also was used to give a qualitative indication of exposure of skin surfaces (Franklin et al. 1981). As can be seen, this study uses both direct (air

Table 9-4 Average urinary excretion of Guthion metabolites (spray day 1 plus postexposure day 2)

Protective clothing	Active agent sprayed (kg)	48-h output (μg)	Active agent ($\mu g/kg$ bw)
Coats and pants	2.3	310	135
Coat	2.5	440	176
None	2.5	440	176

Source: Data from Franklin et al. (1981).

and surface concentrations) and indirect (urinary metabolites, cholinesterase inhibition) estimates of exposure, the latter being estimates of systemic absorption.

Some of the results of this detailed study—the urinary alkyl phosphate excretion—are shown in Table 9-4. All of the workers showed urinary alkyl phosphates, the levels varying and depending upon the type of protective clothing worn. Twenty-four hour urine samples provided a more reliable estimate of exposure than did first morning voids. A high correlation was observed between 48-h alkyl phosphate excretion and the amount of active ingredient sprayed. Urinary excretion of metabolites proved to be a better and more sensitive indicator of exposure. None of the orchard workers showed signs and symptoms of toxicity; this observation was confirmed by depressions of erythrocytic and plasma cholinesterases not exceeding 15% of preexposure levels. Surprising results included the detection of residues of azinophos-methyl on patches beneath the protective clothing.

9-3 CONCLUSIONS

These scenarios, as well as the information in the various chapters, emphasize the need for rigorous, standard methodologies for estimating exposure in pesticide workers. Durham, Wolfe, and Elliott (1972), comparing exposure by dermal and respiratory routes, found that the former accounted for 87% of the total exposure. Although this may be true for many situations, it does not exclude the possibility of inhalation being the predominant route in enclosed situations such as greenhouses, mushroom barns, tractor cabs, and unventilated dwellings, where aerosols do not behave as they do in field situations and tend to remain suspended for extended periods of time. Patch studies on exposed skin surfaces and on clothing only provide information about contact and availability for absorption. Absorption can be quantitated by measuring excretory products (alkyl phosphates, aromatic leaving groups (e.g., phenols), and biological markers such as erythrocytic or plasma cholinesterases). In some instances, correlations between metabolite levels and degree of cholinesterase inhibition have been established but, in other studies, no such correlations have been found. Although many authors would agree that urinary levels of metabolites appear to correlate reasonably well with the amount of agent applied (an indirect index of exposure), this must be accompanied by knowledge of the extent and site of dermal contamination because it is known that the degree of penetration varies considerably through the skin on different parts of the anatomy (Maibach et al. 1971).

It is imperative that a better database be obtained concerning the systemic absorption of various classes of pesticides; these data are key to any estimation of body burden of agents and estimation of possible adverse health effects.

The complex nature of exposure sampling, measuring, and monitoring has led to continued exploration of more sophisticated techniques to quantify absorption and to estimate potential risks to human health. It is our hope that the discussions in this text will provide direction and stimuli to readers, encouraging them to advance this type of research in the future.

REFERENCES

Crabbe, R. S., and M. McCooeye. 1985. Field study of ground deposition and bystander exposure from agricultural aircraft spray emissions. Report LM-UA-203 National Aeronautical Establishment/National Research Council, Ottawa, Canada.

Durham, W. F., H. R. Wolfe, and J. W. Elliott. 1972. Absorption and excretion of parathion by spraymen. *Arch. Environ. Health* 24:381–87.

Fenske, R. A., K. G. Black, K. P. Elkner, C.-L. Lee, M. M. Methner, and R. Soto. 1990. Potential exposure and health risks of infants following indoor residential pesticide applications. *Am. J. Public Health* 80:689–93.

Franklin, C. A., R. A. Fenske, R. Greenhalgh, L. Mathieu, H. V. Denley, J. T. Leffingwell, and R. C. Spear. 1981. Correlation of urinary pesticide metabolite excretion with estimated dermal contact in the course of occupational exposure to Guthion. *J. Toxicol. Environ. Health* 7:715–31.

Maibach, H. I., R. J. Feldmann, T. H. Milby, and W. F. Serat. 1971. Regional variation in percutaneous penetration in man. *Arch. Environ. Health* 23:208–211.

Nolan, R. J., D. L. Rick, N. L. Freshour, and J. H. Saunders. 1984. Chlorpyrifos: Pharmacokinetics in human volunteers. *Toxicol. Appl. Pharmacol.* 73:8–15.

Occupational Safety and Health Administration (OSHA). 1984. Sampling for surface contamination. Chap. 8 in *Industrial hygiene. Technical manual.* Washington, D.C.: U.S. Government Printing Office.

Rodericks, J. V. 1991. Assessing and managing toxicological risks. In *First International Conference, Environmental and Industrial Toxicology. Research and Applications.* ed. M. Ruchirawat and R. C. Shank. 33–61. Bangkok: Chulabhorn Research Institute.

INDEX